TARGET VALIDATION IN DRUG DISCOVERY

TARGET VALIDATION IN DRUG DISCOVERY

Editors
BRIAN W. METCALF AND SUSAN DILLON

AMSTERDAM • BOSTON • HEIDELBERG • LONDON
NEW YORK • OXFORD • PARIS • SAN DIEGO
SAN FRANCISCO • SINGAPORE • SYDNEY • TOKYO

Academic Press is an imprint of Elsevier

Academic Press is an imprint of Elsevier
30 Corporate Drive, Suite 400, Burlington, MA 01803, USA
525 B Street, Suite 1900, San Diego, California 92101-4495, USA
84 Theobald's Road, London WC1X 8RR, UK

This book is printed on acid-free paper. ♾

Library of Congress Cataloging-in-Publication Data
Target validation in drug discovery / Brain W. Metcalf and Susan Dillon, editors.
p. cm.
Includes bibliographical references and index.
ISBN 13: 978-0-12-369393-8 (alk.paper)
ISBN 10: 0-12-369393-4 (alk.paper)
1. Drug development. 2. Drugs—testing. 3. High throughtput screening (Drug development) I. Metcalf, Brain W. II. Dillon, Susan, 1952–
RM301.63.T36 2006
615'.19—dc22

2006049868

British Library Cataloguing-in-Publication Data
A catalogue record for this book is available from the British Library.

ISBN-13: 978-0-12-369393-8
ISBN-10: 0-12-369393-4

For information on all Academic Press publications
visit our Web site at www.books.elsevier.com

Printed in the United States of America
06 07 08 09 10 9 8 7 6 5 4 3 2 1

CONTENTS

PREFACE

The genomic revolution, which involved sequencing of the human genome and subsequent development of a number of brilliant platform technologies, has led to the anticipation that these biomedical advances will lead to the discovery of new medicines. However, owing to a 15-year cycle from initiation of a drug discovery program to market, commercial success emanating from the genomics revolution or from the sequencing of the human genome has not materialized. The report from a recent conference organized by the Association of American Medical Colleges (AAMC) and the Food and Drug Administration (FDA) entitled "The Critical Path to Medical Products" noted that, "[t]hose deciphering the genome had high expectations that these new discoveries would help select better therapeutic targets. Unfortunately, the excitement around genomics has not carried over to validation in humans." In fact, the number of NDAs (New Drug Applications) approved each year by the FDA has declined from 53 in 1996 to 35 in 1999 to 17 in 2002 to 15 in 2005. It is too early to conclude that advances in the numbers and quality of new drugs will not be forthcoming. Nevertheless, the annual decrease in NDAs is a real concern that has prompted most companies to overhaul their drug discovery processes, or their organizational structures, and also is related to an increasing number of mergers and acquisitions. Drugs directed at new targets likely to have a real impact on disease are desperately needed. In fact, with ever-increasing competition from generics, bio-similars, and bio-follow on drugs, the pressure on the pharmaceutical and biotechnology industries for truly novel medicines with added value to patients and to the health care system becomes even more essential.

How then can the pharmaceutical and biotechnology industries respond to the dual needs of increased productivity and novelty? A likely scenario is to take advantage of the sequencing of the human genome by creating new paradigms for selection and validation of targets that are causal to disease. This change of direction presents an enormous new challenge as the sequences of all 30,000 or so human genes are known, but the linkage of gene sequence to human disease is not. In prioritizing which genes to approach as drug targets several issues must be addressed:

- Which genes are causal to, and not just associated with, disease? Which genes are druggable?
- Which technologies does one need to address these issues?
- How does one proceed upon having chosen a disease gene of interest, given that 15 or more years of investment will be required, estimated by some to be at a cost of over $800 million for each new drug?

- How does one mitigate risk in the face of this required investment? Failures in Phase III clinical trials owing to lack of efficacy related to poor choice of target cannot be tolerated.

From our point of view, many of these issues distill down to successful target validation in drug discovery and clinical development. There is no point in providing a potential molecular solution for a drug target that does not have an impact on disease. This book is directed primarily at the overriding issue of target validation. We have chosen to examine the background leading to a number of successful new biologic agents; by definition protein drugs such as antibodies, soluble receptors, or replacement cytokines. The chapters examine how the targets of these drugs were chosen and validated, what technologies were used, and what biological and clinical results were obtained in support of the hypothesis that these targets were linked to human disease.

In general, the following questions must be addressed in the validation of a target:

- Is the target associated with disease pathology?
- Is the target expressed by cells linked to disease pathology?
- Is the target expressed *in vivo*?
- In animal models does efficacy occur at drug concentrations necessary to interact with the target?

Similarly, there are common issues inherent to preclinical target validation studies:

- Animal models frequently do not accurately reflect or mimic human disease processes.
- Species specificity issues sometimes necessitate the use of surrogate molecules to achieve preclinical proof of concept.
- There is occasional lack of clarity around direct human/animal gene homologs.

Nonetheless, animal modeling remains a prerequisite step in target identification and validation prior to investment in clinical development. A number of new technologies have evolved that are deployed to address some of these issues, some of which are described in Section I. Primary among these is the creation of transgenic mice, in which a gene of interest is knocked out and the subsequent mouse phenotype is studied. Chapter one describes the genetic technologies that are deployed in these studies, in which the goal is to knock out the 5000 genes estimated to be druggable over a 5-year period. It is anticipated that a number of genes will be discovered that are causally linked to disease during the course of this program. Indeed, reference to results obtained from the study of transgenic mice appears often in subsequent chapters in this book—Vaddi relies on such data for the choice of CCR-2 as an anti-inflammatory target, whereas Benson references studies on mice deficient in IL-12 or IL-23 subunits as a basis for targeting IL-12p40 for immune-mediated disease.

One can approach targets in the inverse manner by electing a desirable cell phenotype to be attained by drug action, finding a molecule to effect transformation to that phenotype by high throughput screening and then discovering the molecular target or pathway influenced by that compound. This approach called Forward Chemogenomics is described in chapter two, and is illustrated by the discovery of Purmorphamine as a small molecule inducer of osteogenic differentiation, and subsequent delineation of the hedgehog pathway as being the pertinent pathway.

As proteinaceous drugs such as soluble receptors and antibodies do not cross the cell membrane, they generally impact at the extracellular domain of the pertinent receptor by antagonism, agonism, or removal of the endogenous ligand or by increasing its concentration. Their success, however, validates the transduction pathway initiated by agonism at the cell surface as a likely pathway for successful intervention. Conceptually, then, one can move further down that same pathway beyond the surface receptor to other intracellular targets that will also be valid. To do so, the nature of the drug must change from a protein that cannot enter cells and is not orally absorbed, to a small molecule capable of entering cells. The physical characteristics required for cell permeation are also consistent with, but not necessarily completely adequate for, oral absorption. So the paradigm of moving down a validated transduction pathway is also that of transitioning from a proteinaceous drug administered by systemic routes (intravenous, subcutaneous, intramuscular) to a small molecule drug given by oral administration. The themes we have developed by our choice of the chapters solicited are therefore novel technologies required for target identification (Section I), validation of disease-associated pathways by systemically administered biological drugs (Section II), followed by small molecule approaches directed at targets later in the indicated pathways to afford orally available drugs (Section III).

This paradigm is illustrated by the epithelial growth factor (EGF-1) pathway. Cetuximab is an anti-EGF-1 growth factor receptor antibody approved for the treatment of metastatic colorectal cancer. The EGF-1 receptor is a tyrosine kinase, wherein activation at the extracellular domain of the receptor by EGF leads to tyrosine phosphorylation on the cytoplasmic domain of the receptor with subsequent signal transduction. Erlinotib is an orally active inhibitor of the tyrosine kinase associated with the EGF-1 receptor that is now approved for the treatment of non–small cell lung cancer (NSCLC) and pancreatic cancer. Chapter three in Section II describes the role of the EGFR-1 in cancer and subsequent clinical evaluation of cetuximab. In Section III, the cetuximab story is complemented by chapter nine on the discovery and clinical evaluation of erlinotib, which takes advantage of the signal transduction pathway validated by cetuximab. Thus, a systemically administered monoclonal antibody, while a successful drug in its own right, has validated the EGF pathway as a viable collection of novel drug targets, and erlinotib is an orally active small molecule acting in the same pathway as cetuximab.

The paradigm is further enforced by the success of the anti-TNF biologics, in which biologics such as infliximab, which is a monoclonal anti-TNFα antibody, and etanercept, a human p75 TNF receptor-IgG Fc fusion protein, have been demonstrated clinically to have dramatic effects in the treatment of rheumatoid arthritis and other inflammatory conditions. These biologics have demonstrated the validity of antagonizing TNF as an anti-inflammatory target. Many other efforts are now underway to intervene at later points in the TNF pathway, other than the TNF receptor, in the search of small molecule, orally available drugs that might have similar positive effects to the systemically administered TNF biologics. Chapter five in Section II lays the groundwork of the discovery of the anti-TNF biologics. Elliott's emphasis is on the demonstration of linkage of local production of TNF in the gut wall to the etiology of Crohn's disease and to ulcerative colitis. Such basic studies led to a powerful new treatment for patients with IBD. As noted earlier, anti-TNF biologics have also led to powerful new treatments for other inflammatory diseases such as rheumatoid arthritis. Chapter ten in Section III describes p38 kinase as a target that impacts on TNF signaling and production, and the chapter eleven relates to the inhibition of IKK-2 in the NF-KB pathway initiated by TNF-α binding to its receptor as approaches to oral agents.

We present other chapters in Section II that allow us to continue the theme of pathway validation by a biologic agent, and follow in Section III with small molecule approaches directed at the same pathways. In one surprising case, that of antagonizing the integrin GpIIb/IIIa, the protein agent abciximab proved to be successful, whereas the follow-up small molecules failed for unanticipated reasons. These chapters are developed by Jordan and Seiffert and Billheimer, respectively. In comparison, the hematopoietic cytokine thrombopoietin failed in the clinic for reasons of immunogenicity, whereas a small molecule mimetic, eltrombopag is advancing in late phase clinical trials as described by Duffy and Erickson-Miller. The chemokine receptor ligand CCL is described in Section II by Das and Yan, whereas the small molecule approach to antagonism is rationalized by Vaddi in Section III. These studies were also predicated by the success of the anti-TNF biologics, as cytokines are released by infiltrating inflammatory cells, the major one being the macrophage, which undergoes trafficking via the CCR-2 receptor. In some cases only one side is presented (chapter seven), the respective counterpart likely to be developed in the future. Another future counterpart would be represented by the success of the anti-IgE antibody omalizumab, being followed up with a small molecule approach.

We are passionately interested in drug discovery and development. One of us (SD) is a cell biologist with responsibility for the discovery and preclinical development of monoclonal antibodies as drugs, whereas the other (BM) has a chemistry background and leads a small molecule drug discovery effort. We believe that there is a new paradigm to be explored by linking concepts derived from successful biologic drugs that must be administered systemically to the discovery of orally bioavailable small molecules that act in the same signal transduction pathway. Thus the biologic drug points the way by providing clinical proof of concept and demonstrating market success for a later chemical entity that acts in the now-validated pathway. In this book, we attempt to formalize this approach through presentation of several illustrative examples.

Our intended audience includes biotechnology and pharmaceutical executives, those working the laboratories in these industries, academics involved in the basic biology of signal transduction pathways, and students of chemistry and biology. The concepts developed in the chapters presented could be generalized and extrapolated to other new drug discovery efforts. This book should provide an ideal platform to popularize a new paradigm in drug discovery efforts, and hopefully aid in the translation of current revolutionary advancements in the biomedical sciences to new medicines.

Brian W. Metcalf
Moraga, CA

Susan Dillon
Radnor, PA

CONTRIBUTORS

Jerry L. Adams, Ph.D.
GlaxoSmithKline Pharmaceuticals
Collegeville, Pennsylvania

Shannon E. Beard, Ph.D.
OSI Pharmaceuticals
Melville, New York

Jacqueline Benson, Ph.D.
Centocor Research and
Development, Inc.
Radnor, Pennsylvania

Jeffrey T. Billheimer, Ph.D.
University of Pennsylvania
Philadelphia, Pennsylvania

James R. Burke, Ph.D.
Bristol-Myers Squibb Co.
Princeton, New Jersey

Anuk Das, Ph.D.
Centocor Research and
Development, Inc.
Radnor, Pennsylvania

Sheng Ding, Ph.D.
The Scripps Research Institute
La Jolla, California

Kevin J. Duffy, Ph.D.
GlaxoSmithKline Pharmaceuticals
Collegeville, Pennsylvania

Derek E. Eberhart, Ph.D.
Lexicon Genetics Incorporated
The Woodlands, Texas

Michael J. Elliott, M.D., Ph.D.
Centocor Research &
Development, Inc.
Malvern, Pennsylvania

Connie L. Erickson-Miller, Ph.D.
GlaxoSmithKline Pharmaceuticals,
Collegeville, Pennsylvania

Francisco J. Esteva, M.D., Ph.D.
University of Texas M. D. Anderson
Cancer Center
Houston, Texas

John D. Haley, Ph.D.
OSI Pharmaceuticals
Farmingdale, New York

Kathleen H. Holt, Ph.D.
Lexicon Genetics Incorporated
The Woodlands, Texas

Gabriel N. Hortobagyi, M.D.
University of Texas M. D. Anderson
Cancer Center
Houston, Texas

Kenneth K. Iwata, Ph.D.
OSI Pharmaceuticals
Farmingdale, New York

Robert E. Jordan, Ph.D.
Centocor Research and
Development, Inc.
Radnor, Pennsylvania

Laura L. Kirkpatrick, Ph.D.
Lexicon Genetics Incorporated
The Woodlands, Texas

John C. Lee, Ph.D.
GlaxoSmithKline Pharmaceuticals
King of Prussia, Pennsylvania

William J. Pitts, Ph.D.
Bristol-Myers Squibb Co.
Princeton, New Jersey

Dietmar A. Seiffert, M.D.
Bristol-Myers Squibb Co.
Pennington, New Jersey

Kris Vaddi, DVM, Ph.D.
Incyte Corporation
Experimental Station
Wilmington, Delaware

D. Wade Walke, Ph.D.
Lexicon Genetics Incorporated
The Woodlands, Texas

Tom Y-H. Wu, Ph.D.
Merck Frosst, Center for Therapeutic Research
Kirkland, Quebec, Canada

Li Yan, M.D., Ph.D.
Centocor Research and Development, Inc.
Malvern, Pennsylvania;
Peking University
Beijing, China

Brian P. Zambrowicz, Ph.D.
Lexicon Genetics Incorporated
The Woodlands, Texas

Zhenping Zhu, M.D., Ph.D.
ImClone Systems Incorporated
New York, New York

F. Christopher Zusi, Ph.D.
Bristol-Myers Squibb Co.
Pharmaceutical Research Institute
Wallingford, Connecticut

I

PHARMACEUTICAL BIOTECHNOLOGY FOR TARGET VALIDATION

1 GENERATION OF TRANSGENIC ANIMALS

BRIAN P. ZAMBROWICZ, KATHLEEN H. HOLT, D. WADE WALKE, LAURA L. KIRKPATRICK, and DEREK E. EBERHART

Ph.D., Lexicon Genetics Incorporated, The Woodlands, Texas

The development of molecular biological tools for mutant-mouse production signaled the new era of contemporary mammalian genetics. Recent technology advances coupled with completed sequence of the mouse and human genomes are converging synergistically such that today's scientists are exploring their fields of interest in arguably the most productive period for research and discovery of modern biological science. Scientists can now take advantage of an unprecedented ability to manipulate the mouse genome using versatile techniques that afford both custom genetic alteration and random mutagenesis approaches. Together, these strategies can be employed to quickly and comprehensively saturate the genome with a high percentage of gene coverage. Productivity estimates of the leading commercial biotechnology groups that have knockout mouse production and analysis capabilities, together with published knockout reports, suggest that roughly one-fourth of the approximate 25,000 genes in the mouse genome have been knocked out and studied to varying degrees. The international mouse-genomics community has recently begun an initiative to produce knockout alleles for all mouse genes and to develop a system for open access to data and reagents for the broader scientific community. Clearly we can look forward to an escalation of discovery productivity in the upcoming years. It will be even more exciting to track how mouse gene-function discoveries will drive the drug discovery industry's efforts to develop the next generation of therapies for human disease.

I. INTRODUCTION

The demonstration that Mendel's laws of inheritance governing plant traits translated to the inheritance of coat color characteristics in mice launched the classical period of mouse genetics in 1902 (Cuenot, 1902). Over the next several decades, the mouse proved to be a robust experimental model system that afforded unprecedented breakthroughs in understanding mammalian genetics and physiology. In the early 1900s, the first strain of inbred mice—DBA—was derived by C. C. Little, who later became the founder of The Jackson Laboratory (Paigen, 2003). Many of the more than 300 inbred mouse strains in use today were developed during the same time period (Staats, 1980). These mouse strains were the cornerstone of research, which led to the elucidation of fundamental physiological mechanisms, including antibody-mediated tumor transplantation rejection (Gorer, 1937) and retroviral-mediated proto-oncogene activation and neoplastic transformation (Mulbock and Bentvelzen, 1968). They also fueled the beginnings of the contemporary field of biochemical genetics. Mouse geneticists at the time could not have known the degree of conservation between human and mouse genes. Nevertheless, the idea that human and mouse physiology is highly conserved appears to have been a widely held view considering the common theme of comparative physiological and disease studies in mice and humans reported over the years.

Ultimately, classical mouse genetics laid the groundwork for the launch of the new era of modern mouse genetics, which began around 1980 with the abrupt arrival of molecular biological techniques and the coincident development of germline-competent murine embryonic stem (ES) cell lines. A brilliant scheme, known as gene targeting, was devised to engineer precise genetic manipulations by homologous recombination using positive and negative selection in mammalian tissue culture cells (Mansour et al., 1988). In parallel, pluripotent murine ES cell lines were derived from the inner cell mass of blastocysts and found to be capable of contributing to the formation of an embryo (Bradley et al., 1984). Eventually, genome-wide mutational strategies, such as retroviral gene disruption and chemical mutagenesis, were applied to mutant mouse production. These random mutagenesis strategies have proven to be extremely efficient when used in conjunction with viral-insertion site-mapping techniques (Zambrowicz et al., 1998) and robust point mutation mapping techniques (Hrabe de Angelis et al., 2000). Using the technologies of gene trapping and gene targeting by homologous recombination, Lexicon Genetics has generated and completed the phenotypic analysis of more than 2600 knockout mouse lines to date, which represent approximately 2500 distinct genes. Over the last four years, gene function in mice has been explored by focusing efforts on reducing the amount of functional transcript of a given gene of interest. This method, known as RNA interference, takes advantage of the endogenous cellular machinery dedicated to destroying double stranded RNA resulting from viral infection. For the first time in the history of modern science, scientists now have access to a molecular toolbox filled with unique instruments designed to alter the functional expression of genes in a mammal. Based on the physiological consequences of a given genetic alteration in the mouse, scientists can predict the function of the orthologous gene in man. Coupled with drug development capabilities, these inferences can translate to therapeutic treatments for human disease (Zambrowicz and Sands, 2003; Zambrowicz, Turner, et al., 2003).

II. GENERATION OF TRANSGENIC ANIMALS FOR TARGET VALIDATION

A. Gene Targeting by Homologous Recombination

Gene targeting by homologous recombination affords deliberate and precise manipulation of genomic sequence in the mouse. Gene targeting in mammalian cells is an exceedingly rare event and remains a bottleneck in the entire process of knockout mouse generation. Approximately 99% of the targeting vectors introduced into ES cells will insert randomly in the genome, as opposed to integrating homologously into the target sequence (Joyner, 1999). Since the early years of gene targeting, efforts have been focused on identifying key variables that might be optimized to achieve higher gene-targeting rates. For example, the use of positive as well as negative selection protocols to reduce the survival of cells containing random integrants (Mansour et al., 1988), using isogenic DNA within targeting sequences (te Riele et al., 1992), increasing the homology arm length (Deng and Capecchi, 1992; Hasty et al., 1991), and limiting the deletion size (Joyner, 1999) have each contributed to overall improved rates of success.

Previously, Lexicon described a λ phage knockout shuttle (λ KOS) system for streamlined construction of targeting vectors (Wattler et al., 1999). This system employs a murine 129/SvEvBrd λ phage genomic library containing negative selection cassettes flanking the genomic inserts (Figure 1.1A). The genomic inserts are 9 kb average length, which allows for deletion strategies upward of 6 kb or greater, while maintaining conveniently sized arms of homology that allow for efficient targeting as well as ease of screening for

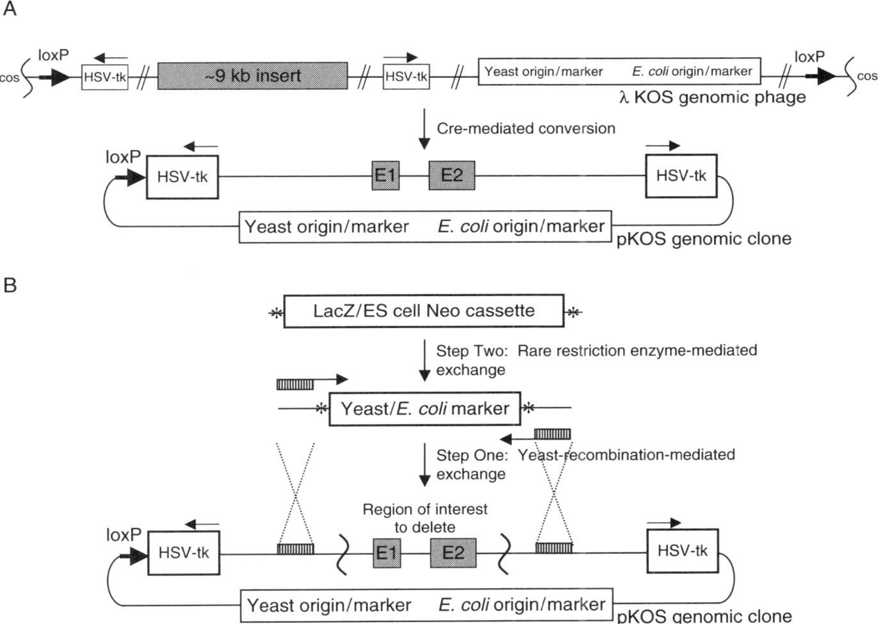

FIGURE 1.1 Construction of gene targeting vectors using the λ KOS system. (A) A schematic of a genomic KOS clone before and after CRE-mediated excision. The positions of the bacterial and yeast origins of replication and selectable markers are shown. The orientation of the HSV-tk genes (negative selection cassettes) and the loxP sites are indicated by arrows. In this illustration, the genomic clone contains exons 1 and 2 of a gene to be targeted. (B) A selection cassette containing both yeast and bacterial selectable markers and flanked by several rare-cutter restriction sites (noted by asterisk*) is PCR amplified using primers containing at their 5′ end 25 to 40 bp of sequence homology flanking the genomic region to be deleted (hatched boxes). In this schematic, a deletion of exons 1 and 2 is planned. Yeast-mediated recombination replaces the genomic region to be deleted with the amplified double-selection cassette. To complete the final targeting vector, the yeast selection cassette is replaced with any desired ES cell-selection cassette using the rare-cutter restriction sites in a single cloning step. The ES selection also contains, in most cases, a reporter gene (LacZ), which can be used to define the expression patterns of the gene targeted (Wattler et al., 1999).

correct recombination events by Southern blot or long-range PCR. The library has a nine-fold genomic coverage and is arrayed in a 96-well format such that any gene of interest can be recovered efficiently by PCR. To date, we have yet to identify a single gene that was not represented in the library, and more than 2500 genes have been successfully retrieved from this resource. The λ phage clone is readily converted to a high copy plasmid via Cre-mediated excision (Figure 1.1A). In our experience, the genomic plasmids are stable and propagation robust in common *E. coli* laboratory strains. Rare base-cutter restriction enzyme sites are placed at the 5′ and 3′ boundaries of the deletion region by introduction of a yeast selection cassette via an efficient homologous recombination event in yeast. Finally, a single directional cloning step replaces the yeast selection cassette with an ES cell selection cassette (Figure 1.1B). Our ES cell selection cassettes usually contain a reporter gene, typically LacZ, which can be used to determine the temporal and spatial expression patterns of the targeted gene. The λ KOS system has proven to be robust in terms of both efficiency and scalability, and we have employed this system to generate by far the world's largest collection of custom mutant mouse lines.

Of note, in the years pre-dating the completion of sequencing the human and mouse genomes, a partial cDNA tag of a given gene of interest was sufficient for engineering virtually any targeting vector utilizing our system. Several overlapping genomic clones would be isolated from the λ KOS library using exonic probes derived from the cDNA of interest. The genomic plasmid clones would be shotgun sequenced to create a large genomic contig, typically greater than 15 kb in length, within which exons could be defined and the mutation strategy designed. Importantly, external probe sequences, restriction maps, and Southern blot strategies could be derived from the large genomic contigs (Figure 1.2A and 1.2B). Meanwhile, the recent completion of the sequencing of the human and mouse genomes (Lander et al., 2001; Venter et al., 2001) has afforded the advantage that we can now upfront design any mutation strategy with a robust screen design *in silico*, and then retrieve an appropriate genomic clone from the λ KOS library to begin vector construction. This has resulted in saving much time that was previously spent troubleshooting Southern blot screening strategies.

Recently, we have optimized a PCR-based targeting vector engineering strategy, taking advantage of the sequenced mouse genome as well as newly available, high-processivity DNA polymerases (Figure 1.3A). This PCR-based approach allows us much greater flexibility in mutation design, as we are no longer limited by the size or the placement of the KOS clones. The strategy utilizes a single directional cloning step whereby the two homology arms, the ES cell selection cassette, and the pKO Scrambler NTKV-1901 vector backbone (Stratagene, La Jolla, Calif.) are ligated together. This approach can significantly decrease the amount of hands-on manipulations and streamlines the overall timeline to final targeting vector construction, from an average of 10 weeks for the λ KOS system to as few as 2 weeks for the PCR-based system. In addition, we have adapted the use of long-range PCR as a method to rapidly screen for targeted ES cell clones in a 96-well format (Figure 1.3B). We screen for targeting events by PCR using one primer outside of the arm of homology and a second primer within the selection cassette. Amplified bands are sequence-confirmed, and the corresponding ES cell clones are expanded for final Southern blot confirmation prior to microinjection into blastocysts. The PCR-based approach for identifying correctly targeted ES cell clones compares

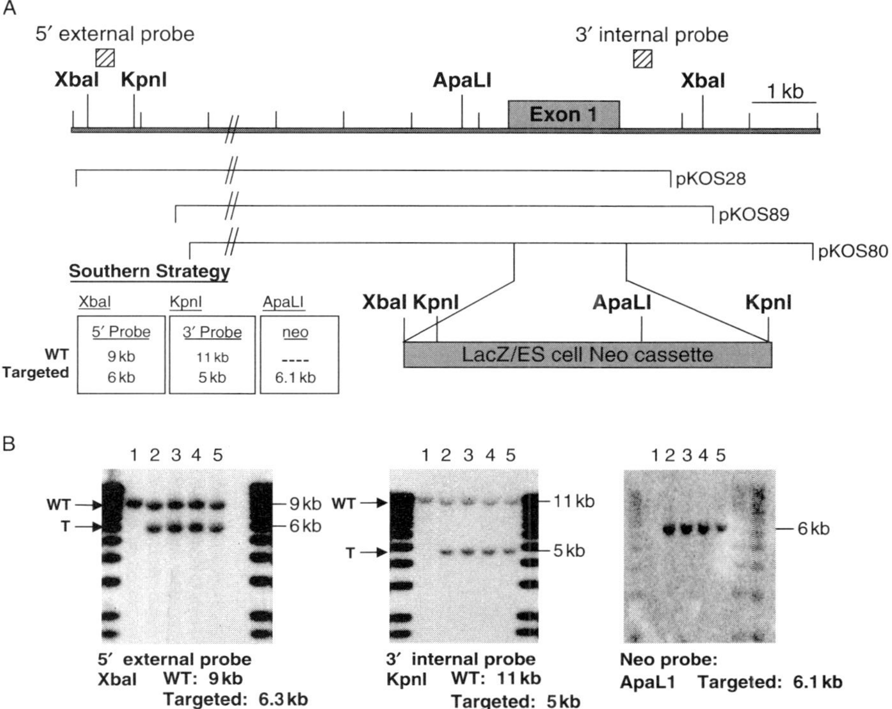

FIGURE 1.2 Construction of a large genomic contig using overlapping KOS clones. (A) A schematic illustrating multiple KOS isolated with an exonic probe derived from the gene of interest. Prior to the sequencing of the mouse genome, these clones would be transposon sequenced to generate a complete working contig in regions where no public sequence existed. External and internal probes could then be amplified and tested on genomic DNA to determine their suitability as probes for the final confirmation of targeted ES cells (B). After electroporation of the target vector, ES cell clones are screened in 96-well format to identify targeted clones. Those clones are then expanded for final confirmation before microinjection into blastocysts.

favorably to Southern blotting as it can be completed quickly and does not require the use of radioactive probes. One disadvantage of the long-range PCR approach for both target vector construction and targeted ES cell clone screening is that time-consuming troubleshooting steps may occasionally be necessary. Based on our experience, the need for PCR troubleshooting often seems to correlate with repetitive sequence stretches or overall high GC content of the target sequences. Nevertheless, having access to the λ KOS and the PCR-based vector construction as well as the targeted clone screening approaches greatly increases our overall throughput.

As of August 2005, almost 2800 gene targeting projects representing over 2400 distinct genes have been initiated at Lexicon, and over 2000 projects have achieved germline transmission. As shown in Figure 1.4, we have focused our gene targeting efforts on druggable gene classes including channels, enzymes, G protein-coupled receptors (GPcRs), kinases, membrane-associated and integral membrane proteins, proteases, secreted proteins, and transporters. We launched a concerted effort to knock out all GPcRs several years ago, and we have virtually completed our phenotypic screen of the GPcR gene class, excluding the odorant and pheromone receptors. Our gene targeting technologies have proven to be equally efficient for mutating all gene classes, with an

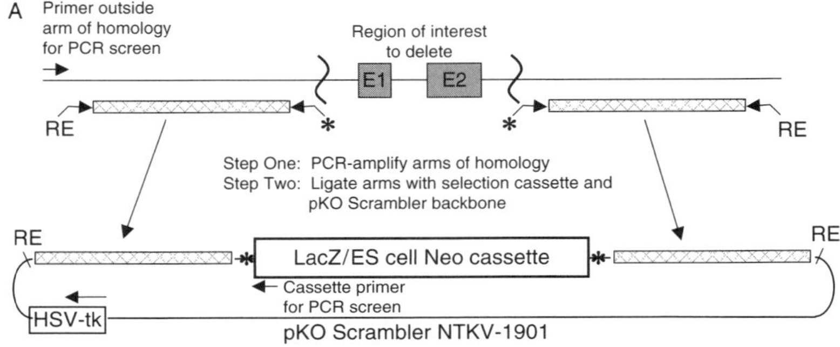

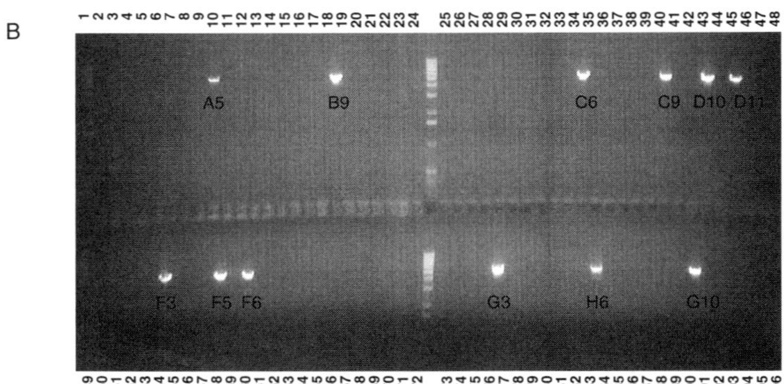

FIGURE 1.3 **Construction of gene targeting vectors and ES cell screening using long-range PCR.** (A) Utilizing multiple bioinformatic tools and the public mouse genome sequence, a target gene is scrutinized to determine the best mutation to create. Once a target region is identified, PCR primers are designed to amplify arms of homology 5′ and 3′ of the deleted region (long hatched boxes). The primers immediately adjacent to the deleted region contain sites for rare-cutter restriction enzymes (noted by asterisk*) and the distal primers contain restriction sites for cloning into the pKO Scrambler NTKV-1901 vector (RE). The arms are typically 3- to 5-kb long and are amplified from genomic DNA prepared from the same ES cells used for electroporation to ensure isogenicity. After amplification, the homology arms are ligated in a four-piece reaction to the ES cell-selection cassette and the Scrambler backbone. (B) After electroporation of the target vector, ES cell clones are screened in 96-well format using long-range PCR to identify targeted clones. A primer is placed outside one of the homology arms and used with a primer inside the selection cassette. Positive bands are sequence-confirmed and the clones expanded for final confirmation as shown in Figure 1.2 prior to microinjection.

average of 6.43% targeting efficiency (Figure 1.5). This compares favorably to published results, including a recent report describing an average 3.8% targeting efficiency for 200 genes (Valenzuela et al., 2003). Surprisingly, utilizing lengthy homology arms derived from bacterial artificial chromosomes (BACs) did not afford any advantage in terms of targeting efficiency compared to our system. While the use of large homology arms does allow for very large deletions, it also mandates that a great deal of additional scrutiny is applied to the mutant locus to ensure deletions and rearrangements have not occurred across the entire homology arm length.

We have built a sizable gene targeting dataset compiled from 1860 targeting vectors, consisting of 5′ and 3′ homology arm sequence, the sequence of the deleted region, and the percent of targeting events based on an average of 400 ES cell clones screened by 96-well Southern blot or PCR per targeting vector. As previously reported by others, whose data was based on a

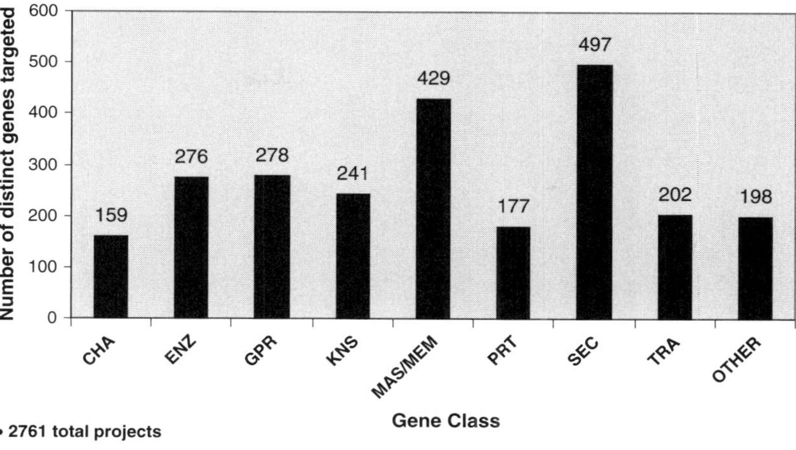

FIGURE 1.4 Genes targeted by homologous recombination at Lexicon. As of August 2005, 2761 total gene targeting projects have been initiated representing 2457 distinct genes. The number of total projects initiated includes null mutations, conditionals, humanizations, and point mutations. More than 2000 gene targeting projects have achieved germline transmission at this time. The total number of unique genes is broken down by gene class here. CHA = channels; ENZ = various types of enzymes, including phophodiesterases and phosphatases; GPR = G protein-coupled receptors; KNS = kinases; MAS = membrane and secreted proteins; MEM = membrane proteins; PRT = proteases; SEC = secreted proteins; TRA = transporters; OTHER = other gene classes not considered "druggable."

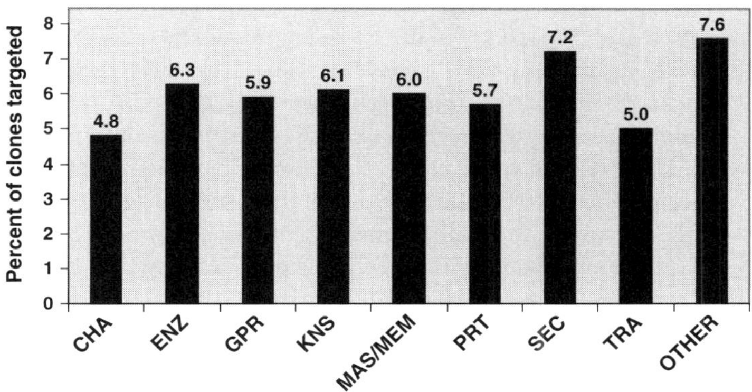

FIGURE 1.5 Targeting efficiency by homologous recombination broken down by gene classes. A dataset comprising 1860 target vectors was complied and used to identify variables that may or may not correlate with targeting efficiency. The dataset is 86% KOS vectors and 14% Scrambler vectors for standard null and conditional projects only. The overall average efficiency in this dataset is 6.43%. The efficiency across gene class is broken down and depicted here. CHA = channels; ENZ = various types of enzymes, including phophodiesterases and phosphatases; GPR = G protein-coupled receptors; KNS = kinases; MAS = membrane and secreted proteins; MEM = membrane proteins; PRT = proteases; SEC = secreted proteins; TRA = transporters; OTHER = other gene classes not considered "druggable."

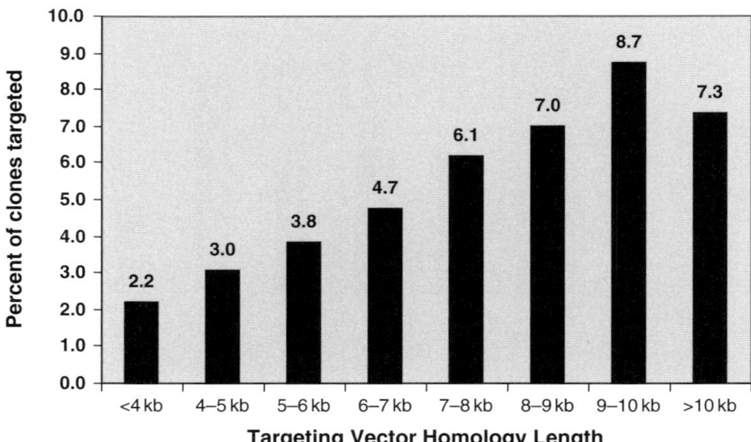

FIGURE 1.6 Total homology positively correlates with targeting efficiency. Total homology refers to the lengths of the 5′ and 3′ arm combined. The average length of homology in the dataset vectors is 7.6 kb, with a high of 12.3 kb and a low of 2.4 kb.

significantly smaller number of targeting vectors, we also observed a positive correlation between increasing overall homology sequence length and gene targeting efficiency (Figure 1.6). Not surprisingly, given that total homology length and deletion size are inversely dependent variables in our λ KOS system, we observed that deletion size negatively correlates with gene targeting efficiency (Figure 1.7). We have scrutinized several variables within our gene targeting dataset with the idea that certain criteria, such as sequence composition or relative position of the intended deletion within a genomic context, might correlate with gene targeting efficiency and therefore elucidate additional core principals to improve the efficiency of gene targeting. Disappointingly, we were unable to find any correlation with targeting efficiency between multiple variables including repeat content, distance from ATG or transcription start site, whether or not the gene was expressed in ES cells, or distance from the nearest gene. Nevertheless, targeting efficiency by homologous recombination is undoubtedly influenced by locus-specific variables we do not yet understand. For example, small changes in homology arms can dramatically increase targeting efficiency (Lexicon unpublished data).

While there have been no major recent advances resulting in significant increases in gene targeting rates since the introduction of isogenic DNA and positive negative selection, many groups have reported new techniques aimed at streamlining the steps of target vector construction and screening for targeting events (Lee et al., 2001; Valenzuela et al., 2003; Wattler et al., 1999). For the most part, these new vector construction approaches have employed either yeast or bacterial recombination systems to avoid tedious cloning steps otherwise dependent on native restriction sites. In addition, new high-fidelity DNA polymerases and real-time PCR have been deployed to speed the screening steps. Despite the many challenges involved with engineering gene knockouts,

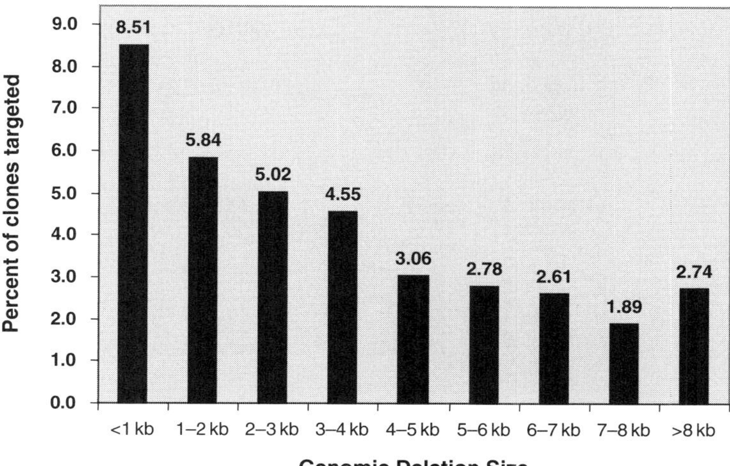

FIGURE 1.7 Deletion size negatively correlates with targeting efficiency. The average deletion size in the dataset is 2.0 kb, with a high of 20.6 kb and a low of 41 bp.

targeted knockouts for approximately 10% of the known 25,000 murine genes have already been reported in the literature (Austin et al., 2004). To date, Lexicon has achieved germline transmission for more than 2000 lines created by gene targeting, and a sizeable portion of the over 2600 knockouts having completed phenotypic analysis to date were generated by gene targeting. We have demonstrated that it is possible to use gene targeting to create knockout lines on a large scale and in a high-throughput manner.

In addition to serving as the workhorse technology to create standard null alleles for our large-scale screen for valid drug targets (see Section IIE. Genome5000™), the λ KOS system has also proven to be extremely adaptable to designing diverse genetic alterations, such as conditional floxed alleles, knock-ins, and humanizations (Figure 1.8). We have created custom intermediate cloning vectors that allow for directional cloning of the floxed or knock-in cassettes in conjunction with rare-cutter restriction enzyme sites introduced into the genomic clone by a recombination step in yeast (Wattler et al., 1999). A minor variation of this approach also allows for the introduction of point mutations into exons.

We routinely generate a conditional floxed allele and a humanized allele for our advanced drug discovery programs. The conditional allele is typically crossed to a protamine-Cre transgenic line, which excises the floxed region at an early stage of spermatogenesis. Importantly, this provides us with an independent confirmation of the function and potential clinical utility of the target. In addition, the conditional allele can be crossed to various tissue-specific or inducible Cre lines as appropriate. The humanized lines are subjected to in-depth biochemical studies to confirm that the human transgene supports functional expression of active protein. The humanized lines are also submitted to our comprehensive clinical evaluation to confirm that mice are normal

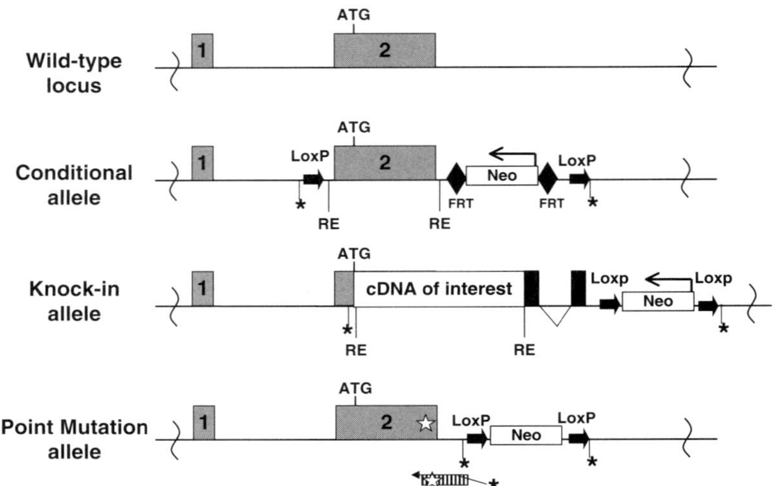

FIGURE 1.8 Other types of mutations can be made by gene targeting. Although the majority of gene targeting vectors are designed to create null alleles, our technology allows for the generation of other mutations as well. **Conditional alleles:** A custom universal cloning cassette has been created that allows for the directional cloning of a genomic region to be flanked by loxP sites. To make this type of mutation, the KOS clone is first subjected to yeast-mediated recombination to introduce the rare-cutter restriction sites (noted by asterisk*) in the appropriate location, (Wattler et al., 1999), or arms of homology are PCR-amplified carrying the rare-cutter sites. The region to be floxed is cloned into the intermediate vector using other restriction sites (RE). Then that intermediate cassette is cloned together with the arms of homology utilizing the rare-cutter sites. The positions of the loxP sites are indicated with arrows. The neo selection cassette is flanked with FRT sites (diamonds) to allow for FLIP recombinase-mediated excision later if desired. **Knock-in or humanization alleles:** The open reading frame of a cDNA of interest can be knocked-in to either a ubiquitously expressing locus (e.g., HPRT or ROSA-26) or to another locus. When a human cDNA is knocked-in to and disrupts the corresponding mouse locus, the mutation is termed a humanization. This allele again begins by PCR-amplifying arms of homology with rare-cutter restriction sites in the desired genomic positions. The cDNA of interest is cloned into another custom universal vector using other directional RE sites. This intermediate vector includes two exons of a minigene (black boxes) to provide a splicing event (enhancing expression) and also a floxed neo selection cassette (loxP sites as arrows). The intermediate cassette is then cloned together with the arms of homology utilizing the rare-cutter sites. **Point mutations:** Individual amino acid residues (white star) can be mutated by incorporation of the desired base changes into the primer used to PCR the 5′ arm of homology. This primer also carries the rare-cutter site (noted by asterisk*) that, when coupled with the rare-cutter site incorporated into the 3′ arm of homology, allows for the cloning of an ES cell-selection cassette into the adjoining intron. The selection cassette can be subsequently excised using CRE-recombinase.

in every measurement as compared to wild-type controls. The generation of humanized mouse lines can be challenging because the human transgene must be able to respond to normal regulatory signals governing transcription, splicing, translation, and protein processing, within the context of the partially deleted mouse genomic locus. Furthermore, the human protein must be able to perform all normal functions within the murine cellular context. A fully validated humanized mouse line can be used as a preclinical model in the evaluation of lead compounds that are meant to agonize or antagonize the human target, providing important information at an early stage of drug development. When successful, this approach can result in tremendous time and cost savings for a preclinical program.

B. Gene Trapping

Gene trapping provides a high-throughput approach for producing insertional mutations in genes (Stanford et al., 2001). When gene trapping is carried out in mouse ES cells, these cells can subsequently be used to produce mutant mice. Gene trapping vectors are delivered to ES cells most often by infection with a retrovirus or by electroporation. Upon entry into the cell, the vector inserts randomly into the genomic DNA, in some cases inserting into and mutating genes. The vectors are designed so that insertion events in genes result in the expression of a selectable marker gene. Subsequent selection allows the growth of ES cell colonies from clones containing a single gene trap insertion event. The gene trap event not only mutates the gene; it also marks the gene, making it possible to identify the sequence of the mutated gene using a variety of techniques.

The advantage of the gene trapping approach over gene targeting by homologous recombination is that a single vector can be used to mutate many thousands of genes, thereby eliminating the molecular biology steps required to make a targeting vector for each gene. Screening of ES cells for a rare targeted event is also largely eliminated since with gene trapping every clone of ES cells that is isolated has a mutation in a gene. The primary disadvantages of gene trapping are that 1) it is not possible to choose ahead of time what gene will be mutated, and 2) while it is an excellent approach for producing null mutations, it is currently not yet possible using gene trapping to produce more subtle mutations such as point mutations or conditional alleles, though several groups are actively exploring the possibilities of conditional trap vectors (Schnutgen et al., 2005; Xin et al., 2005). Most commonly, gene trapping vectors contain a promoterless selectable marker gene preceded by a strong splice acceptor sequence and followed by a polyadenylation sequence (Figure 1.9). The selectable marker gene will not be expressed unless it inserts within the exons or introns of a gene that is expressed in ES cells. Upon inserting into an expressed gene, the selectable marker gene is transcribed, allowing for the selection of the gene trap event. The expression results from upstream exons of the trapped gene splicing into the splice acceptor sequence of the gene trap vector. This also is the mechanism that results in disruption of the gene as the trapped gene's exons upstream of the gene trap insertion preferentially splice into the strong splice acceptor sequence of the gene trap vector. The strong splice acceptor prevents these upstream exons from splicing into the endogenous exons downstream of the gene trap insertion, thereby abolishing the formation of the normal gene transcript. Gene trapping vectors are very efficient at producing null mutations. We have found, based on the analysis of more than 1200 lines of mice produced using our gene trap library, that as long as the gene trap vector inserts within the exons or introns of a gene, the trapping event results in severe transcriptional disruption. Ninety-six percent of the events result in complete loss of the endogenous transcript (functional knockout) and the other 4% result in greater than 95% reduction in transcript abundance (hypomorph) (Zambrowicz, Abuin, et al., 2003).

One of the limitations of this typical gene trapping approach is that it relies on expression of the trapped gene in ES cells in order to select for the trapping event. This prevents the trapping of genes that are not expressed in ES cells. We have, however, trapped more than 60% of mouse genes using a 5′ trapping strategy (Zambrowicz, Abuin, et al., 2003), and others have used similar 5′ trapping strategies to trap an additional 6% of genes not found in

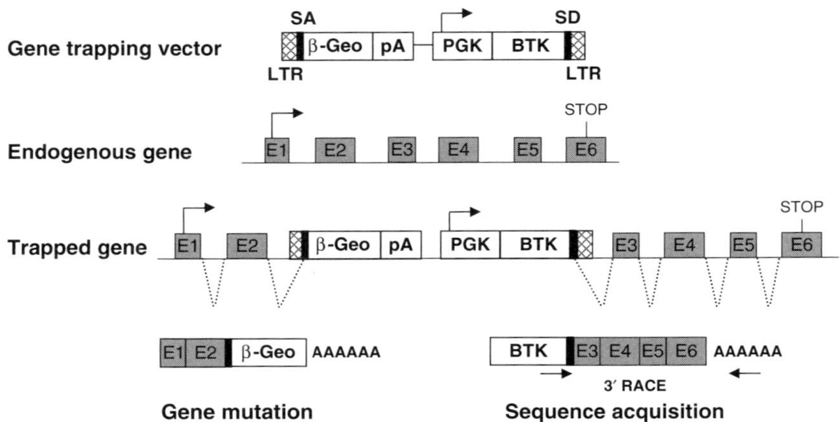

FIGURE 1.9 **Gene trapping using a vector that allows simultaneous gene mutation and insertion site identification.** A retroviral vector contains a splice acceptor sequence (SA, heavy black line) followed by a promoter-less selectable marker such as β-Geo, a functional fusion between the β galactosidase and neomycin resistance genes with a polyadenylation signal (pA). Insertion of the retroviral vector into an expressed gene (exons E1-E6 in this illustration) leads to the splicing of the endogenous upstream exons (E1-E2) into this cassette to generate a fusion transcript. The vector also contains a promoter that is active in ES cells, such as that of the mouse phosphoglycerate kinase (PGK) gene, followed by a first exon (such as that of the Bruton's Tyrosine Kinase gene, BTK) upstream of a splice donor (SD, heavy black line) signal. Splicing from this signal to the exons downstream of the insertion gives rise to a fusion transcript that can be used to generate a sequence tag of the trapped gene by rapid amplification of cDNA ends (RACE) (Zambrowicz, Friedrich, et al., 1998). The BTK exon contains termination codons in all reading frames to prevent translation of downstream fusion transcripts. Within the inserted proviral element, the LTRs (hatched boxes) are 590 bp of sequence that form the terminus of the retroviral element. They contain the promoter and enhancer elements for the retrovirus, which allow it to transcribe the retroviral RNA (that is packaged within viral coat proteins and infects the ES cells). The splice acceptor is placed slightly within the retroviral backbone; the splice donor is within the BTK cassette. (This figure was adapted from Zambrowicz, Abuin, et al., 2003.)

our library (Skarnes et al., 2004). This suggests that a minimum of 66% of all genes are transcribed at a sufficient level in ES cells to allow for selection of the trapping event using 5′ trapping vectors.

We and others have used a 3′ trapping and selection component in our trapping vectors in order to trap genes that are not expressed in ES cells (Zambrowicz, Friedrich, et al., 1998). In this strategy, the vector still contains the 5′ trapping component, but a 3′ trapping component is also added. The 3′ component consists of a strong ubiquitous promoter directing the expression of a selectable marker gene that lacks a polyadenylation sequence and instead contains a splice donor sequence. The 3′ trapping component is expressed only if it inserts within a gene and splices into exons downstream of the insertion that provide a polyadenylation sequence. Because the 3′ trapping component carries its own promoter it does not rely on insertion into an expressed gene in order to be expressed and provide selection for the gene trap event. We have used this type of vector to demonstrate that it is possible to trap genes not normally expressed in ES cells. The 3′ trapping strategy, however, leads to the recovery of a more limited number of trapped genes. We demonstrated this by testing vectors containing both a 5′ and 3′ trapping component in ES cells. The 5′ component contained a neomycin resistance gene as a selectable marker, and the 3′ component contained a puromycin resistance gene as the selectable marker. After infection of ES cells with these vectors and selecting with either G418 or puromycin, respectively, we found that the number of colonies obtained by 5′ trapping and G418 selection was always about 10- to 30-fold greater than the number of colonies obtained by 3′ trapping and puromycin selection. This suggests that the 5′ trapping component is much more efficient at trapping and allowing selection for a gene trap event, which could be because of better splicing of the 5′ trapping component or instability of the 3′ trapped transcript resulting in rapid degradation and an inability to select.

Shigeoka et al. (2005) suggested that the poor efficiency of the 3′ trapping event may be due to instability of the fusion transcript as a result of nonsense mediated decay, a process in which mRNA degradation is triggered by the presence of stop codons in the 5′ end of the transcript. They observed a bias in trapping events in that their 3′ trapping vector only allowed selection for integrations into the last introns of genes. They modified their vector and were able to prevent both nonsense-mediated decay of the 3′ fusion transcripts and the bias in vector integration sites. These investigators also suggested that their vectors permitted the trapping of genes that are transcriptionally silent in ES cells. However, we have trapped all three of the genes they provided as examples of transcriptionally silent genes using a 5′ trapping selection protocol that is dependant on expression of the trapped genes in ES cells. It remains to be determined whether these vectors have any effect on the target size of genes that can be trapped and selected using 3′ trapping cassettes. If 3′ trapping could be made more efficient, it could allow the trapping of all genes in ES cells.

Gene trapping has been used to make mutant mice and subsequently identify the mutated gene. Because this can be a laborious process, there have also been a number of strategies developed to pre-select gene trap events *in vitro* for specific types of genes before making mice. These include elegant approaches to screen for trap events in genes encoding secreted proteins (Skarnes et al., 2004), genes induced by a variety of factors (Forrester et al., 1996), genes expressed only in certain lineages (Stanford et al., 1998), and genes involved in specific biochemical pathways, such as glycosylation (Hubbard et al., 1994).

In the early to mid 1990s, the sequencing of large numbers of mouse and human ESTs empowered investigators in the gene trapping field to employ a nucleotide-based approach to identify what gene trap events to study in mice. With ESTs as the reference sequence, it was now possible to match a gene trap sequence tag to a gene sequence when exonic or coding sequence was obtained to identify the trapped gene. We and others have developed libraries of ES cells containing gene trap events in thousands of genes (Skarnes et al., 2004; Wiles et al., 2000; Zambrowicz, Friedrich, et al., 1998; Zambrowicz, Abuin, et al., 2003). These libraries are catalogued by a sequence tag that is used to identify what gene has been mutated in each clone. We developed an automated process using robotics and a 96-well format to produce a library of more than 272,000 sequence-tagged ES cell clones covering approximately 60% of all genes (Zambrowicz, Friedrich, et al., 1998). Our method involved a modified 3′ trapping cassette containing a promoter, a short 5′ noncoding exon, and a splice donor sequence. We selected for trapping events using a 5′ trapping cassette contained within the gene trap vector, but the modified 3′ trapping cassette allowed us to use a 3′ RACE approach for obtaining sequence tags to identify the trapped genes. We collected more than 522,000 ES cell clones and obtained an overall sequence acquisition success rate of about

52%. Others have used 5′ RACE or cloning of DNA flanking the insertion for identification of the trapped gene. An academic consortium has pooled their combined gene trap clones with sequence tags and these total about 27,000 (Skarnes et al., 2004).

Because sequencing of the mouse genome has been completed, other efficient methods can now be contemplated for characterizing gene trap clones. Exonic sequence information is no longer required since intronic sequence surrounding a gene trap insertion can now be mapped to the mouse genome and used to identify what gene has been trapped. We have developed a highly efficient automated 96-well approach to obtain genomic sequence flanking gene trap insertions using inverse PCR (Figure 1.10). This process is highly efficient and requires very little DNA from a trapped ES cell clone. Unlike RNA-based sequence acquisition approaches, the DNA from ES cells grown in a single well of a 96-well plate is sufficient to provide material for multiple attempts at cloning flanking sequence, allowing one to use multiple enzyme combinations for inverse PCR and increasing the success rate of sequence tag acquisition for a given gene trap event.

There is currently a great deal of interest in developing libraries of ES cells containing mutations in all mouse genes and establishing standardized phenotyping methodologies (Austin et al., 2004). To date, fewer than 25,000 protein-encoding genes have been identified in the mouse genome. The combined approaches of gene targeting and gene trapping provide the tools to create a comprehensive library of gene knockouts. The gene trap method is cost effective and has already been used to mutate at least 66% (Austin et al., 2004) of all genes based on the current published data. The enhancements

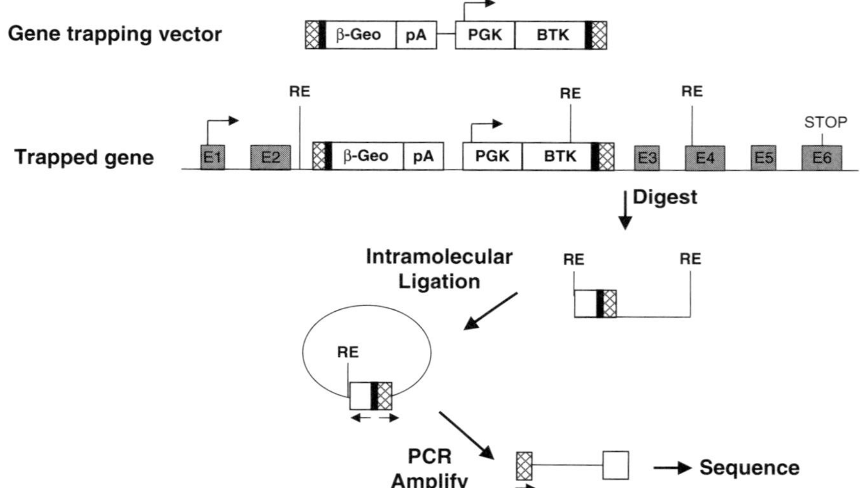

FIGURE 1.10 Inverse PCR allows sequence acquisition from genomic regions adjacent to an inserted gene trap vector. See Figure 1.9 for a detailed description of the components of the gene trap vector. Genomic DNA from individual gene trap mutant clones is cleaved with one or more restriction enzymes (RE) to produce a vector-genomic junction fragment that, when ligated, produces a circular template. Two vector-specific primers can be used to amplify the flanking genomic DNA using PCR. Products are sequenced using a vector-specific primer, and the resulting genomic sequences are then mapped to the mouse genome to pinpoint the exact insertion site of the vector.

in gene trap sequence tag acquisition, the clear data on the frequency of null mutations obtained with gene traps, and the ability to cover a significant portion of the genome make gene trapping an attractive approach for producing null mutations en masse. Genes not covered by gene trapping or genes for which a more sophisticated type of mutation is desired can be mutated by homologous recombination. Adding the approximately 66% of genes mutated in gene trap libraries worldwide, the over 2800 genes that have been knocked out by homologous recombination and published (Austin et al., 2004), the more than 2600 genes we have knocked out by homologous recombination in ES cells and other unpublished knockouts, it is likely that, regardless of some overlap, mutations in the vast majority of genes may already exist in ES cells.

Gene trapping has provided an excellent approach for producing mice with null or strong hypomorphic mutations in thousands of genes. Within the last couple of years, there has been speculation about whether gene trap vectors can be developed that would provide the opportunity to produce conditional knockouts. Published data related to this topic, however, is severely limited. In one of the few reports published to date on conditional gene inactivation using gene trapping, Hong-Bo et al. (Xin et al., 2005) describe vectors that enable inversion of a portion of the gene trap vector using Cre recombinase to control loss-of-function of the trapped gene. This work is based on the concept that in most intronic gene trap insertion events, it is the strong splice acceptor fused to a reporter gene and polyadenylation sequence (the gene inactivation cassette) that causes the loss of gene function. With this in mind, the investigators hypothesized that by flipping the orientation of the gene inactivation cassette, normal splicing could take place and the gene would not be inactivated. The group developed constructs that would begin with the gene inactivation cassette in the reverse orientation within an artificial intron of an expression vector for DsRED. The gene inactivation cassette was flanked by lox sites that would allow only unidirectional inversion in response to Cre recombinase. They demonstrated that in cell culture the addition of a Cre-expressing plasmid to cells containing the DsRED expression cassette resulted in inversion of the gene inactivation cassette and loss of DsRED expression. They also demonstrated that a transgene with a similar inversion cassette could be driven to inversion when mice containing the inversion cassette were crossed with a Cre-expressing mouse line. The vectors were not tested in any endogenous genes either in cell culture or *in vivo*, however. Schnutgen et al. (2005) have also shown promising results. Their conditional gene trap vectors employ two directional site-specific recombination systems (Cre & Flip) that, when activated in succession, invert the gene trap from its mutagenic orientation to a nonmutagenic orientation. This inversion is done *in vitro* ("repairing" the mutation with Flip) prior to microinjection, and then it can be reintroduced *in vivo* (with Cre) for a conditional mutation.

In both cases, conditional trapping has not been tested *in vivo*. The question remains whether conditional gene trap vectors will be reliable and of general utility. Although this approach may work in some cases, it is unlikely to work in all cases because many introns contain regulatory elements required for proper gene expression that could be disrupted by the gene trap insertion. In addition, it is unclear if insertions with an inverted gene inactivation cassette will be benign in all introns, even if they do not contain regulatory elements. To more reliably produce predictable conditional knockouts, it may be wiser to use a more standard gene targeting approach to flank a critical gene sequence with lox sites for subsequent elimination with Cre recombinase.

C. RNAi for Studying Mammalian Gene Function *In Vivo*

A technology that is currently of high interest for studying gene function *in vivo* is RNA interference (RNAi). The RNAi approach relies on an innate cellular surveillance mechanism used to combat viral infection (Dykxhoorn et al., 2003). RNAi was first discovered in *Caenorhabditis elegans* when dsRNAs were found to induce specific gene silencing. Subsequently, this cellular surveillance mechanism has been demonstrated to be widespread and present in a range of other organisms including plants and mammals. RNAi provides a post-transcriptional method of suppressing or knocking-down gene function through a sequence-specific degradation of messenger RNA (mRNA). In this process, double-stranded mRNAs are recognized and cleaved by the dicer RNase resulting in 21–23 nucleotide short interfering dsRNAs (siRNAs). These siRNAs are incorporated into and unwound by the RNA-inducing silencing complex (RISC). The single antisense strand then guides the RISC to mRNA containing the complementary sequence, resulting in endonucleolytic cleavage of the mRNA.

The potential advantages of the RNAi approach over traditional transgenic and knockout approaches are the increased speed, decreased cost, and possibility of dynamic knockdown in the adult animal. Delivery has been one of the bigger challenges, but advances have been made in the *in vivo* delivery of naked siRNAs as well as in delivery of constructs that express hairpin mRNA for viral delivery or to produce transgenic mice. Typically the hairpin RNA is expressed from a ubiquitous promoter, such as an RNA polymerase III promoter. The resulting animals express the hairpin RNAs that are recognized and processed by dicer into siRNAs that cause gene silencing. In this chapter, we will focus on the use of hairpin RNA constructs to produce transgenic mice to study gene function.

An alternative to direct delivery of RNAi *in vivo* has been to engineer DNA constructs that express short hairpin mRNA either delivered in viral vectors or used to produce transgenic mice. The vectors can be delivered by adeno- or adeno-associated virus or used to create transgenic animals by standard techniques, such as transfection of ES cells, pronuclear injection, or viral transduction of early embryos. Direct delivery of shRNA-expressing constructs by adenovirus has been used to knockdown gene function specifically in the liver of infected animals. One group was able to use adenovirus containing a construct expressing hepatitis B virus (HBV)-specific shRNAs to knockdown HBV gene expression and replication to almost undetectable levels in a line of HBV transgenic mice (Uprichard et al., 2005). A second group used adenovirus to suppress the aryl hydrocarbon receptor (AhR) expression in the liver of mice, resulting in protection from the lethal effects of Fas-induced apoptosis. The greater tropism of adeno-associated virus allows AAV to be used to suppress gene expression in a broader range of tissues. Haibin et al. used intracerebroventricular injection of AAV expressing shRNs specific for a mutant form of ataxin to inhibit polyglutamine-induced neurodegeneration in a mouse model of spinocerebellar ataxia (Xia et al., 2004). These viral delivery methods can allow one to test specific hypotheses related to organs and tissues susceptible to viral infection.

The creation of transgenic mice expressing shRNAs ubiquitously has provided a promising means of studying gene function throughout the organism. The transgenic approach was described in a publication by Kunath et al. (2003). They transfected mouse ES cells with constructs expressing shRNAs for GTPase-activating protein (RasGAP) and characterized the suppression of RasGAP mRNA in several cell lines. By choosing ES cell lines exhibiting dramatic reductions in RasGAP mRNA, they were able to produce mice that died during embryogenesis with defects similar to those of RasGAP knockout mice produced by homologous recombination. The pre-screening of ES cell lines for suppression of RasGAP mRNA was an important step in producing transgenic mice with a high degree of gene suppression. One might imagine that this would limit the approach to genes that are expressed in ES cells, but recently Lickert et al. (2004) demonstrated that the pre-screening approach may also be used for genes not expressed in ES cells. They produced ES cells containing a vector to express Baf60c-specific shRNA. Although Baf60c is a transcription factor expressed specifically in the heart and somites of early mouse embryos, the group was able to successfully pre-screen for ES cell clones that would produce a high level of suppression by differentiating the ES cells into cardiomyocytes and choosing those clones that exhibited the greatest suppression of Baf60c in cardiomyocytes for use in transgenic mouse production. This work suggests that it may be possible to pre-screen ES cells for maximum suppression of many genes not expressed in ES cells by using this *in vitro* differentiation strategy.

Lentiviral vectors provide an alternative method for the production of hairpin RNA-expressing transgenic mice in a process that bypasses the ES cell steps by using direct infection of early embryos. One example is the demonstration of an 88% to 94% reduction in CD8 mRNA levels in T cells from eight lines of transgenic mice produced using lentiviral delivery (Hemann et al., 2003; Rubinson et al., 2003). In both transgenic examples, potential variability in gene knockdown due to selection of gene sequences or sites of construct integration were overcome by pre-screening in cell lines. The variability in gene knockdown can provide the potential for creating an allelic series of mutations in a gene of interest, as was done with the p53 gene in hematopoietic stem cells (Hemann et al., 2003). Variable knockdown, however, indicates that pre-screening of siRNAs for their level of gene knockdown in cell culture may be an important step for proceeding on a gene-by-gene basis to produce knockdown mice.

Although this transgenic RNAi methodology provides advantages of speed in producing animals, it is not yet as well characterized as the knockout approach. To date, there remain only a limited number of reports of the use of RNAi to suppress gene function *in vivo* throughout the entire mammalian organism. There have not been detailed analyses done to determine how reliably these techniques can knock down gene function in all tissues at all times throughout the life of the animal. It remains unclear how much characterization would be required for each gene knockdown line to understand how well the technology has worked, and it is difficult to know what level of knockdown would be required for each gene to detect a phenotype. The lentiviral approach often results in multiple integration events (Hemann et al., 2003; Rubinson et al., 2003), and like any transgenic approach, mosaic expression of the shRNA may be expected. Although the production of knockdown mice by RNAi provides a potential for increased speed in production of animals, it must be balanced with the requirement for further data to determine if the technology is robust enough to represent a general strategy for *in vivo* studies. Importantly, recent data suggests that even short siRNAs may induce an interferon response in mammalian cells (Bridge et al., 2003; Sledz et al., 2003) and has raised questions about the absolute sequence specificity of gene silencing (Jackson et al., 2003; Scacheri et al., 2004). Because of these issues, the null allele produced at the DNA level remains an important benchmark for any study of gene function.

Perhaps the greatest theoretical advantage of using *in vivo* RNA knockdown would be to more accurately mimic the dynamics of drug action by inducing the knockdown in the adult. Dynamic knockdown could be accomplished using RNAi methods coupled with inducible promoters, such as those responsive to tetracycline or ecdysone. Inducible vectors have been developed for use in cell culture to regulate the knockdown event. If these technologies can be developed to produce *in vivo* induction of gene knockdown in the mouse, they would have great potential as a dynamic method for modeling drug action that would be highly valued by the pharmaceutical industry. Two recent reports indicate progress in this approach. Ventura et al. (2004) used a Cre-lox–regulated shRNA expression system to produce conditional knockdown mice. The expression of the shRNA was turned on in mice both tissue-specifically and ubiquitously by breeding the shRNA vector line with different Cre-expressing lines. Using shRNA vectors for CD8, they demonstrated a reduction in CD8+ splenic T cells in response to Cre expression. A second group used a combination of shRNA vectors dependant on Cre recombinase and the tet-inducible system to develop an inducible, tissue-specific knockdown approach in mice (Chang et al., 2004). Using this approach they knocked down the ABC1 gene and produced a Tangier's disease-like phenotype.

As the siRNA technology continues to advance, it may provide an excellent means of studying gene function *in vivo* as well as a new therapeutic approach for disease treatment. Last year, two reports indicated that by using longer siRNA duplexes, a more extended and potent suppression of gene expression could be obtained without the activation of the interferon response (Kim et al., 2005). The results suggest that by using siRNAs of 25 to 30 nucleotides rather than 21-mers, an up to 100-fold increase in potency can be obtained. It is important to note that these studies were carried out *in vitro* in cell culture; it will be interesting to see how these potential improvements translate for the *in vivo* use of siRNA.

D. Chemical Mutagenesis

Chemical mutagenesis has been used as a means of accelerating forward genetics, moving from phenotypic trait to identified gene, and can be used to generate mutant alleles that are useful for elucidating gene function (Balling, 2001; O'Brien and Frankel, 2004; Guenet, 2004). *N*-Ethyl-*N*-nitrosourea (ENU) is the principal chemical mutagen employed in mice and is used primarily as an *in vivo* mutagen of the male germline, although ENU has also been used to effectively mutagenize ES cells. Conditions for ENU mutagenesis in mice have been well established (Guenet, 2004; Justice et al., 1999). ENU induces lesions in DNA resulting in point mutations that can lead to complete loss of function (null), partial loss of function (hypomorph), increased function (hypermorph), and altered or gain of function (neomorph) alleles (O'Brien and Frankel, 2004). This variety of alleles can be useful in understanding the full spectrum of gene function.

Large-scale phenotype-driven approaches using ENU have been utilized to identify novel phenotypes in mice (Hrabe de Angelis et al., 2000; Nolan et al., 2000). Phenotypic screens to identify dominant traits have been most commonly employed as screening for recessive traits, requires breeding and analysis of considerably more mice. Utilization of mouse strains carrying balancer or marker chromosomes or strains with defined chromosomal deletions and inversions can help facilitate recessive screens (Balling, 2001; Hentges and Justice, 2004).

The primary advantages of a phenotype-driven ENU mutagenesis approach are the variety of alleles generated by ENU and that no prior knowledge of the target gene or pathway is required. The main drawback in using this approach to identify drug targets lies in the hurdles encountered in moving from the observed phenotypic trait to the underlying genetic mutation. Advances in high-throughput sequencing and mutation detection technologies have the potential to increase the efficiency of identifying the mutation responsible for the observed phenotype. Given the current state of ENU mutagenesis technology, however, gene targeting and gene trapping offer significant advantages over chemical mutagenesis as a means to identify tractable drug targets.

E. Genome5000™

Since the completion of the sequencing of the human genome in 2003, there has been a large effort to determine the biological function of all 20,000 to 25,000 protein-encoding genes contained in human DNA. These efforts have included work in proteomics, gene expression arrays, and mutational studies. Proteomics, which focuses primarily on the structure and function of the "protein universe" as well as interactions between proteins, offers the closest large-scale look at protein function, which ultimately defines the phenotype of a particular organism. By its very nature, proteomics is quite complicated and often involves the study of hundreds or thousands of protein-protein interactions. This level of complexity can make it difficult to accurately predict *in vivo* biological function in a systemic context.

Large-scale analysis of gene expression via microarrays also promises a way to classify gene function by observing when and where genes are turned on or off. Unfortunately, spatial and temporal gene activity does not always correlate with *in vivo* gene function. Consequently, large-scale gene expression analysis may not be a good predictor of *in vivo* gene function. This is frequently the case with secreted proteins that function as paracrine hormones. The functions of these secreted proteins are more accurately reflected by the expression of their cognate receptors. Large-scale mutational studies in mice offer one of the best ways to determine gene function in relation to the entire organism. Moreover, the high degree of genomic conservation between human and mouse—99% of genes in mouse have homologs in human (Waterston et al., 2002)—ensures that, more often than not, the biology between mouse and human directly correlate.

Recent reviews of the productivity of the pharmaceutical industry have overwhelmingly demonstrated that using knockout mice for drug target validation is a proven means to elucidate utility of drugs directed against the human target and predict clinical outcomes in patients (Giaever et al., 2002; Zambrowicz and Sands, 2003; Zambrowicz, Turner, et al., 2003). Lexicon

Genetics has established a program for knocking out and systematically discovering the physiological and behavioral functions of the 5000 genes encoding druggable proteins in the mammalian genome. This program, termed the "Genome5000™", aims to define the function of these genes via mutational analysis using both gene trapping and gene targeting approaches in mice. Genes encoding druggable proteins present in our gene trap library are knocked out by gene trapping, and those genes not in the library—or for which a more sophisticated type of mutation is desired—are knocked out by homologous recombination. We have made and analyzed more than 2600 knockout projects to date and, at our current run rate of 1000 knockouts analyzed per year, will complete the project over the next 2 years. The comprehensive nature of this *in vivo* approach allows for the discovery of gene function within the context of mammalian physiology at a scale that is unprecedented.

Lexicon is in a position to catalogue the physiological functions of 20% to 25% of all known mammalian genes. Ninety-seven percent of all yeast genes were knocked out by directed gene targeting in the single-celled organism *Saccharomyces cerevisiae* (Giaever et al., 2002). Although *S. cerevisiae* possesses only 25% to 30% of the number of genes compared to mouse or humans, it is interesting to note that 19% of the yeast genes were absolutely essential for life. More recently, the majority of the predicted 19,427 genes in the metazoan *C. elegans* have been inhibited by RNA interference using bacteria expressing dsRNA in feeding studies (Kamath et al., 2003). These investigators observed that 5.5% of genes were required for viability. They also noted that *C. elegans* genes known to have an orthologue in other eukaryotic species were more likely to have an observable phenotype when knocked down by RNAi, as compared to nonconserved genes. In particular,

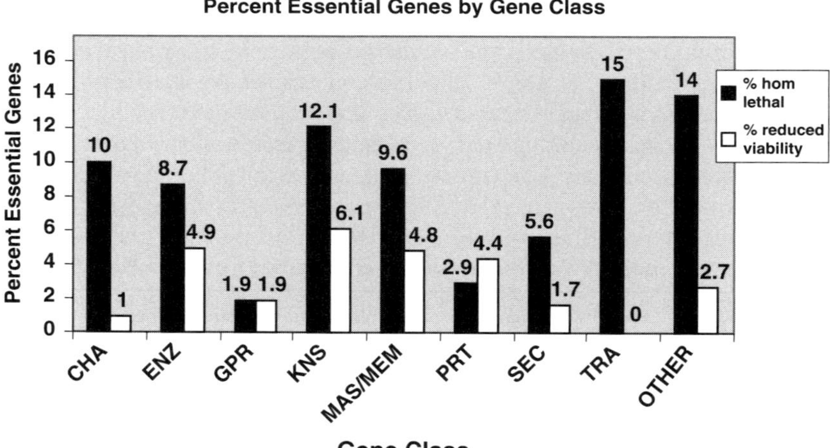

FIGURE 1.11 Percent essential genes by gene class. Homozygous lethal and reduced viability projects in a dataset consisting of 1250 completed gene targeting projects are broken down by gene class. Reduced viability lines are defined as those showing a total number of homozygous animals at 12.5% or less of the total F2 generation. (The expected value is 25% homozygotes in the total F2 generation.)

the essential *C. elegans* genes as a class are the most evolutionarily conserved. These observations from global gene ablation studies in yeast and worm are clearly relevant to human physiology because 50% of human genes have homologues in yeast and worms (Lander et al., 2001) and serve as a reminder that all eukaryotes share basic cellular machinery. In comparison to the percentage of yeast and *C. elegans* genes found to be required for life, we have observed an average 8% embryonic lethality rate and 3% reduced viability rate across all gene classes in a subset of 1250 targeted mouse gene knockouts (Figure 1.11). In agreement with observations from the global yeast and worm genome knockout studies, genes such as the GPcR class—which have relatively less time accumulated on their evolutionary clock—appear to encode for nonessential functions in large part, based on the relatively low number of lethal phenotypes we have observed in over 270 unique GPcR knockouts. Nevertheless, we have catalogued an expansive list of clinically and commercially relevant phenotypes amongst our GPcR knockout resource, which is not surprising given the fact that GPcRs are the largest drug target class based on currently marketed medicines.

III. CONCLUSION

The Genome5000™ project clearly demonstrates that the combined gene trapping and gene targeting approach provides a feasible means to knock out any desired set of genes or to contemplate a project encompassing all genes within the genome. By utilizing gene knockout technology on a genome-wide scale, we have the unprecedented capability to identify novel drug targets that are highly tractable and therapeutically efficacious. By focusing on the physiological function and pharmaceutical utility of genes at the outset of the drug discovery process, we increase the likelihood of success in discovering breakthrough treatments for human disease. Moreover, this approach is likely to reveal side effects caused by inhibiting or otherwise modulating the drug target. Such target-related side effects might limit the utility of potential therapeutics directed at the drug target or prove to be unacceptable in light of the potential therapeutic benefit. These advantages will contribute to better target selection and, therefore, to the success of our drug discovery and development efforts.

ACKNOWLEDGMENTS

Special thanks go to the following: To Ken Platt, Margaret Allen, Bill Paradee, Adisak Suwanichkul, Mike Donoviel, Emily Cullinan, Indrani Rajan, Jean-Pierre Revelli, Anne Phillips, Ragini Shankar, Chunmei Yang, Melissa Yang, Mike Kelly, Guixian Chai, Ken Coker, Mark Potter, Glenn Winnier, and Claire Gelfman for data collection necessary to generate the gene targeting dataset. To Buckley Kohlhauff and Qichao Zhu for data manipulation and statistical analysis of the dataset. To Alex Abuin and Gwenn Hansen for providing help with Figures 1.9 and 1.10. To Bill Sonnenburg, Peter Kipp, Mike Kelly, and J.D. Wallace for their comments on all or parts of the manuscript. To Kristi Lilleberg and Karly Birkenfeld for administrative assistance in preparing the final manuscript.

RECOMMENDED RESOURCES

MGI: http://www.informatics.jax.org/
Mouse Genome Informatics (MGI) provides integrated access to data on the genetics, genomics, and biology of the laboratory mouse.
UCSC Genome Bioinformatics: http://genome.ucsc.edu/
Ensembl: http://www.ensembl.org/index.html
NCBI: http://www.ncbi.nlm.nih.gov/

REFERENCES

Austin, C. P., J. F. Battey, et al. (2004). The knockout mouse project. *Nat. Genet.* 36(9): 921–924.

Balling, R. (2001). ENU mutagenesis: analyzing gene function in mice. *Annu. Rev. Genomics Hum. Genet.* 2: 463–492.

Bradley, A., M. Evans, et al. (1984). Formation of germ-line chimaeras from embryo-derived teratocarcinoma cell lines. *Nature* 309(5965): 255–256.

Bridge, A. J., S. Pebernard, et al. (2003). Induction of an interferon response by RNAi vectors in mammalian cells. *Nat. Genet.* 34(3): 263–264.

Chang, H. S., C. H. Lin, et al. (2004). Using siRNA technique to generate transgenic animals with spatiotemporal and conditional gene knockdown. *Am. J. Pathol.* 165(5): 1535–1541.

Cuenot, L. (1902). Notes et revues. *Arch Zool. Exp. Gen.* xxvii.

Deng, C. and M. R. Capecchi (1992). Reexamination of gene targeting frequency as a function of the extent of homology between the targeting vector and the target locus. *Mol. Cell Biol.* 12(8): 3365–3371.

Dykxhoorn, D. M., C. D. Novina, et al. (2003). Killing the messenger: Short RNAs that silence gene expression. *Nat. Rev. Mol. Cell Biol.* 4(6): 457–467.

Forrester, L. M., A. Nagy, et al. (1996). An induction gene trap screen in embryonic stem cells: Identification of genes that respond to retinoic acid in vitro. *Proc. Natl. Acad. Sci. U. S. A.* 93(4): 1677–1682.

Giaever, G., A. M. Chu, et al. (2002). Functional profiling of the Saccharomyces cerevisiae genome. *Nature* 418(6896): 387–391.

Gorer, P. Z. (1937a). Further studies on antigenetic differences in mouse erythrocytes. *Br. J. Exp. Pathol.* 18: 31–36.

Gorer, P. Z. (1937b). The genetic and antigenetic basis of tumour transplantation. *J. Pathol. Bacteriol.* 44: 691–697.

Guenet, J. L. (2004). Chemical mutagenesis of the mouse genome: an overview. *Genetica* 122(1): 9–24.

Hasty, P., J. Rivera-Perez, et al. (1991). The length of homology required for gene targeting in embryonic stem cells. *Mol. Cell Biol.* 11(11): 5586–5591.

Hemann, M. T., J. S. Fridman, et al. (2003). An epi-allelic series of p53 hypomorphs created by stable RNAi produces distinct tumor phenotypes in vivo. *Nat. Genet.* 33(3): 396–400.

Hentges, K. E. and M. J. Justice (2004). Checks and balancers: balancer chromosomes to facilitate genome annotation. *Trends Genet.* 20(6): 252–259.

Hrabe de Angelis, M. H., H. Flaswinkel, et al. (2000). Genome-wide, large-scale production of mutant mice by ENU mutagenesis. *Nat. Genet.* 25(4): 444–447.

Hubbard, S. C., L. Walls, et al. (1994). Generation of Chinese hamster ovary cell glycosylation mutants by retroviral insertional mutagenesis. Integration into a discrete locus generates mutants expressing high levels of *N*-glycolylneuraminic acid. *J. Biol. Chem.* 269(5): 3717–3724.

Jackson, A. L., S. R. Bartz, et al. (2003). Expression profiling reveals off-target gene regulation by RNAi. *Nat. Biotechnol.* 21(6): 635–637.

Joyner, A. L. (1999). *Gene targeting: a practical approach.* Oxford University Press, New York.

Justice, M. J., J. K. Noveroske, et al. (1999). Mouse ENU mutagenesis. *Hum. Mol. Genet.* 8(10): 1955–1963.

Kamath, R. S., A. G. Fraser, et al. (2003). Systematic functional analysis of the *Caenorhabditis elegans* genome using RNAi. *Nature* 421(6920): 231–237.

Kim, D. H., M. A. Behlke, et al. (2005). Synthetic dsRNA Dicer substrates enhance RNAi potency and efficacy. *Nat. Biotechnol.* 23(2): 222–226.

Kunath, T., G. Gish, et al. (2003). Transgenic RNA interference in ES cell-derived embryos recapitulates a genetic null phenotype. *Nat. Biotechnol.* 21(5): 559–561.

Lander, E. S., L. M. Linton, et al. (2001). Initial sequencing and analysis of the human genome. *Nature* 409(6822): 860–921.

Lee, E. C., D. Yu, et al. (2001). A highly efficient *Escherichia coli*-based chromosome engineering system adapted for recombinogenic targeting and subcloning of BAC DNA. *Genomics* 73(1): 56–65.

Lickert, H., J. K. Takeuchi, et al. (2004). Baf60c is essential for function of BAF chromatin remodelling complexes in heart development. *Nature* 432(7013): 107–112.

Mansour, S. L., K. R. Thomas, et al. (1988). Disruption of the proto-oncogene int-2 in mouse embryo-derived stem cells: A general strategy for targeting mutations to non-selectable genes. *Nature* 336(6197): 348–352.

Mulbock, O. and P. Bentvelzen (1968). The transmission of the mammary tumor viruses. *Perspect Virol.* 6: 75–87.

Nolan, P. M., J. Peters, et al. (2000). A systematic, genome-wide, phenotype-driven mutagenesis programme for gene function studies in the mouse. *Nat. Genet.* 25(4): 440–443.

O'Brien, T. P. and W. N. Frankel (2004). Moving forward with chemical mutagenesis in the mouse. *J. Physiol.* 554(Pt 1): 13–21.

Paigen, K. (2003). One hundred years of mouse genetics: An intellectual history. II. The molecular revolution (1981–2002). *Genetics* 163(4): 1227–1235.

Rubinson, D. A., C. P. Dillon, et al. (2003). A lentivirus-based system to functionally silence genes in primary mammalian cells, stem cells and transgenic mice by RNA interference. *Nat. Genet.* 33(3): 401–406.

Scacheri, P. C., O. Rozenblatt-Rosen, et al. (2004). Short interfering RNAs can induce unexpected and divergent changes in the levels of untargeted proteins in mammalian cells. *Proc. Natl. Acad. Sci. U. S. A.* 101(7): 1892–1897.

Schnutgen, F., S. De-Zolt, et al. (2005). Genomewide production of multipurpose alleles for the functional analysis of the mouse genome. *Proc. Natl. Acad. Sci. U. S. A.* 102(20): 7221–7226.

Shigeoka, T., M. Kawaichi, et al. (2005). Suppression of nonsense-mediated mRNA decay permits unbiased gene trapping in mouse embryonic stem cells. *Nucleic Acids Res.* 33(2): e20.

Skarnes, W. C., H. von Melchner, et al. (2004). A public gene trap resource for mouse functional genomics. *Nat. Genet.* 36(6): 543–544.

Sledz, C. A., M. Holko, et al. (2003). Activation of the interferon system by short-interfering RNAs. *Nat. Cell Biol.* 5(9): 834–839.

Staats, J. (1980). Standardized nomenclature for inbred strains of mice: Seventh listing for the International Committee on Standardized Genetic Nomenclature for Mice. *Cancer Res.* 40(7): 2083–2128.

Stanford, W. L., G. Caruana, et al. (1998). Expression trapping: Identification of novel genes expressed in hematopoietic and endothelial lineages by gene trapping in ES cells. *Blood* 92(12): 4622–4631.

Stanford, W. L., J. B. Cohn, et al. (2001). Gene-trap mutagenesis: Past, present and beyond. *Nat. Rev. Genet.* 2(10): 756–768.

te Riele, H., E. R. Maandag, et al. (1992). Highly efficient gene targeting in embryonic stem cells through homologous recombination with isogenic DNA constructs. *Proc. Natl. Acad. Sci. U. S. A.* 89(11): 5128–5132.

Uprichard, S. L., B. Boyd, et al. (2005). Clearance of hepatitis B virus from the liver of transgenic mice by short hairpin RNAs. *Proc. Natl. Acad. Sci. U. S. A.* 102(3): 773–778.

Valenzuela, D. M., A. J. Murphy, et al. (2003). High-throughput engineering of the mouse genome coupled with high-resolution expression analysis. *Nat. Biotechnol.* 21(6): 652–659.

Venter, J. C., M. D. Adams, et al. (2001). The sequence of the human genome. *Science* 291(5507): 1304–1351.

Ventura, A., A. Meissner, et al. (2004). Cre-lox-regulated conditional RNA interference from transgenes. *Proc. Natl. Acad. Sci. U. S. A.* 101(28): 10380–10385.

Waterston, R. H., K. Lindblad-Toh, et al. (2002). Initial sequencing and comparative analysis of the mouse genome. *Nature* 420(6915): 520–562.

Wattler, S., M. Kelly, et al. (1999). Construction of gene targeting vectors from lambda KOS genomic libraries. *Biotechniques* 26(6): 1150–1156, 1158, 1160.

Wiles, M. V., F. Vauti, et al. (2000). Establishment of a gene-trap sequence tag library to generate mutant mice from embryonic stem cells. *Nat. Genet.* 24(1): 13–14.

Xia, H., Q. Mao, et al. (2004). RNAi suppresses polyglutamine-induced neurodegeneration in a model of spinocerebellar ataxia. *Nat. Med.* 10(8): 816–820.

Xin, H. B., K. Y. Deng, et al. (2005). Gene trap and gene inversion methods for conditional gene inactivation in the mouse. *Nucleic Acids Res.* 33(2): e14.

Zambrowicz, B. P., A. Abuin, et al. (2003). Wnk1 kinase deficiency lowers blood pressure in mice: A gene-trap screen to identify potential targets for therapeutic intervention. *Proc. Natl. Acad. Sci. U. S. A.* 100(24): 14109–14114.

Zambrowicz, B. P. and G. A. Friedrich (1998). Comprehensive mammalian genetics: History and future prospects of gene trapping in the mouse. *Int. J. Dev. Biol.* 42(7): 1025–1036.

Zambrowicz, B. P., G. A. Friedrich, et al. (1998). Disruption and sequence identification of 2,000 genes in mouse embryonic stem cells. *Nature* 392(6676): 608–611.

Zambrowicz, B. P. and A. T. Sands (2003). Knockouts model the 100 best-selling drugs—will they model the next 100? *Nat. Rev. Drug. Discov.* 2(1): 38–51.

Zambrowicz, B. P., C. A. Turner, et al. (2003). Predicting drug efficacy: Knockouts model pipeline drugs of the pharmaceutical industry. *Curr. Opin. Pharmacol.* 3(5): 563–570.

2 TARGET VALIDATION IN CHEMOGENOMICS

TOM Y-H. WU* **and SHENG DING**[†]

**Ph.D., Merck Frosst, Center for Therapeutic Research, Kirkland, Quebec, Canada*
[†]Ph.D., The Scripps Research Institute, La Jolla, California

Target validation in drug discovery generally involves characterizing the pharmacological response resulted from modulating the activity of a target protein. The chemogenomic approach utilizes small molecules as tools to establish the relationship between the target and the phenotype. This can be begun by investigating the biological activity of enzyme inhibitors (reverse chemogenomics) or by identifying the relevant target(s) of a pharmacologically active small molecule (forward chemogenomics). In either directionality, there are various methods that can be applied to better understand the target's functional role in a disease model. This chapter illustrates some of the strategies to identify and validate targets by chemogenomics.

I. INTRODUCTION

With the completion of several genome-sequencing projects, including that of humans (Venter et al., 2001), researchers now have an immense amount of genetic data. Coupled with modern informatics and molecular biology techniques, it has become standard laboratory practice to identify, quantify, clone, and modify any gene desired in a straightforward manner. Nevertheless, systematically and effectively establishing the link between genotype and phenotype still remains a major challenge. In drug discovery, this involves identifying genes or gene products (proteins) whose modulation will cause a favorable pharmacological response with minimal side effects. Recent advances in genetic manipulations (e.g., recombineering or RNA interference/RNAi technology) have allowed scientists to modulate a specific gene and evaluate the phenotypic consequences *in vitro* and *in vivo* in a relatively straightforward fashion (Lee and Threadgill, 2004). To develop a potent and selective small molecular tool for target validation, on the other hand, can be time- and resource-consuming. The molecular pharmacology approach to target validation, however, is often still the ultimate method of choice because chemical treatment is conditional, is compatible with virtually any *in vitro* and *in vivo* pharmacological models, and can be used directly as drug leads. Facilitated by the advances in high-throughput screening and

Target Validation in Drug Discovery

molecular profiling technologies, identifying functional small molecules and using them to discover/validate biological targets constitutes a basic practice for many pharmaceutical companies. Oftentimes, this type of research is broadly referred to as "chemical genomics" or "chemogenomics," especially in academic settings (Bleicher, 2002; Bredel and Jacoby, 2004; Savchuk et al., 2004).

As the traditional genetic methods of linking genotype to phenotype are classified into forward and reverse approaches, chemogenomics could also be fitted into these two categories (Figure 2.1). In forward genetics, the phenotype of interest is first observed and then the underlying genotype is traced out through genetic mapping. In contrast, reverse genetics starts out by manipulation of a particular gene, followed by examination of the phenotypic outcomes. Likewise, screening compound libraries for small molecules that can generate a biological phenotype of interest and then identifying the target(s) of the active small molecule is referred to as forward chemogenomics, whereas reverse chemogenomics is the use of chemical probes with known molecular targets to study the pharmacological response of modulating their targets in cellular or *in vivo* models. Although differing by their directionality, both forward and reverse chemogenomics make uses of chemical compounds as tools for a common purpose: to establish the link between the molecular target and the phenotype. This chapter will be divided into two sections to address the different approaches to identify and validate drug targets using chemogenomics. Because this field of research is broad, only selected representative examples will be highlighted.

II. REVERSE CHEMOGENOMICS: TARGET VALIDATION USING COMPOUNDS WITH KNOWN MOLECULAR TARGET (AND/OR MECHANISM OF ACTION)

Small molecules can serve as useful chemical tools to facilitate the process of target validation. In most small molecule-based drug discovery programs

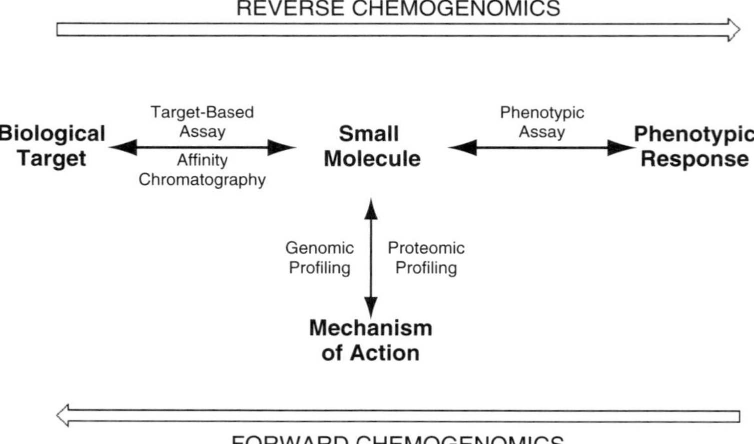

FIGURE 2.1 Overview of target validation in chemogenomics. Small molecules are used as tools to discover and validate drug targets by establishing the functional relationship between the biological target, the mechanism of action, and the phenotypic response through various approaches.

today, the target is initially identified/validated through genetics and then further validated using drug-like molecules. For example, if the genetic manipulation (e.g., gain or loss of function) of a certain target led to a favorable outcome in a disease model, then this would prompt the development of a small-molecule modulator of that target to further validate the therapeutic effect of such pharmacological modulation *in vivo*. Often these chemical modulators are enzyme or receptor agonists/antagonists that could be either designed from structural modification of the target's natural ligand or substrate or identified from screening diverse compound libraries. These small molecules would then have to be optimized for affinity, selectivity, and ADMET (absorption, distribution, metabolism, excretion, and toxicity) properties, so that they could be used to conditionally modulate the activity of their targets under biologically relevant settings (e.g., in cells or animals). In this sense, drug development can be viewed as a process of target validation, from the molecular target to its pharmacological validations in cells, animals, and eventually controlled human clinical trials.

A. Development of Potent and Selective Small Molecules via Target-Based Assays

Target-based assay establishes the functional link between target and compound by measuring the compound's ability to modulate the target, which often, but not always, is a purified protein. Often, screening against several protein members in the same family is required to determine compound selectivity. High selectivity of a compound is generally needed to avoid any undesired off-target effect.

Using the kinase drug-discovery program as an example, the selectivity profile of a kinase inhibitor often determines its potential therapeutic window. While conventional means of selectivity determination relying on enzymatic assay of individual kinase is still widely used, researchers have developed an interesting new strategy to test compounds against a panel of hundreds of kinases simultaneously (Fabian et al., 2005). This novel experimental approach involves the expression of kinases on T7 bacteriophage particles as a way to tag each protein with its own genetic information (Rodi and Makowski, 1999). A series of ATP-competitive probes (e.g., nonspecific kinase inhibitor, staurosporine) are immobilized through a flexible linker to act as "baits" for the phage-tagged kinases. The compound of interest then competes with the probes for the ATP-binding sites of its target kinases. The higher the compound affinity toward a particular kinase, the more that kinase will be competed off from the immobilized probes. Analysis of the kinases bound to the probes by traditional phage plaque assays or by quantitative PCR determines the binding constant of the compound for each kinase. As a test set for this method, 20 well-characterized kinase inhibitors were profiled against a panel of 119 kinases (this comprises about one fifth of the total estimated kinases in the human genome). The result not only confirmed the previously known targets of these inhibitors, it also revealed some novel and unexpected interactions. For example, the authors showed that a p38 MAP kinase inhibitor BIRB-796 could bind to a Gleevec-resistant variant of the ABL kinase. In addition, a new *in vitro* target of Gleevec was identified as the Src-family kinase Lck. These results also show the current lack of understanding in the true kinome-wide selectivity of even some of the most extensively studied kinase inhibitors. Hence, it raises the notion that even though an inhibitor is developed against a single target, the pharmacological response or the side effect of that molecule may not be solely due to inhibition of that target. This prompts the need to develop more comprehensive target-based assays to address the true selectivity of individual molecules. Nevertheless, it is also important to take into account the relative expression level of the target proteins when carrying out target-validation experiments with small molecule modulators. For instance, if a compound is found to be nonselective among several kinases but only one of them is expressed and/or active/involved in the model under which the compound is being examined, then the compound should still be considered a sufficient target validation tool in that particular model. Ultimately, integration of global selectivity profiles and relative protein expression/activity levels with *in vivo* response of the inhibitor should provide a better understanding of the consequences of inhibiting a particular combination of targets.

B. Validation of Drug Targets Using Cell-Based Phenotypic Assays

With potent and selective small-molecule modulators, cell-based phenotypic assays could then be used to evaluate compounds' biological effects in a more relevant setting before taking them into animal studies, which are usually more time and resource consuming. In contrast to target-based assays, phenotypic assays measure the ability of a small molecule to modulate specific phenotypic changes in cells, thereby establishing the functional link between compound/target and cellular phenotype. Easily obtained and manipulated (relative to *in vivo* experimental subjects), cells can be isolated directly from relevant models or engineered to possess specific properties. Even more versatile are the phenotypic readouts, which may simply involve counting the number of viable cells, examining cell morphologies, quantifying a reporter gene activity, or immunostaining a specific biomarker, and so on. Through careful design and interpretation, phenotypic assays can be applied in various ways for target validation in chemogenomics.

In reverse chemogenomics, where the small molecule has a defined target, phenotypic assays are used to validate the outcome of target modulation by the small molecule. This concept was nicely illustrated in the target validation of XIAP by small molecule inhibitors (Schimmer et al., 2004; Wu et al., 2003). XIAP (X-chromosome linked inhibitor of apoptosis protein) is a negative regulator of caspase-3, a protease that initiates apoptosis (Deveraux et al., 1997). Elevated XIAP expression prevents apoptosis by directly binding to and inhibiting caspase-3, and this is hypothesized to contribute to the reduced cell death commonly observed in certain cancer cells (Salvesen and Duckett, 2002). From a high-throughput protein-protein interaction screen, a class of compounds bearing sulfonamide cores was found to effectively block the association of XIAP and caspase-3 *in vitro*, thereby preventing the inhibition of caspase-3 proteolytic activity by XIAP (Wu et al., 2003). To validate that the disruption of the XIAP/caspase-3 interaction is effective in sensitizing cellular apoptosis under relevant conditions, the molecules were tested in a phenotypic assay involving overexpression of XIAP (in 293 cells), which prevented apoptosis induced by added soluble fas ligand in the medium. Under these conditions, additional treatment with a water-soluble sulfonamide analog, TWX024, induced apoptosis of 293 cells. At the same concentration, TWX024 had little effect in normal 293 cells, indicating that the cytotoxicity it exhibits is specifically from removing he XIAP blockade on fas-promoted cell death. In addition, a structurally distinct series of XIAP/caspase-3 inhibitors bearing a polyphenylurea scaffold were shown, in an independent experiment, to display greater toxicity against transformed MEF cells derived from wild-type (*xiap*$^{+/+}$) than from knockout (*xiap*$^{-/-}$) mice (Schimmer et al., 2004). These target-validation studies using engineered phenotypic cell-based assays confirmed the antagonistic role of XIAP in apoptosis and provided further evidence for targeting the XIAP/caspase-3 interaction as a potential strategy aimed at sensitizing cancer cells to conventional therapy.

C. Elucidation of Drug Mechanism Using Genomic Profiling

Besides characterizing the pharmacological outcome/phenotype of drug treatment (whether in cells or *in vivo*), drug target validation also involves elucidation of drug mechanism, that is, how modulation of the intended target leads to the desired phenotype. Often, it is informative to analyze the effect of drug treatment on the transcriptional, translational, and/or post-translational levels. The development of DNA microarrays, or more colloquially known as gene chips, has allowed rapid determination of gene expression profiles across the entire genome (Gabig and Wegrzyn, 2001). In drug discovery research, such transcriptome profiling is often used to determine which genes are affected by a given small molecule on a genome-wide level. Through the comparison of altered gene expression patterns, transcriptome profiling may provide a better understanding of the signaling pathways associated by drug treatment.

Perhaps one of the first examples of using DNA microarrays to validate/elucidate drug mechanism is demonstrated with the immunosuppressants FK506, and cyclosporine A (Marton et al., 1998). FK506 and cyclosporine A (CsA) can both bind to and inhibit the peptidyl-proline isomerase activity of their intracellular immunophilins, FK506 binding proteins (FKBP) and cyclophilins (CyP), respectively. The drug-immunophilin complexes, in turn, can bind to and inhibit calcineurin A, which is a key regulator of T-cell activation genes (Cardenas et al., 1995). Using DNA microarrays, comparative expression analysis revealed large correlations in the altered gene-expression patterns between FK506 treated wild-type yeast and CsA treated wild-type yeast, as well as calcineurin A deleted yeast strain. This experiment established that genetic or pharmacological inhibition of the same pathway may often result in closely correlated gene expression profiles. Furthermore, it was found that treatment of a calcineurin deletion yeast strain with FK506 (1 μg/mL) did not afford significant transcriptional changes relative to the untreated calcineurin deletion strain, suggesting that FK506 was unable to elicit its normal transcriptional response without functional calcineurin. When treated with a higher concentration of FK506 (50 μg/mL), however, the calcineurin deletion strain yielded a much more complex expression profile that is not associated with calcineurin-regulated genes, indicating that FK506 can affect additional targets at higher dose. Altogether, the authors demonstrated that genomic profiling tools can be used to confirm a potential drug target and possibly identify additional targets by using a deletion mutant defective in the gene encoding the putative target. This method, called the "decoder" strategy, and other applications of DNA microarrays may prove useful in elucidating drug mechanism and validating drug targets.

D. Elucidation of Drug Mechanism Using Proteomic Profiling

Because signaling pathways are mainly mediated by proteins, whose expression and activity are not always correlated with their mRNA expression, global analysis of the complex proteome would provide a better picture of drug-induced response and could also facilitate the process of target validation in chemogenomics. Whereas the former is based on quantification of mRNA transcripts to establish the gene expression at the transcriptional levels, proteomic profiling is more complicated in that it deals with two parameters: abundance and activity. In other words, two proteomic samples may have some similar protein expression levels, but the activity of individual proteins may differ substantially due to various post-translational modifications (e.g., phosphorylation, glycosylation, ubiquination), which would direct different signaling pathways/phenotypic outcomes. One of the most widely used techniques for protein abundance profiling is 2DGE (two-dimensional gel electrophoresis) (Patton et al., 2002). Although this method provides an unbiased view of the entire proteome, it lacks the ability to resolve and detect low abundance proteins. Other proteomic profiling tools such as yeast two-hybrid (Uetz et al., 2000) and protein microarrays (Zhu et al., 2001) require the proteins be studied in somewhat artificial environments that may not be relevant in the endogenous settings in mammalian cells. Hence, new technologies that detect global changes in the expression and activity of proteins are highly desirable. The remainder of this section describes emerging methods that attempt to fulfill these needs and discusses how they can be applied to profile the changes within the functional proteome caused by chemical treatments.

Activity-based protein profiling (ABPP) is a technique that uses chemistry to address the functional state of selected protein families in the complex proteome (Jessani and Cravatt, 2004). Reactive chemical probes are designed to covalently bind the active site of different enzyme classes via reactions with proximal catalytic or non-catalytic nucleophilic residues. Only active enzymes react to the probes because inactive enzymes would have a different conformation and/or inaccessible active sites. The other end of the probe consists of a tag for the detection or purification of these enzyme-probe adducts. Separation of fluorescent-tagged proteins on gel electrophoresis allows the visualization and comparison of enzyme activities across different proteomic samples. Using the biotinylated probes, the target proteins can be purified and identified through mass spectrometry. One example uses sulfonate ester probes, which can covalently react with a number of distinct enzyme classes, to profile enzyme activities in estrogen-receptor positive (ER^+) and negative (ER^-) human breast cancer cell lines (Adam et al., 2002). Because ER expression inversely correlates with several metastatic phenotypes in breast carcinomas, identification of enzyme activities that are upregulated in ER^- cells may provide a better understanding or even a therapeutic target of breast cancer aggression. Using a phenyl sulfonate probe, an omega-class glutathione *S*-transferase (GSTO 1-1) was found to be active only in ER^- cells relative to the control ER^+ cells. This preliminary analysis of cancer tissues using reactive chemical probes highlights the ability of activity-based profiling to identify previously unrecognized enzyme activities associated with pathological conditions.

ABBP has also been applied in the determination of protein-activity changes associated with small-molecule treatment, as means of identifying and validating protein function. In one such case study, several reversible serine hydrolase inhibitors were profiled using a serine hydrolase activity probe (Leung et al., 2003). The probe, a rhodamine-tagged fluorophosphonate (FP-rhodamine), was known to react non-specifically with members of the serine hydrolase family. Incubation of FP-rhodamine with mouse heart or brain membrane proteome resulted in the labeling of several active enzymes (e.g., FAAH, KIAA1363, and TGH). Addition of inhibitors to the proteome caused the differential disappearance of several protein bands, indicating the diminishing of those enzyme activities in the presence of these compounds. Further investigation is required to determine if the reduction in each protein activity is the result of direct competition by the inhibitors or secondary downstream effect. Nevertheless, ABPP provides an alternative focused approach to profile protein activity associated with the treatment of specific inhibitors. Not only can this information be used to facilitate elucidating drug mechanism on the proteomic level, the discovery of downstream enzyme activities may be used as biomarkers for pharmacodynamic readout. Although the activity profile is currently limited to the proteins that the probe can label, the development of diverse and cell-permeable activity probes (Speers and Cravatt, 2004) should facilitate ABPP in the process of target validation by establishing functional relationship between protein activities and pharmacological response.

Many chemical treatments in cells have direct effects on post-translational modifications of proteins. Therefore, one proteomic approach of target validation is to identify and monitor these protein modifications to provide a better understanding of the signaling pathways perturbed by the small molecule. Mass spectrometry and peptide mapping has now become the method of choice for protein identification (Reinders et al., 2004). Given the ultrahigh sensitivity, it can even detect certain post-translational modifications. One case study illustrates the use of mass spectrometry and peptide mapping to identify protein tyrosine phosphorylation caused by Gleevec (Salomon et al., 2003). In this experiment, cellular extracts were obtained from Gleevec-treated and non-treated CML cell lines, and proteins containing phosphotyrosine were immunoprecipitated by monoclonal anti-phosphotyrosine antibodies, subjected to tryptic digestion, and further enriched by a combination of methyl esterification and immobilized metal affinity chromatography. The resulting samples were then analyzed by reversed phase high performance liquid chromatography (HPLC) and tandem mass spectometry (MS) methods, enabling the unambiguous assignment of tyrosine phosphorylation sites in proteins. Examination of changes in the sites of phosphorylation over time helped to clarify the temporal orchestration of protein tyrosine phosphorylation within signaling pathways. Consistent with previous reports, Gleevec treatment inhibited the phosphorylation of Tyr-177 and Tyr-393 on BCR-ABL. Furthermore, nearly 30 other previously unrecognized sites of tyrosine phosphorylation in other proteins were identified using this method. Future technological improvements should be able to extend this method to detect other protein modifications, such as serine/threonine phosphorylation and protein acetylation.

III. FORWARD CHEMOGENOMICS: TARGET IDENTIFICATION/VALIDATION USING COMPOUNDS WITH UNKNOWN MECHANISM OF ACTION

Perhaps the most significant aspect of forward chemogenomics is that it holds the potential to identify novel targets. Before the advent of molecular biology, drugs were discovered largely by *in vivo* testing of compounds in animals with little knowledge concerning the molecular mechanism of the disease or the therapy—a "phenotype first" approach that is later termed forward chemogenomics. Today, mainstream drug discovery follows the reverse chemogenomics trend, where drug candidates are selected on the basis of modulating a disease-related target (protein or nucleic acid) and then evaluated in model systems of higher complexity (cells or organism). Though reverse chemogenomics have delivered many successful drugs in the past decade, the concept of forward chemogenomics is slowly re-emerging, especially in academic research, because it gives the opportunity to evaluate novel phenotypes and discover novel targets that are poorly understood or less obvious at the current state of knowledge. For example, how would one control a specific stem cell fate by reverse chemogenomics if there were no defined biological targets yet for such process? Or, how can more effective cancer therapeutics be developed if cancer eradication requires targeting multiple proteins? These types of problem can potentially be addressed in forward chemogenomics by screening for compounds that directly produce the desired phenotypes in cells or animals and then using these compounds to discover the responsible target(s).

A. Target Identification Using Genomic Profiling

Target identification of compounds with unknown mechanism of action often begins by characterizing the biological consequences of the small-molecule treatment. Because transcriptome profiling can reliably provide unbiased snapshots of the global gene expression pattern at various time points post compound treatment, many researchers have thus relied on expression analysis as a first step to examine small-molecule treatments (Butcher and Schreiber, 2005). Using various bioinformatics tools, the transcriptional changes are clustered by relevance to produce gene expression patterns from which the affected signaling pathways can be deduced. Hence, comparative transcriptome profiling may provide the missing link between the compound-induced phenotype and the affected biological pathway.

Purmorphamine is a potent small-molecule inducer of osteogenic differentiation of mesenchymal progenitor cells (Wu et al., 2002). It was discovered through a high-throughput cell-based assay that detects the enzymatic activity of alkaline phosphatase, a bone-specific marker, in C3H10T1/2 cells (a mouse mesenchymal progenitor cell line capable of differentiating into bone, fat, cartilage, and muscle cells). Using DNA microarrays, the transcriptional profile of C3H10T1/2 cells was examined with and without purmorphamine or BMP-4 (a protein known to induce osteogenesis of mesenchymal stem cells) treatments in different time courses. Based on the gene expression of bone biomarkers, purmorphamine was confirmed to be a more specific inducer of bone differentiation compared to BMP-4, which appears also to activate adipogenic (fat formation) pathways (Wu et al., 2004). Furthermore, Ingenuity Pathway analysis suggests that the Hedgehog (Hh) signaling is the primary affected (activated) biological pathway. To confirm this mechanistic insight by chemical epistasis, the authors showed that two known antagonists of Hg signaling, cyclopamine (targeting Smo at the membrane level) and forskolin (inhibiting downstream Gli transcriptional activity), can abolish the Hg-pathway activating and osteogenesis inducing effects of purmorphamine in C3H10T1/2 cells. This result also implies that purmorphamine may target Smo or proteins upstream of Smo in activating Hh pathway. This case study illustrates the use of genome-wide expression analysis to reveal the biological pathways affected by small molecules with unknown mechanism of action, though additional genetic and biochemical experiments are required for confirmation.

B. Classification of Drug Actions Using Phenotypic Profiling

In forward chemogenomics, compound hits can be categorized based on their generated phenotypic responses (e.g., cytotoxicity pattern among a panel of cancer cell lines), and such information may facilitate elucidation of the compounds' mechanism of action. This information-intensive approach to molecular pharmacology has been extensively applied in the field of cancer-cell biology (Weinstein et al., 1997).

In order to better understand the response of cancer chemotherapy with the hope of finding more effective and even personalized treatments, the Developmental Therapeutics Program (DTP) of the National Cancer Institute (NCI) has initiated large-scale compound screens of cell proliferation using a panel of 60 representative cancer cell lines (NCI60) (Shoemaker et al., 2002). The anti-proliferative effect of every molecule on every cell line was recorded and the accumulated dataset was analyzed to differentiate various classes of anti-proliferative compounds. This allows researchers to quickly make hypotheses about the mechanism of action of novel cytotoxic agents. For example, molecules that are cytotoxic across the entire panel may be general DNA-modifying agents or cell-cycle inhibitors, whereas molecules that selectively kill specific cancer cells may act on certain signaling pathways that the cancer cells depend on. Coupled with the transcriptome profiling of these cancer cell lines, the screening results can be used to investigate gene-to-drug relationship, one potential application of which lies in predictive chemogenomics. To exemplify this concept, the activity pattern of 5-fluorouracil (5-FU)—a drug commonly used in the clinic to treat colorectal cancer—was examined in the NCI60 screen (Scherf et al., 2000). It was found that the cytotoxic effect of 5-FU showed a significant inverse correlation with the expression of dihydropyrimidine dehydrogenase (DPYD) in these cell lines. Perhaps not coincidentally, all of the seven colon-derived cancer cell lines in the NCI60 fall into the panel with low DPYD expression. This study provides evidence that DYPD may serve as a clinical biomarker to select patients that would respond to 5-FU. Predictive cancer chemogenomics will potentially allow researchers to design more specific clinical trials for achieving better efficacy of novel anti-cancer drug candidates and may ultimately lead to future personalized chemotherapy.

C. Target Identification/Validation Using Target-Based Assays

While most drug-discovery programs now utilize target-based assays as the engine to drive the preclinical development of chemical entities against single targets, there are a few exceptions where the drug candidate's target is unknown until later in the development stage, such as FTY720. FTY720 is a novel class of immunosuppressant (Kiuchi et al., 2000). It functions by reducing the circulating level of lymphocytes in peripheral blood, but the precise target that this drug acts on was not clear to begin with. Nevertheless, the novel phenotype exhibited by this compound prompted further investigation and development.

Analysis of the circulating FTY720 metabolite led to the finding that the drug existed largely as the phosphoester form in the blood (Mandala et al., 2002). Based on structural similarity shared by the phosphorylated form of FTY720 (phospho-FTY720) and the lysophospholipid sphingosine, it was hypothesized that FTY720 was a prodrug whose phosphorylated form targets the sphingosine-1-phosphate (S1P) receptors. To validate this hypothesis, several binding assays were set up to evaluate the affinity of FTY720 and phospho-FTY720 against the different S1P receptor isoforms. The assay involved competition of radioligand binding using [^{33}P]-labeled S1P on transfected CHO cells expressing each of the five S1P-binding receptors ($S1P_1$, $S1P_2$, $S1P_3$, $S1P_4$, and $S1P_5$). It was found that while FTY720 itself has no binding activity, phospho-FTY720 is a high affinity binder to all of the five S1P receptors except $S1P_2$ (Mandala et al., 2002). Furthermore, Ca^{2+} mobilization assays demonstrated the agonistic effect of phospho-FTY720. In summary, these simple binding assays provided strong evidence linking the target (SIP receptor), the molecule (FTY720), and the phenotype (lymphocyte trafficking). This serves as an excellent illustration of using target-based assays to validate the proposed target of a drug with unknown mechanism of action.

D. Target Identification/Validation Using Affinity Chromatography

Perhaps the most straightforward method of target identification/validation in forward chemogenomics—though not necessarily the easiest—is to pull down the targets that directly interact with the small molecule of interest using affinity-based approaches. In affinity chromatography, the compound is attached onto a solid matrix and exposed to cellular extracts. After a series of washing steps to elute non-specific binding species, the protein(s) retained by the affinity matrix is(are) thought to be the putative target(s). Another type of affinity-based approach uses a radio-labeled compound with a photo-labile cross-linking functionality to covalently tag the compound to its macromolecular binder(s) followed by identification of the adduct. While all of these methods have limitations in identifying low-abundance proteins, there are cases that have led to the discovery of novel target and mechanism.

As the molecular mechanism of regulating stem cells is still poorly understood currently, discovery of small molecules that can control stem cells' fates will provide useful chemical tools for target identification and validation (Ding and Schultz, 2004). A disubstituted pyrrolopyrimidine, named TWS119, was identified as a potent inducer of neuronal differentiation in pluripotent mouse embryonic stem cells (Ding et al., 2003). A panel of affinity matrices, prepared from representative TWS analogs, was used to pull down target proteins from the cell extracts. Proteins specifically bound to all positive resins derived from active molecules but not to the negative resins derived from inactive molecules were considered to be the putative targets of TWS119. Consequently, glycogen synthase kinase 3β(GSK-3β) was identified as one target of TWS119, and was further confirmed by biochemical and cellular assays (e.g., surface-plasmon resonance, kinase-inhibition assay, Western blot, and reporter assay). GSK-3β is known to be a negative regulator of the canonical Wnt pathway, which is involved in embryonic patterning and cell behaviors. Through chemogenomics, the discovery of the phenotype caused by TWS119 and its intrinsic activity provide yet another link between neuronal differentiation and the Wnt signaling pathway.

Affinity chromatography will often reveal non-obvious interacting protein partners of well-studied compounds, as in the case of purvalanol, a class of purine-based small molecules designed to selectively inhibit cyclin-dependent kinase (CDK; $IC_{50} = 4–6\,nM$) (Gray et al., 1998). To determine the selectivity of purvalanol in cells, affinity matrices derived from purvalanol analogs were prepared. In addition to CDK, an isoform of casein kinase I (CK1) was unexpectedly pulled down from protozoan parasite but not mammalian cell extracts (Knockaert et al., 2000). Given the ability of purvalanol to inhibit protozoan parasite growth, the protozoan CK1 isoform was cloned and further investigated as a potential anti-parasitic target. It was found that Purvalanol B and aminopurvalanol inhibited the α form of *T. gondii* parasitic CK1 (TgCK1α; $IC_{50} = 120$ and 42 nM) but not the mammalian CK1 (Donald et al., 2005). In *T. gondii* whole-cell growth inhibition assay, the efficacy of aminopurvalanol ($IC_{50} = 0.36\,\mu M$) is approaching that of coccidostat ($IC_{50} = 0.2\,\mu M$), a known anti-parasitic compound whose mechanism of action is thought to be associated with the inhibition of *T. gondii* cGMP-dependent kinase (TgPKG). Because aminopurvalanol shows no inhibition on TgPKG, it would be reasonable to propose that TgCK1α could be another significant intracellular anti-parasitic target. Hence, through the use of affinity chromatography, novel uses for a well-characterized enzyme inhibitor were identified, which led to the discovery of a potential therapeutic target.

IV. CONCLUSION

Target validation by small molecules is a well-established concept in pharmaceutical research. Recent discovery and technology advancements—such as completion of the several genome projects and developments of various genomic, proteomic, and HTS tools—have enabled researchers to design more effective strategies to better understand the relationship between the molecular target and the pharmacological response for a given drug. While theoretically most strategies seem straightforward and promising, in practice, target validation in chemogenomics is a complex process. Each drug-discovery program is unique; therefore, different approaches will be employed in target validation, which often integrate many disciplines (biology, chemistry, pharmacology, genomics, proteomics, informatics, etc). This chapter has illustrated some of the approaches from different target validation/identification case studies. It is no doubt that future works will involve more advanced and effective biotechnological methods to improve the efficiency of drug-discovery research.

REFERENCES

Adam, G. C., E. J. Sorensen, and B. F. Cravatt (2002). Proteomic profiling of mechanistically distinct enzyme classes using a common chemotype. *Nat. Biotechnol.* 20: 805–809.

Bleicher, K. H. (2002). Chemogenomics: Bridging a drug discovery gap. *Curr. Med. Chem.* 9: 2077–2084.

Bredel, M. and E. Jacoby (2004). Chemogeomics: An emerging strategy for rapid target and drug discovery. *Nat. Rev. Genet.* 5: 262–275.

Butcher, R. A. and S. L. Schreiber (2005). Using genome-wide transcriptional profiling to elucidate small-molecule mechanism. *Curr. Opin. Chem. Biol.* 9: 25–30.

Cardenas, M. E., D. Zhu, and J. Heitman (1995). Molecular mechanisms of immunosuppression by cyclosporine, FK506, and rapamycin. *Curr. Opin. Nephrol. Hypertens.* 4: 472–477.

Deveraux, Q. L., R. Takahashi, et al. (1997). X-linked IAP is a direct inhibitor of cell-death proteases. *Nature* 388: 300–304.

Ding, S., T. Y. Wu, et al. (2003). Synthetic small molecules that control stem cell fate. *Proc. Natl. Acad. Sci. U S A* 100: 7632–7637.

Ding, S. and P. G. Schultz (2004). A role for chemistry in stem cell biology. *Nat. Biotechnol.* 22: 833–840.

Donald, R. G., Zhong, T., et al. (2005). Characterization of two *T. gondii CK1* isoforms. *Mol. Biochem. Parasitol.* 141: 15–27.

Fabian, M. A., W. H. Biggs 3rd, et al. (2005). A small molecule-kinase interaction map for clinical kinase inhibitors. *Nat. Biotechnol.* 23: 329–336.

Gabig, M. and G. Wegrzyn (2001). An introduction to DNA chips: Principles, technology, applications and analysis. *Acta. Biochim. Pol.* 48: 615–622.

Gray, N. S., L. Wodicka, et al. (1998). Exploiting chemical libraries, structure, and genomics in the search for kinase inhibitors. *Science* 281: 533–538.

Jessani, N. and B. F. Cravatt (2004). The development and application of methods for activity-based protein profiling. *Curr. Opin. Chem. Biol.* 8: 54–59.

Kiuchi, M., K. Adachi, et al. (2000). Synthesis and immunosuppressive activity of 2-substituted 2-aminopropane-1,3-diols and 2-aminoethanols. *J. Med. Chem.* 43: 2946–2961.

Knockaert, M., N. Gray, et al. (2000). Intracellular targets of cyclin-dependent kinase inhibitors: Identification by affinity chromatography using immobilised inhibitors. *Chem. Biol.* 7: 411–422.

Lee, D. and D. W. Threadgill (2004). Investigating gene function using mouse models. *Curr. Opin. Genet. Dev.* 14: 246–252.

Leung, D., C. Hardouin, et al. (2003). Discovering potent and selective reversible inhibitors of enzymes in complex proteomes. *Nat. Biotechnol.* 21: 687–691.

Mandala, S., R. Hajdu, et al. (2002). Alteration of lymphocyte trafficking by sphingosine-1-phosphate receptor agonists. *Science* 296: 346–349.

Marton, M. J., J. L. DeRisi, et al. (1998). Drug target validation and identification of secondary drug target effects using DNA microarrays. *Nat. Med.* 4: 1293–1301.

Patton, W. F., B. Schulenberg, and T. H. Steinberg (2002). Two-dimensional gel electrophoresis: Better than a poke in the ICAT? *Curr. Opin. Biotechnol.* 13: 321–328.

Reinders, J., U. Lewandrowski, et al. (2004). Challenges in mass spectrometry-based proteomics. *Proteomics* 4: 3686–3703.

Rodi, D. J. and L. Makowski (1999). Phage-display technology: Finding a needle in a vast molecular haystack. *Curr. Opin. Biotechnol.*, 10, 87–93.

Salomon, A. R., S. B. Ficarro, et al. (2003). Profiling of tyrosine phosphorylation pathways in human cells using mass spectrometry. *Proc. Natl. Acad. Sci. U. S. A.* 100: 443–448.

Salvesen, G. S. and C. S. Duckett (2002). IAP proteins: Blocking the road to death's door. *Nat. Rev. Mol. Cell. Biol.* 3: 401–410.

Savchuk, N. P., K. V. Balakin, and S. E. Tkachenko (2004). Exploring the chemogenomic knowledge space with annotated chemical libraries. *Curr. Opin. Chem. Biol.* 8: 412–417.

Scherf, U., D. T. Ross, et al. (2000). A gene expression database for the molecular pharmacology of cancer. *Nat. Genet.* 24: 236–244.

Schimmer, A. D., K. Welsh, et al. (2004). Small-molecule antagonists of apoptosis suppressor XIAP exhibit broad antitumor activity. *Cancer Cell* 1: 25–35.

Shoemaker, R. H., D. A. Scudiero, D. A., et al. (2002). Application of high-throughput, molecular-targeted screening to anticancer drug discovery. *Curr. Top. Med. Chem.* 2: 229–246.

Speers, A. E. and B. F. Cravatt (2004). Profiling enzyme activities in vivo using click chemistry methods. *Chem. Biol.* 11: 535–546.

Uetz, P., L., L. Giot, et al. (2000). A comprehensive analysis of protein-protein interactions in *Saccharomyces cerevisiae*. *Nature* 403: 623–627.

Venter, J. C., M. D. Adams, et al. (2001). The Sequence of the Human Genome. *Science* 291: 1304–1351.

Weinstein, J. N., T. G. Myers, et al. (1997). An information-intensive approach to the molecular pharmacology of cancer. *Science* 275: 343–349.

Wu, T. Y., K. W. Wagner, et al. (2003). Development and characterization of nonpeptidic small molecule inhibitors of the XIAP/caspase-3 interaction. *Chem. Biol.* 10: 759–767.

Wu, X., S. Ding, et al. (2002). A small molecule with osteogenesis-inducing activity in multipotent mesenchymal progenitor cells. *J. Am. Chem. Soc.* 124: 14520–14521.
Wu, X., J. Walker, et al. (2004). Purmorphamine induces osteogenesis by activation of the hedgehog signaling pathway. *Chem. Biol.* 11: 1229–1238.
Zhu, H., M. Bilgin, et al. (2001). Global analysis of protein activities using proteome chips. *Science* 293: 2101–2105.

II

TARGET VALIDATION FOR BIOPHARMACEUTICAL DRUG DISCOVERY

3

CETUXIMAB (ERBITUX®), AN ANTI-EPIDERMAL GROWTH FACTOR RECEPTOR ANTIBODY FOR THE TREATMENT OF METASTATIC COLORECTAL CANCER

ZHENPING ZHU

M.D., Ph.D., ImClone Systems Incorporated, New York, New York

Compelling experimental and clinical evidence suggests that epidermal growth factor receptor (EGFR) plays an important role in pathogenesis of a variety of human cancers, thus providing a strong rationale for the development of receptor antagonists as effective and specific therapeutic strategies for the treatment of EGFR-expressing cancers. Cetuximab (Erbitux®, IMC-C225), a chimeric monoclonal antibody directed against EGFR, is the first and the only FDA-approved anti-EGFR antibody for the treatment of patients with refractory metastatic colorectal cancer (mCRC). Cetuximab has demonstrated significant antitumor activity, both as a single agent and in combination with irinotecan, in mCRC patients refractory to irinotecan therapy. Cetuximab, via blocking ligand/receptor interaction, exerts its biological activity via multiple mechanisms, including inhibition of cell-cycle progression, potentiation of cell apoptosis, inhibition of DNA repair, inhibition of angiogenesis, tumor-cell invasion and metastasis, and potentially induction of immunological effector mechanisms. Cetuximab is currently being tested in multiple advanced clinical trials in various cancer patients, including those with squamous cell carcinoma of head and neck, non–small cell lung cancer, pancreatic cancer, and ovarian cancer.

I. INTRODUCTION

A large number of genes in eukaryotic cells encode for proteins that function as trans-membrane cell-surface receptors, of which many are endowed with intrinsic protein tyrosine kinase activity. These receptor tyrosine kinases

(RTKs) play important roles in the control of some of the most fundamental cellular processes including cell cycle, proliferation, and differentiation (Schlessinger, 2000; Ullrich et al., 1990). In recent years, it has become increasingly evident that the developmental fate of individual cells is often established by activation of the trans-membrane receptors of the RTK family. Abnormal cell growth resulted from aberrant signal transduction of RTKs has been implicated in the initiation and progression of a variety of human cancers (Schlessinger, 2000; Ullrich et al., 1990). Tyrosine kinase activity has been detected in a major proportion of oncoproteins, including both the transmembrane receptors and the cytoplasmic proteins, and increase in kinase activity of these oncoproteins has been shown to correlate with the degree of malignant transformation (Chiao, 1990; Hunter, 1991). In addition, overexpression of wild-type or various mutated RTK in normal cells usually confers the neoplastic phenotypes, resulting in abnormal growth of the transformed cells (Chiao, 1990; Hunter, 1991). This transformation can be reversed by specific RTK inhibitors such as neutralizing antibodies and synthetic small molecule compounds in experimental systems (Bianco et al., 2005; Krause and Van Etten, 2005; Nooberg et al., 2000; Tibes et al., 2005). Further, there is accumulating evidence to suggest that in tumors, over-expression of certain types of RTKs, such as epidermal growth factor receptor (EGFR) and HER-2/neu, directly correlates with poor response to standard treatment protocols and shorter survival times of the patients (Hynes and Lane, 2005; Klapper et al., 2000). Taken together, all this knowledge suggests that various RTK pathways may represent excellent targets for effective cancer intervention (Krause and Van Etten, 2005; Nooberg et al., 2000).

In the past years, many efforts have been made by both academic laboratories and pharmaceutical/biotechnology companies to identify and develop effective RTK inhibitors for cancer therapy (Nooberg et al., 2000; Bianco et al., 2005; Krause and Van Etten, 2005; Tibes et al., 2005). Our understanding of the complex series of events that are involved in the abnormal signaling pathways has improved dramatically during the past several decades. New information regarding how each of the signal transduction pathways participate in abnormal cell growth and the molecules responsible for these events has led to a variety of novel and increasingly mechanism-based approaches for the development of RTK inhibitors for cancer therapy, including those that specifically interfere with growth factor ligand/receptor interaction, thus blocking the activation of the receptors and the subsequent signaling cascade. For example, a number of anti-RTK monoclonal antibodies (mAb) have been developed to specifically block growth factors, such as epidermal growth factor (EGF) and vascular endothelial growth factor (VEGF), from binding to their respective receptors (Bianco et al., 2005; Starling and Cunningham, 2004; Paz and Zhu, 2005). Additional approaches (e.g., the development of small molecule compounds that inhibit phosphorylation of tyrosine residues of the receptors, and/or inhibit activation of other signaling molecules) are also being rigorously pursued (Albanell and Gascon, 2005; Paz and Zhu, 2005).

MAb, owing to their high specificity towards a given target, represent a unique class of novel therapeutics as inhibitors to RTKs. Recent commercial success with antibody-based therapeutics, for example, rituximab (Rituxan®) and trastuzumab (Herceptin®), has led to an upsurge in the clinical development of these agents (Glennie and van de Winkel, 2003; Stern and Herrmann, 2005; von Mehren et al., 2003). In fact, antibodies have rapidly evolved from the ideal of "magic bullet" to practical and efficacious cancer therapeutics in the last 10 years. Traditional obstacles in antibody therapy, such as immunogenicity of rodent-derived antibodies and difficulty in producing antibodies in sufficient quantity and quality for commercial application, are being rapidly superseded by the advancement in antibody engineering technologies. These include antibody chimerization, humanization, and direct identification of fully human antibodies from either phage display libraries or human transgenic mice (Jones et al., 1986; Lonberg et al., 1995; Morrison and Huszar, 1984; Winter et al., 1994), as well as the development of efficient manufacturing processes for high level production of mAb at costs that are more economical than ever (Chadd and Chamow, 2001). Since 1994, the Food and Drug Administration (FDA) has approved 18 therapeutic mAb for clinical use in the United States, including 8 for oncology indications. In addition, there are several hundred mAb currently being tested in clinical trials worldwide for various indications.

II. EPIDERMAL GROWTH FACTOR RECEPTOR AND ITS ROLE IN HUMAN CANCER

EGFR is a trans-membrane receptor encoded by the *c-erb*B1 proto-oncogene with a molecular weight of approximately 170 kDa (Normanno et al., 2005; Ullrich et al., 1984; Wells, 1999). EGFR belongs to subclass I family of RTKs and is the receptor to at least six distinct ligands including EGF, transforming growth factor α (TGF-α), heparin-binding EGF, amphiregulin, betacellulin, and epiregulin (Normanno et al., 2005; Yarden, 2001). The subclass I family of RTK consists of EGFR (also known as HER1), Her2/neu (*erbB-2*), HER3 (*erbB-3*), and Her-4 (*erbB-4*) (Carpenter 2000; Yarden, 2001; Yarden and Sliwkowski, 2001). Much evidence suggests that these receptors function in various homodimeric and heterodimeric pairs, depending on their density on cell surface, the concentrations of a particular ligand, and intrinsic dimerization preference between the receptors (Gullick, 1998). EGFR is normally expressed in a wide variety of epithelial tissues as well as in the central nervous system. Binding of a ligand to the extracellular domain of EGFR leads to receptor dimerization, followed by activation of the intrinsic RTK activity and autophosphorylation of specific residues within the receptor's cytoplasmic domain. These phosphorylated residues serve as docking sites for other molecules involved in the regulation of intracellular signaling cascades. The major signaling cascades activated by EGFR include the Ras/MAP kinase, PLC-gamma, PI-3 kinase/Akt and STAT3 pathways. The integrated biological responses to EGFR signaling are pleiotropic, including mitogenesis or apoptosis, enhanced cell motility, protein secretion, cell adhesion, invasion, differentiation or dedifferentiation, and increased neovascularization (Carpenter, 2000; Grant et al., 2002; Normanno et al., 2005; Wells, 1999; Yarden, 2001; Yarden and Sliwkowski, 2001). In tumor cells, EGFR activation provides the cells with growth and survival advantages through these same regulatory pathways.

A large body of experimental evidence supports a role for EGFR activation and signaling in the pathogenesis of human cancers. In 1984, analyses of the EGFR gene and protein revealed that its sequence was homologous to the v-erbB proto-oncogene (Downward et al., 1984; Ullrich et al., 1984). Direct preclinical evidence for a role of EGFR in malignant transformation emerged from studies in which the transfection of EGFR or TGF-α cDNA was associated with cellular transformation (Di Fiore et al., 1987). EGFR is expressed in a variety of human solid tumors, including head and neck squamous cell cancer and carcinomas of cervix, renal cell, lung, prostate, bladder, colorectal, pancreatic, and breast, as well as melanoma, glioblastoma, and meningioma. Accumulating evidence suggests that the level of EGFR overexpression is an important factor that directly correlates with active proliferation of malignant cells and poor prognosis of patients (Brandt et al., 2000; Dassonville et al., 1993; Humphreys et al., 2000; Klijn et al., 1992; Salomon et al., 1995). A literature survey of more than 200 studies published between 1985 and 2000 showed EGFR expression is a strong adverse prognostic factor in cancers of the head and neck, ovary, cervix, bladder, and esophagus, and was correlated with reduced recurrence-free and overall survival in 70% of studies (Nicholson et al., 2001). EGFR expression on tumor cells in lymph nodes, distant metastases, and at the invasion front of colorectal lesions showed the strongest negative correlation with patient survival (Goldstein and Armin, 2001). In addition, several tumor types, including colorectal cancers, have been demonstrated to co-express EGFR and its ligands, leading to an autocrine activation of the receptor and poor outcome in the clinic (Ekstrand et al., 1991; Grandis et al., 1998; Salomon et al., 1995; Yonemura et al., 1992; Yamanaka et al., 1993). Finally, mutants of EGFR, because of gene rearrangement that results in in-frame deletion of portions of the extracellular domain of the receptor, have been found in a significant fraction of EGFR-expressing tumors. For example, the most common mutation (EGFRvIII)—with a deletion of amino acids 6-273—that is frequently found in brain tumors, such as glioblastoma, results in a protein with defective ligand binding capacity but is constitutively activated and its tumorigenicity *in vivo* is enhanced (Nishikawa et al., 1994; Pedersen et al., 2001). Taken together, these data indicate that expression of EGFR in human cancers has a significant effect on their biological behavior, thus providing the rationale for the development of EGFR antagonists as potentially useful therapeutic strategies for the treatment of EGFR-expressing cancers. In this regard, a number of EGFR targeted therapeutics, including mAb and small-molecule tyrosine kinase inhibitors (TKI), have been developed for preclinical and clinical testing (Arteaga, 2003a, 2003b; Bianco et al., 2005; Krause and Van Etten, 2005).

III. CETUXIMAB (ERBITUX®, IMC-C225)

Cetuximab (Erbitux®, IMC-C225) is a monoclonal anti-EGFR antibody being developed for the treatment of EGFR-expressing human cancers via inhibition of the function of EGFR. Cetuximab is a chimeric version of the murine anti-EGFR mAb M225 (Sato et al., 1983). M225 efficiently competes with EGF and TGF-α for binding to the receptor, inhibits both ligand stimulated activation of the receptor and downstream signaling molecules, and inhibits tumor cell mitogenesis *in vitro*. Binding of M225 to EGFR induces rapid receptor internalization, hence effectively stripping the receptor from tumor cell surface. M225 also induces apoptosis in some EGFR-overexpressing tumor cell lines. The anti-proliferative effects *in vitro* in cultured cells as well as tumor inhibition *in vivo* in animal models were observed in numerous human tumor cell lines treated with M225 (Masui et al., 1984; Mendelsohn, 1997).

Further, the antibody was found to enhance the anti-tumor activity of doxorubicin *in vitro* as well as both doxorubicin and cisplatin in several human xenograft models (Baselga et al., 1993; Fan et al., 1993a). In a small phase I trial, administration of ^{111}In-labeled M225 significantly improved tumor visualization (Divgi et al., 1991).

A. Preclinical Studies on Cetuximab

M225 was subsequently engineered into a chimeric form by genetic fusion of the murine variable regions of the antibody with a human IgG1 (κ) constant domain. The resultant antibody, cetuximab (IMC-C225), was more efficacious than the parent M225 in inhibiting tumor growth in animal models, at least partly because of an improved binding affinity (K_d, 0.2 to 0.3 nM) of the chimeric antibody (Goldstein et al., 1995).Extensive preclinical studies using a variety of assays, including ligand competition, receptor phosphorylation, cell proliferation, and anchorage-independent growth in soft agar, demonstrated that cetuximab inhibits EGFR activation and the growth of several different EGFR-expressing human tumor cell lines *in vitro*, including those carcinomas of the bladder, breast, colon, epidermoid carcinoma, kidney, ovary, pancreas, and prostate (Waksal, 1999). The extent of cell growth inhibition varies among different tumor cell lines used in each study, and is dependent on the type of assay utilized, EGFR expression levels, the presence of autocrine stimulatory pathways and the intrinsic biology of the tumor cell lines. Cetuximab treatment under optimal *in vitro* culture conditions results in suppression of tumor cell growth ranging from 10% to 90%. Some tumor cells are uniquely sensitive to EGFR blockade *in vitro*, such as the DiFi colon tumor cell line, which undergoes extensive apoptosis when treated with saturating concentrations of cetuximab (Wu et al., 1995).

The tumor growth inhibitory activities of cetuximab were confirmed *in vivo* in a number of human tumor xenograft models that include bladder, breast, colon, epidermoid, lung, pancreas, prostate, and renal carcinomas (Waksal, 1999). In these studies, cetuximab was administered by intraperitoneal injection two or three times per week at doses of 0.2 to 1 mg/injection and treatment was carried out from 2 weeks to more than 3 months. The anti-tumor activity of cetuximab therapy in these preclinical studies ranged from no effect to regression of established tumors depending on the model system. Data from human tumor xenograft studies demonstrated that the antitumor efficacy of cetuximab is markedly enhanced *in vivo* compared to effects on tumor cells *in vitro*. For example, cetuximab treatment of A431 cells in culture generally results in an inhibition of cell proliferation of 20% to 50% depending on the assay conditions. In contrast, cetuximab treatment of mice bearing A431 xenografts led to complete tumor growth arrest and regression of well-established tumors (Goldstein et al., 1995; Petit et al., 1997). The enhanced efficacy *in vivo* is likely the result of increased growth factor dependence of tumor cells under stress conditions when grown in mice (e.g., decreases in availability of nutrients, growth factors, oxygenation), as well as inhibition of other events associated with tumor growth and progression, such as tumor cell invasion, migration and neovascularization, that cannot be readily modeled in cell culture systems. Moreover, immune effector mechanisms, including antibody-dependent cellular cytotoxicity (ADCC) and complement-mediated cytotoxicity (CMC), may also contribute to the enhanced antitumor efficacy observed *in vivo*, as has been demonstrated for other anti-tumor antibodies (Clynes et al., 2000).

It has been known that tumor cells often upregulate growth factor and its receptors, for example, EGF and EGFR, in response to cellular stress or cytotoxic insult in order to activate their survival mechanisms (Dent et al., 1999, 2003; Hagan et al., 2000; Kari et al., 2003). Consequently, an increased ability of tumor cells to survive is inherently associated with resistance to cytotoxic agents (Baselga et al., 1993; Kari et al., 2003; Talapatra et al., 2001). Several *in vivo* studies have demonstrated direct correlation between increased EGFR expression and decreased response (i.e., the development of resistance) to chemotherapy and radiotherapy regimens for several different tumor types (Nicholson et al., 1989; Sheridan et al., 1997; Volm et al., 1992). To this end, a number of studies have shown that cetuximab could augment the anti-tumor activity of various anti-cancer agents, including cisplatin, doxorubicin, fluorouracil, gemcitabine, paclitaxel, and topotecan, both *in vitro* and *in vivo* in human tumor xenograft models in mice (Aboud-Pirak et al., 1988; Baselga et al., 1993; Fan et al., 1993a; Ciardiello et al., 1999; Prewett et al., 1996a, 1996b). In these studies, combination treatment with both cetuximab and cytotoxic drugs resulted in markedly enhanced tumor inhibition over treatment with either agent alone, and in some models, led to tumor regression and eradication of established tumors. It is pertinent to note that the tumor cells used in some of these studies were poorly responsive to the cytotoxic agents alone, but were sensitized to these agents by the antibody treatment (Baselga et al., 1993; Fan et al., 1993a). For example, cetuximab has been shown to significantly potentiate—in an additive or synergistic manner—the antitumor activity of irinotecan (CPT-11) in a variety of preclinical models, including several human colorectal cancer xenografts, such as HT-29, DLD-1, HCT-8, and SN-38 (an active metabolic derivative of irinotecan)-resistant tumor models, in which irinotecan and/or cetuximab exhibit poor efficacy as mono therapies (Prewett et al., 2002). Similar enhancement of anti-tumor activity has been observed in studies in which cetuximab is used in combination with radiation (Huang et al., 1999; Milas et al., 2000; Raben et al., 2005; Saleh et al., 1999). These observations suggest that simultaneous inhibition of EGFR-mediated biological activities may improve (or sensitize) tumor response to conventional cytotoxic therapy. Similar phenomena have been observed in HER2 over-expressing tumors with the use of the anti-HER2 mAb, trastuzumab, and cytotoxic agents or radiations (Hancock et al., 1991; Pegram et al., 1999; Pietras et al., 1994, 1998, 1999; Shepard et al., 1991).

B. Early Clinical Development of Cetuximab as Antitumor Agent

The anti-tumor activity and lack of significant toxicity of cetuximab observed in animal models encouraged the clinical evaluation of the antibody in patients with EGFR-expressing tumors. Of the clinical trials carried out to date, cetuximab has been administrated both as a single agent and in combination with chemotherapeutic agents or radiation treatment (Ng and Cunningham, 2004; Waksal, 1999). Initial trials were performed in patients with head and neck cancers but have expanded to other patients with EGFR-expressing tumors including colorectal, lung, pancreatic, prostate, and renal cell carcinomas. A number of earlier phase I/II studies were carried out in patients with advanced epithelial malignancies of head and neck, lung, prostate, breast, and renal cell, with various cetuximab dose schedules (Mendelsohn, 2000, 2001; Waksal, 1999). Clinical experience to date suggests a favorable toxicity profile (Needle, 2002; Thomas, 2005). The most common toxicity associated with cetuximab, either alone or in combination with chemotherapy or radiation, is an acne-like rash in a majority of patients (with grade 3 folliculitis in approximately 10% of patients). The rash appears during the first 2 to 3 weeks of therapy and resolves spontaneously upon cessation of antibody treatment without scarring. Approximately 7% of the patients experienced various grades of allergic reaction, with 2% of the patients developing grade 4 allergic anaphylactic reactions that occurred during the test dose or during the first infusion. All the allergic reactions responded to standard outpatient treatment. Out of 196 patients tested, 11 patients (5.6%) developed antibodies to cetuximab, but sera from two patients with the highest antibody titer did not neutralize the biological activity of cetuximab in an *in vitro* assay, nor did the human antibodies have any effect on the pharmacokinetics of the repeated infusions of cetuximab (Khazaeli et al., 2000). It is important to point out that the combination of cetuximab with conventional chemotherapy or radiation did not appear to influence or exacerbate the side effects of the cytotoxic therapy. Pharmacokinetic studies showed dose-dependent saturable kinetics, with no change in the clearance of the antibody after repeated doses (Baselga et al., 2000). In one study, tumor tissues from treated patients were removed upon completion of the therapy and assessed for EGFR receptor saturation by immunohistochemistry (IHC) analysis. Combining results from both the pharmacokinetic and the IHC analysis, the recommended dosing schedule of cetuximab in later clinical studies is 400 mg/m^2 loading dose followed by a 250 mg/m^2 weekly maintenance dose (Baselga et al., 2000). This dosing schedule was confirmed to achieve sustained circulation levels of cetuximab that block EGFR activation and downstream signaling in biopsy specimens from patients (Albanell et al., 2001; Shin et al., 2001).

IV. CETUXIMAB IN CLINICAL STUDIES IN PATIENTS WITH mCRC

EGFR is expressed in a significant percentage (from 25% to 80%) of human colorectal tumors, and its overexpression is usually associated with advanced diseases. Clinical trials have been conducted using both cetuximab alone and cetuximab in combination with irinotecan in the treatment of mCRC patients who have failed prior irinotecan-based regimens. These trials were aimed to demonstrate that treatment with cetuximab, either as single agent or in combination with irinotecan, would lead to significant antitumor activities as those observed in the preclinical studies. That is, not only does cetuximab have substantial clinical activity as a monotherapy, but when combined with irinotecan, results in further improvement in antitumor efficacy even in the setting of prior irinotecan failure.

EMR-007 (BOND), a randomized phase II study, is designed to evaluate the activity of cetuximab monotherapy and of cetuximab in combination with irinotecan in a prospective manner (Cunningham et al., 2004). In this design, patients with EGFR-positive tumors who had recently failed irinotecan therapy were randomized in a 2:1 fashion to receive the combination of cetuximab and irinotecan or cetuximab alone. All patients were determined by the investigators to have progressed on their most recent prior irinotecan therapy at time of randomization. The radiographic and clinical data collected during the study was reviewed by an Independent Review Committee (IRC) to determine the patients' refractoriness to prior irinotecan at baseline and their response to treatment on-study. A total of 329 patients were randomized, 218 to receive the combination of irinotecan and cetuximab and 111 to receive cetuximab alone. The objective response (partial response, PR) was achieved in 22.9% of the patients who received both cetuximab and irinotecan, and in 10.8% of the patients treated with cetuximab alone ($P = 0.007$). The time-to-progression was also significantly different between the two groups in favor of the combination regimen; the median time-to-progression for the combination arm is 4.1 months versus that of 1.5 months for the monotherapy group ($P < 0.001$). The most common toxicity associated with cetuximab treatment was acne-like skin rash, which occurred in 80% of treated patients (grade 3/4, 9.4% in the combination group and 5.2% in the monotherapy group). Severe anaphylactic reactions to cetuximab were developed in four (1.2%) patients requiring the discontinuation of the treatment. Other grade 3 or 4 toxicities in the combination group included diarrhea (21.2%), asthenia (13.7%), neutropenia (9.4%), nausea and vomiting (7.1%), and dyspnea (1.4%), compared to those of 1.7%, 10.4%, 0%, 4.3%, and 13%, respectively, in the antibody monotherapy group. The tumor response rate in this trial was consistent with that observed in an earlier trial in a similar patient population reported by Saltz et al. (study CP02-9923). In the latter trial, a single group of 120 mCRC patients who had failed irinotecan was treated with combination of cetuximab and irinotecan. PR was observed in 22.5% of treated patients (Saltz et al., 2001).

Two non-randomized, single arm studies provided additional data on the antitumor activity of cetuximab as monotherapy. In a small single-arm study (CP02-0141), 57 patients with documented progression on irinotecan or an irinotecan-based regimen were treated with cetuximab. Five patients (9%) achieved a PR. Twenty-one additional patients had stable disease (SD) or minor responses. The median survival time was 6.4 months (Saltz et al., 2004). Eighty-three percent of patients developed skin rash (18% with grade 3), 3.5% (two patients) had grade 3 allergic reactions, and 56% of patients experienced asthenia, fatigue, malaise, or lethargy (9% grade 3). In another trial (IMCL-0144) reported by Lenz et al. (2004), cetuximab as a single agent yielded a 11.6% PR rate in 346 patients refractory to both irinotecan and oxaliplatin, with another 31.8% of patients experiencing SD for at least 6 weeks. Median overall survival was 6.7 months. The most common adverse events were very similar to those observed in the CP02-0141 trial, including acne-like skin rash (90%, with 6% of grade 3/4), and fatigue/malaise (48%, grade 3/4, 10%). Detailed analysis of the response rate among patient subgroups, who had received two to nine regimens (median, 4) of prior chemotherapy (including 259 patients had received oxaliplatin after irinotecan failure and 87 patients had received irinotecan after or with oxaliplatin), revealed that cetuximab was equally active in all patient subgroups regardless of the numbers of prior therapy or the sequence of prior agents (Lenz et al., 2005).

Of special interest regarding observations across the clinical trials using cetuximab were: 1) the degree of EGFR expression in the tumor, either as the percentage of EGFR-positive tumor cells or as the staining intensity determined by standard IHC, did not seem to correlate with the clinical response of patients to cetuximab treatment. In EMR-007 study, patients with tumors that stained faint, weak, or moderate, and strong for EGFR expression, demonstrated response rates to cetuximab treatment of 20.8%, 24.7%, and 22.7%, respectively ($P = 0.64$). The response rates were also similar among patients with tumors that had either $\leq 10\%$ EGFR-positive cells or $>35\%$ EGFR-positive cells (Cunningham et al., 2004). 2) Patients with skin rash or any skin reactions had a higher response rate than those not showing any skin reactions, and the severity of skin reactions seemed to correlate well with the patient's response and survival to the antibody therapy (Cunningham et al., 2004; Saltz et al., 2003). For example, in EMR-007 study, patients with any degree of skin reaction to cetuximab therapy yielded response rates of 25.8% (the combination group) and 13% (the antibody alone group), compared to those of 6.3% and 0% in patients without any skin reactions, respectively. Further, patients who developed grade 3/4 skin reactions also showed higher response rates than those with grade 1/2 reactions; 55.2% versus 20.4% (in the combination group), and 33.3% versus 11.6% (in the antibody alone group), respectively. The median survival time among patients with skin reactions and those without skin reactions were 9.1 and 3.0 months, respectively, in the combination group, and 8.1 and 2.5 months, respectively, in the antibody alone group.

Cetuximab is currently being evaluated for safety and efficacy in mCRC patients in both first-line and second-line settings in combination with various chemotherapy regimens. In one phase II study, cetuximab was given in combination with FOLFOX-4 (oxaliplatin/5-fluorouracil/folinic acid) to patients with non-resectable mCRC (Rubio et al., 2005). In a preliminary analysis of 42 patients, 10% had a complete response, 71% had a PR, and 17% had SD. Median progression-free survival was 12.3 months. Nine patients subsequently underwent surgery of their metastases. The major grade 3/4 toxicities were acne-like rash (30%), neurotoxicity (30%), diarrhea (26%), neutropenia (21%), and stomatitis/mucositis (16%). Currently, a number of phase III clinical trials with cetuximab in mCRC patients are being carried out, including: a first-line phase III FOLFRI ± cetuximab in chemotherapy naïve patients (CRYSTAL trial in Europe, 1080 patients); a second-line phase III irinotecan ± cetuximab in oxaliplatin-failure patients (EPIC trial, 1300 patients); a first-line phase III continuous FOLFOX ± cetuximab versus FOLFOX (intermittent) (COIN trial, 2400 patients); and in adjuvant setting using cetuximab plus FOLFOX in CRC patients with high likelihood of recurrence.

V. MECHANISMS OF ACTION OF CETUXIMAB

The biochemical action of cetuximab, following binding to the extracellular domain of EGFR, is the immediate inhibition of ligand binding that leads to the blockade of EGFR activation and inhibition of downstream signal transduction pathways. Several mechanisms of action may play a role in the anti-tumor effect of cetuximab (Fan et al., 1993b; Goldstein et al., 1995; Mendelsohn, 1997, 2000; Naramura et al., 1993; Waksal, 1999). These mechanisms include inhibiting cell-cycle progression, inducing cell apoptosis, inhibiting angiogenesis, inhibiting invasion and metastasis, inhibiting DNA repair and recovery after chemotherapy and/or radiation, and inducing immunological effector responses. It appears that the main mechanism of action for cetuximab is because of its ability in disrupting EGFR-mediated signal transduction pathway (Mendelsohn, 1997; Waksal, 1999). Binding of cetuximab to cell surface EGFR not only efficiently blocks the association of the receptor to its ligands, EGF and TGF-α, but also induces a rapid internalization of the receptor, preventing further exposure of the receptor (Sunada et al., 1986). The down-regulation of cell surface receptor, combined with the direct competition effect of cetuximab for receptor/ligand interaction, effectively blocks EGF and TGF-α stimulated receptor activation and the initiation of the subsequent signal transduction cascade. As a result, the cellular process necessary for cell survival and proliferation does not ensue properly. In the case of combinational therapy, blockade of EGFR pathway by cetuximab inhibits DNA synthesis and repairing processes following the cytotoxic insults, such as chemotherapeutics or radiation, thereby enhancing the anti-tumor effects of the cytotoxic agents (Mendelsohn, 1997, 2000; Waksal, 1999).

A. Inhibition of Cell-Cycle Progression

Cetuximab blocks ligand from binding to EGFR, leading to inhibition of autophosphorylation of the receptor and activation of downstream signaling events that lead to cell proliferation. Inhibition of EGFR activation by cetuximab *in vitro* and *in vivo* leads to increased expression of the cyclin-dependent kinase inhibitors—including $p27^{KIP1}$, $p21^{CIP1}$, and $p16^{INK}$—resulting in cell-cycle arrest in the G1 phase and inhibition of tumor cell proliferation. In addition, a decrease in the cell-cycle regulators such as cyclin A, D, and E, an accumulation of hypophosphorylated retinoblastoma (Rb) protein and a decrease in the cell-cycle proliferating-cell nuclear antigen (PCNA) have also been observed in tumor cells following exposure to cetuximab (Fan et al., 1997; Kiyota et al., 2002; Liu et al., 2001; Peng et al., 1996; Prewett et al., 1996a, 1996b; Wu et al., 1996).

B. Potentiation of Apoptosis

Inhibition of signaling pathways in tumor cells can result in the modulation of pro- and anti-apoptotic regulators leading to decreased cell survival. Cetuximab has been shown to increase apoptosis in tumor cell lines by several mechanisms, including upregulating the pro-apoptosis regulator Bax, decreasing expression of the anti-apoptotic molecule Bcl-2, inducing hyperphosphorylation of Bcl-2 (leading to its functional inactivation), inhibiting the Akt survival pathway, and activating the pro-apoptotic caspases-3, -8, and -9 (Huang et al., 1999; Liu et al., 2000, 2001; Mandal et al., 1998; Wu et al., 1995).

C. Inhibition of Angiogenesis

There is evidence to suggest that cetuximab may affect tumor growth through inhibition of the production of mediators of angiogenesis. Cetuximab treatment causes a dose-dependent inhibition of VEGF mRNA and protein expression in tumor cells and consequently, a significant reduction in tumor neovasculature. Down regulation of the angiogenic factors, such as VEGF, basic fibroblast growth factor, interleukin 8, and TGF-α has been observed following cetuximab treatment of colon, pancreatic, and transitional cell tumor cells *in vitro*. Similar down regulation of these angiogenic factors and decreased microvascular density (a surrogate marker of angiogenesis) has been observed following cetuximab treatment of xenografts of these tumors in mouse models (Bancroft et al., 2002; Bruns et al., 2000; Ciardiello et al., 2000; Huang and Harari, 2000; Huang et al., 2002; Karashima et al., 2002; Petit et al., 1997; Perrotte et al., 1999).

D. Inhibition of Tumor Cell Invasion and Metastasis

Cetuximab has been shown to inhibit the expression and the activity of several metalloproteinases (MMPs) that play important roles in tumor invasion and metastasis. Several preclinical models have demonstrated that cetuximab treatment significantly inhibits the growth of disseminated metastatic tumors. In addition, a marked reduction in MMP-9 expression has been observed in cetuximab treated tumors *in vivo* (Huang et al., 2002; Inoue et al., 2000; Perrotte et al., 1999; Prewett et al., 1998).

E. Inhibition of DNA Repair

There is preclinical evidence demonstrating that the ability of tumor cells to repair DNA damage is reduced following EGFR blockade with cetuximab. It has been demonstrated that cellular damage by cytotoxic agents, radiation, and other stresses not only upregulate the EGFR/TGF-α autocrine loop in tumor cells initiating compensatory survival pathways involving MAPK or Akt signaling, but also activate NF-κB signaling in numerous cancer cell lines, including colorectal carcinomas and in tumor xenograft models (Bottero et al., 2001; Cusack et al., 2000; Huang et al., 2000; Li and Karin, 1998; Wang et al., 1999; Wang et al., 2000). Increased NF-κB activity is found in many tumors and is responsible for both promoting growth and survival of the malignant cells (Amit and Ben-Neriah, 2003; Baldwin et al., 2001), and for increased resistance to chemotherapeutic agents, including irinotecan (Xu and Villalona-Calero, 2002). Several studies have shown that EGFR activation enhances NF-κB activity in various cell types (Biswas et al., 2000; Habib et al., 2001; Sun and Carpenter, 1998). It is plausible that when cetuximab is combined with other cytotoxic agents (e.g., irinotecan), blockade of EGFR activity by the mAb would inhibit both the autocrine and paracrine EGFR/TGF-α (EGF) growth loops and NF-κB signaling provoked by the concurrent cytotoxic insult leading to impaired DNA repair mechanism, resulting in restoration of apoptosis potential and enhancement of the overall sensitivity of tumor cells to the cytotoxic agents (Sclabas et al., 2003). Similarly, anti-HER2 antibodies have also been shown to be capable of attenuating DNA repair pathways following exposure to DNA damaging agents such as platinum salts, topoisomerase II inhibitors, alkylating agents, and ionizing radiation (Pegram et al., 1999; Pietras et al., 1998, 1999). Taken together, these data suggest that inhibition of DNA repair is a potentially important mechanism responsible for the observed additive or synergistic antitumor activity between anti-erbB antibodies (EGFR or HER-2) and conventional cytotoxic agents.

F. Immunological Mechanisms

Both ADCC and CMC have been shown to play important roles in the efficacy of certain antitumor antibodies in animal models, including the anti-HER mAb, trastuzumab, and the anti-CD20 mAb, rituximab (Clynes et al., 2000). Cetuximab is an IgG1 antibody that is capable of eliciting both ADCC and CMC on tumor cells *in vitro*. In theory, either of these immune effector mechanisms may contribute to the antitumor activity of cetuximab. Data supporting an antitumor immune effector function of cetuximab *in vivo* are very limited, however. In contrast, there is evidence to suggest that the biological activity of cetuximab is derived, in large part, from its ability to inhibit EGFR signaling rather than from its theoretical immune effector functions. For example, a study investigating the anti-tumor effects of M225 and the corresponding $F(ab')_2$ fragment (Fan et al., 1993b) showed comparable *in vitro* activities for the intact antibody and $F(ab')_2$ fragment and for inhibition of A431 xenograft tumor growth in a mouse model. Moreover, no reduction in cetuximab efficacy has been observed in human tumor xenograft studies carried out in various genetic backgrounds of immunodeficient mice and/or in animals treated with myelosuppressive doses of cytotoxic agents, suggesting that immune effector functions unlikely play a major role in these models. Further, an aglycosylated version of cetuximab produced in transgenic corn plant, which had minimal ADCC activity in *in vitro* assays, was almost equally efficacious in inhibiting the growth of human xenografts in nude mice (Ludwig et al., 2004). Based on these observations, it is not clear at this time to what extent ADCC and CMC may contribute to the anti-tumor activity of cetuximab in both preclinical models and in clinical settings, although they certainly cannot be ruled out.

VI. CONCLUSIONS AND PERSPECTIVES

The biological activity of cetuximab, either as single agent or in combination with other cytotoxic agents, has been adequately proven in the clinic in patients with refractory mCRC. Recent clinical trials also demonstrated that cetuximab may have significant antitumor activity in patients with other malignancies, including those of squamous cell carcinoma of head and neck (SCCHN), pancreatic carcinoma (Mochlinski et al., 2005), non–small cell lung carcinoma (NSCLC) (Lilenbaum et al., 2005; Rosell et al., 2004) and ovarian carcinoma (Aghajanian et al., 2005) (for reviews see Kim et al., 2004; Ng and Cunningham, 2004). For example, in a recently reported phase III trial (IMC-9815), 424 SCCHN patients were randomized to receive high-dose radiation alone or in combination with cetuximab. The addition of cetuximab resulted in a significantly prolonged overall survival (54 versus 28 months, $P = 0.02$) and improved locoregional control of the tumors at 2 years (56% versus 48%, $P = 0.02$), without exacerbating the general side effects associated with high dose radiation (Bonner et al., 2004). In several other trials in SCCHN patients with advanced disease refractory to platinum-based regimen, cetuximab—either alone or in combination with the same dose and schedule of platinum (that patients had failed)—yielded a response rate of 10% to 13% (Baselga et al., 2005; Herbst et al., 2005), which compares favorably to the expected response rate in a similar patient population treated with the more toxic second-line chemotherapeutic agents. Despite the overall clinical success, there are a number of biological and clinical questions associated with cetuximab therapy that remain to be addressed. Answers to these questions will, undoubtedly, greatly facilitate further clinical development of the antibody in a more rational and efficient manner.

A. EGFR Expression and Efficacy

Unlike the case of trastuzumab in the treatment of HER2-expressing breast cancer where overexpression of HER2 in tumors is positively associated with patients' response, it seems apparent that the levels of EGFR expression in tumors (as determined by IHC staining) do not correlate with tumor response to cetuximab therapy. It is even more intriguing that in a recent study, in which 16 refractory EGFR-negative (by IHC) mCRC patients were treated with either cetuximab alone (two patients) or in combination with irinotecan (14 patients), 4 PR (including one patient received cetuximab alone) and two minor responses were achieved (Chung et al., 2005). Despite the small numbers of patients studied, this observation may suggest that the current IHC method used for EGFR staining is not sensitive enough for detecting minimal receptor expression and therefore may represent a poor screening protocol for both identifying mCRC patients who are eligible for cetuximab therapy and for predicting clinical outcome of the treated patients. It is possible that there is a threshold of EGFR expression, which is below the sensitivity of the current IHC detection, that is required for response to cetuximab therapy. There are a number of technical issues (variations) associated with the IHC protocol that can significantly affect the outcome of EGFR staining, such as the choice of the reagents used for the staining and specimen fixation methods, the tissue storage time, as well as the scoring systems used in different reference laboratories. Further, expression of EGFR in primary tumors may not necessarily correlate with receptor status in metastatic sites. In a retrospective study, primary tumors and related metastatic sites from 99 mCRC patients were examined for EGFR expression by IHC. EGFR expression was seen in the primary tumors in 53 (53%) patients, but the corresponding metastatic sites from 19 (36%) of these patients were found to be EGFR-negative. On the other hand, seven patients (15%) showed positive EGFR staining in the metastatic sites but not in the primary tumors (Scartozzi et al., 2004). This result suggests that screening of EGFR expression by IHC in primary CRC tumors may not be adequate for patient selection for anti-EGFR–based therapy.

As an alternative approach for assessing EGFR status in tumors, a recent study examined the gene copy numbers of the receptor by fluorescence *in situ* hybridization (FISH) in 31 patients (10 responders and 21 non-responders) treated by cetuximab or panitumumab (an anti-EGFR antibody currently being developed by Abgenix/Amgen). Eight of nine patients with objective response had an increased EGFR gene copy number, compared to that in 1 of 20 non-responders, suggesting that gene amplification of EGFR may serve as a better criterion for patient selection and indicator of patients' response to therapy (Moroni et al., 2005). Another report examined the correlation between EGFR gene amplification and protein expression by IHC revealed that only a small percentage of EGFR-positive (by IHC) tumors also harbor gene amplification (>5 copies/nucleus). In 158 primary or metastatic CRC tumors studied, positive IHC staining was detected in 85% of primary and 79% of metastatic tumors, whereas gene amplification was only seen in 12% of primary and 8% of metastatic tumors (Shia et al., 2005). Taken together, all these findings suggest that large perspective clinical trials are in clear need to further delineate the true relevance of the levels of EGFR expression (or gene amplification) in tumors and their response to cetuximab therapy.

B. Biomarker for Cetuximab Efficacy

Apart from EGFR expression, there are also significant efforts aiming to identify other biomarkers, or surrogate markers, for patient stratification and for predicting patients' response to anti-EGFR therapies. One obvious phenomenon that is positively associated with patients' response to cetuximab is the development of skin rash or other types of skin reactions during cetuximab therapy. As noted, patients who experienced skin rash are more likely to respond to the antibody than those who do not, and the severity of skin rash seems to correlate well with patients' response and overall survival. This observation suggests that skin rash may be used as a pharmacodynamic marker (i.e., an indication of EGFR inhibition) for biological activity of the anti-EGFR antibody, although one should keep in mind that activity seen in skin would not likely be an accurate indication of tumor inhibition since the downstream consequences of EGFR blockade are clearly different in the skin and tumors (Perez-Soler, 2003). Nevertheless clinical studies are currently being conducted in which the patients are given escalating doses of cetuximab until the development of skin rash in a hope to further enhance the biological activity of the antibody by achieving maximum receptor saturation/blockade.

On the molecular level, several preclinical and clinical studies have been performed in an attempt to identify molecular markers as predictive indicators for the outcome of anti-EGFR therapy. One preclinical study examined the proteome profile of two CRC cell lines with high expression of EGFR but a different response to cetuximab. Using two-dimensional electrophoresis and subsequent mass spectrometry, 14 proteins were identified that expressed differentially among the two cell lines: the responder Caco-2 and the non-responder HRT-18. While all the proteins are involved in the metabolic pathways and malignant growth, expression of certain proteins such as fatty acid–binding protein and heat shock protein 27 were implicated in the anti-apoptotic activity responsible for the non-responsiveness to cetuximab treatment by the HRT-18 cells (Skvortsov et al., 2004). In a retrospective clinical study, tumor specimens from 39 patients enrolled in the IMCL-0144 trial were examined for intratumoral mRNA levels of EGFR, VEGF, cyclin D1, cyclooxygenase 2, and interleukin 8 by real-time RT-PCR following laser-capture microdissection. High gene expression of VEGF was associated with resistance to cetuximab, whereas the combination of low gene expression of EGFR, cyclooxygenase 2, and interleukin 8 was significantly associated with longer overall survival (Vallbohmer et al., 2005). Both findings were independent of skin rash, which is itself correlated to survival. Finally, in several early clinical studies using EMD72000 (an anti-EGFR antibody being developed by Merck KGaA), or small-molecule EGFR kinase inhibitors, it was noted that while there was complete inhibition of tumor phosphorylated EGFR and MAP kinases in all antibody treated patients, phosphorylated Akt and Ki67 (a cell proliferation marker) were only inhibited in patients who responded to the therapy (Cappuzzo et al., 2004; Baselga and Arteage, 2005; Pao et al., 2004b). Taken together, these observations suggest that in addition to the target EGFR itself, careful and detailed analysis of downstream signaling pathways may yield molecular biomarkers that can be used as useful indicators for prediction of the efficacy of anti-EGFR–based therapies. Research advancements and successful implementation of new technologies, including high-throughput cDNA and protein arrays, will further enhance our ability to identify a series of genes/proteins that are differentially regulated by anti-EGFR treatment, thus providing good guidance in both patient selection and prediction of treatment outcome in future clinical studies.

C. EGFR Mutation and Efficacy of Anti-EGFR Therapy

It has been shown that, in a number of clinical trials in NSCLC patients involving the use of small-molecule TKI to EGFR including both gefitinib and erlotinib, a subset of patients including those with bronchioalveolar carcinoma, females, never-smokers, and Asian patients, had higher response rates and better clinical outcomes. Also it was later discovered that NSCLC patients with somatic mutations in the EGFR kinase domain yielded much better response to gefitinib and erlotinib (Lynch et al., 2004; Paez et al., 2004; Pao et al., 2004a. For review, see Pao and Miller, 2005). For example, together in three separate studies, 25 of 31 (81%) patients experiencing objective response to gefitinib or erlotinib have been shown to contain mutations in the exons 18 through 21 encoding EGFR kinase domain (Lynch et al., 2004; Paez et al., 2004; Pao et al., 2004a). On the other hand, none of the 29 tumor specimens from patients who are refractory to these agents showed such mutation. In a more recent study, EGFR kinase mutations were analyzed in 90 consecutive NSCLC patients who have received gefitinib (Han et al., 2005). EGFR kinase mutations were found in 17 (18.9%) patients. Eleven out of these 17 patients (64.7%) achieved objective response, compared to that of 10 out of 73 patients (13.7%) without mutations. In addition, these 17 patients had significantly prolonged time-to-progression (21.7 versus 1.8 months, $P < 0.001$) and overall survival (30.5 versus 6.6 months, $P < 0.001$) compared with the other 73 patients without mutations. Similar observations were also reported by Taron et al. (2005). In the latter study, EGFR kinase mutations were found in 17 out of 68 (25%) gefitinib-treated patients, and radiographical response to therapy was achieved in 16 out of these 17 (94.1%) patients compared to that of six out of 51 (12.6%) patients with wild-type EGFR. In a study of a Japanese population, EGFR mutations were found in 33 out of 59 (56%) NSCLC patients treated with gefitinib. Objective response was observed in 24 out of 29 (82.8%) patients with EGFR kinase mutations compared to 2 out of 21 patients (9.5%) without mutations ($P < 0.0001$) (Mitsudomi et al., 2005). *In vitro*, tumor cells harboring EGFR mutants demonstrated enhanced tyrosine kinase activity in response to EGF and increased sensitivity to inhibition by gefitinib (Lynch et al., 2004; Paez et al., 2004; Pao et al., 2004a). Early reports proposed that this increased activity of gefitinib and erlotinib toward tumor cells with EGFR kinase mutations is probably because of a greater sensitivity of the mutated kinases to the inhibitors, but more recent studies suggest that it is more likely the result of alterations in EGFR-dependent signaling pathways in tumor cells than that of direct changes in kinase-inhibitor interaction (Fabian et al., 2005).

Subsequent sequencing studies of a large panel of lung cancer specimens (Lynch et al., 2004; Paez et al., 2004; Pao et al., 2004a; Shigematsu et al., 2005) revealed that the overall rate of EGFR TK mutations was appropriately 19.6% (149 mutations out of 759 tumors studies), and the mutations are more frequent in tumors from female (37.5%) than male (13%), never-smokers (50.8%) than smokers (9%), adenocarcinomas (31.3%) than tumors of other histology (2.3%), and from patients of Asian origin (29.1%) than those of non-Asian origin (7.9%). Overall, the incidence of EGFR kinase mutations in patients from the United States is 9.5%. This observation is rather intriguing in that the subgroups of patients harboring higher EGFR TK mutation rate in fact correlates well to the same population of patients who had better response to gefitinib and erlotinib. Detailed analysis of the mutation sites (192 in total identified so far) demonstrated that the majority of the mutations occur in two hot spots—one (55.8%) is an in-frame deletion of four highly conserved amino acids (LREA) encoded by exon 19, and the other (44.2%) point mutation in exon 21 that lead to an amino acid substitution at position 858 (L858R) (Lynch et al., 2004; Paez et al., 2004; Pao et al., 2004a; Sordella et al., 2004; Shigematsu et al., 2005). Amid all these findings, perspective clinical trials to correlate these mutations with actual patients' response to either gefitinib or erlotinib are, however, yet to be conducted.

Rather than in NSCLC patients, earlier studies failed to positively identify any EGFR TK domain mutations in other tumors (Barber et al., 2004; Lee et al., 2005b; Shigematsu et al., 2005). Two recent reports have, however, revealed that EGFR kinase mutations may also be present in other human malignancies including CRC and SCCHN. In the study by Nagahara et al. (2005), while none of the 11 CRC cell lines examined exhibited somatic mutations, 4 out of 33 clinical tumors (12%) exhibited mutations in the EGFR kinase domain. Similarly, Lee et al. (2005a) observed three mutations (7.3%), all the same in-frame deletion mutation in exon 19 (del746-750), in 41 SCCHN patients analyzed. In another recent report, EGFR kinase domain in tumor specimens from 38 NSCLC and 39 mCRC patients participating in two separate cetuximab monotherapy studies were sequenced. Three mutations were identified in the 38 NSCLC patients: two del746-750 mutations in 13 patients experiencing SD, and one L861Q mutation in 21 patients with progressive disease. No mutations were found in one patient achieving a PR. In the 30 mCRC patients, including 20 experiencing PR and one CR, no mutations were identified (Tsuchihashi, et al., 2005). Further sequencing analysis of 160 biopsy samples of previously untreated CRC tumors from patients outside of cetuximab trials did not reveal any mutations in exons 18, 19, and 21 in the EGFR kinase domain. Taken together, these results suggest that the presence of EGFR kinase mutations may not represent a major predictive and prognostic factor for the efficacy of cetuximab therapy in CRC patients.

D. Cetuximab in Combination with Other Targeted Agents

One of the hallmarks of effective cancer treatment is the use of combinational therapeutic regimens comprising several cytotoxic or cytostatic agents that target cancer cells via different mechanisms. To this end, cetuximab has demonstrated significant enhanced antitumor activity in combination with either chemotherapeutics or radiation in the clinic, for example, with irinotecan in mCRC and with radiation in SCCHN. The side effects of these combination therapies are usually associated with the cytotoxic components in the regimens. Based on these observations, it is plausible that combination of anti-EGFR antibodies with other targeted therapeutic agents, including small-molecule TKI and antibodies directed against different tumor-associated targets, may yield enhanced therapeutic activity without adding severe unwanted toxicity. A number of preclinical studies have shown that combination of cetuximab with a small-molecule TKI, either gefitinib or erlotinib, resulted in enhanced tumor growth inhibition both *in vitro* and *in vivo* of a number of different tumor cell lines (Huang et al., 2004; Matar et al., 2004). There was, however, at least one study showed that the combination of cetuximab and gefitinib was rather antagonistic (Fischel et al., 2005). Taken together, these observations suggest that the concept of combining an anti-EGFR antibody with a small-molecule TKI for double-hitting the same target in tumor cells. While encouraging, more preclinical validation is needed and caution should be used in the clinical trials. Similarly, additive or synergistic antitumor activity has also been observed when cetuximab was used in combination with antisense oligonucleotides targeting other molecules such as protein A kinase, VEGF and CRIPTO (Ciardiello et al., 1998, 2000; Normanno et al., 1999; Tortora et al., 1999).

Cetuximab has also demonstrated additive or synergistic antitumor activities in various xenograft models when used in combination with mAb targeting other growth factor receptors, including those directed against HER2 (Ye, et al., 1999), VEGF receptor 2 (Jung et al., 2002) and insulin-like growth factor receptor (Lu et al., 2005). In a recent phase II clinical trial (NCI-6444), patients with mCRC refractory to irinotecan were given both cetuximab and bevacizumab (an anti-VEGF antibody being developed by Genentech, Inc., approved by the FDA for the first-line treatment of mCRC in combination with irinotecan), with or without irinotecan. Of 41 patients who received cetuximab/bevacizumab plus irinotecan, 37% had a PR, and the median time to progression was 7.9 months. Of 40 patients who received cetuximab/bevacizumab alone, 20% had a PR, and the median time to progression was 5.6 months. The most commonly reported adverse events in the cetuximab/bevacizumab/irinotecan arm were skin rash (grade 2, 60%; grade 3, 17%), diarrhea (grade 2, 29%; grade 3/4, 24%), fatigue (grade 2, 32%; grade 3, 10%), and neutropenia (grade 3/4, 22%), and the most commonly reported adverse event in the cetuximab/bevacizumab alone arm was skin rash (grade 2, 65%; grade 3, 20%) (Saltz et al., 2005). As a historical control, in the EMR-007 trial, patients who received cetuximab alone or cetuximab plus irinotecan had PR rate of 10.8% and 22.9%, respectively (Cunningham et al., 2004). Based on this encouraging result, a randomized phase III trial is being conducted by CALGB/NCI in which the chemotherapy naïve patients will be randomized into three groups to receive cetuximab/bevacizumab plus chemotherapy (FOLFOX or FOLFIRI depending on the choice of the individual investigators), cetuximab plus chemotherapy, or bevacizumab plus chemotherapy in the first-line settings.

RECOMMENDED RESOURCES

Erbitux prescription information: www.erbitux.com
Reviews on preclinical and clinical development of Erbitux and its adverse event profile:

1. Harding, J. and B. Burtness (2005). Cetuximab: An epidermal growth factor receptor chimeric human-murine monoclonal antibody. *Drugs Today (Barc)*. 41: 107–127.
2. Thomas, M. (2005). Cetuximab: Adverse event profile and recommendations for toxicity management. *Clin. J. Oncol. Nurs.* 9: 332–328.

Scientific literature reviews on EGFR and clinical development of EGFR inhibitors:

1. Arteage, C. (2003). Targeting HER1/EGFR: A molecular approach to cancer therapy. *Semin. Oncol. Suppl.* 7: 3–14.
2. Baselga, J. and C. L. Arteage (2005). Critical update and emerging trends in epidermal growth factor receptor targeting in cancer. *J. Clin. Oncol.* 23: 2445–2459.

REFERENCES

Aboud-Pirak, E., E. Hurwitz, et al. (1988). Efficacy of antibodies to epidermal growth factor receptor against KB carcinoma in vitro and in nude mice. *J. Natl. Cancer Inst.* 80: 1605–1611.

Aghajanian, C., P. Sabbatini, et al. (2005). A phase II study of cetuximab/paclitaxel/carboplatin for the initial treatment of advanced stage ovarian, primary peritoneal, and fallopian tube cancer. *Proc. Am. Soc. Clin. Oncol.* 24: abstract #5047.

Albanell, J., J. Codony-Servat, et al. (2001). Activated extracellular signal-regulated kinases: Association with epidermal growth factor receptor/transforming growth factor alpha expression in head and neck squamous carcinoma and inhibition by anti-epidermal growth factor receptor treatments. *Cancer Res.* 61: 6500–6510.

Albanell, J. and P. Gascon (2005). Small molecules with EGFR-TK inhibitor activity. *Curr. Drug Targets* 6: 259–274.

Amit, S. and Y. Ben-Neriah (2003). NF-κB activation in cancer: A challenge for ubiquitination- and proteasome-based therapeutic approach. *Semin. Cancer Biol.* 13: 15–28.

Arteage, C. (2003a). ErbB-targeted therapeutic approaches in human cancer. *Exp. Cell Res.* 284: 122–130.

Arteage, C. (2003b). Targeting HER1/EGFR: A molecular approach to cancer therapy. *Semin. Oncol.* Suppl. 7: 3–14.

Baldwin, A. S. (2001). Control of oncogenesis and cancer therapy resistance by the transcription factor NF-κB. *J. Clin. Invest.* 107: 241–246.

Bancroft, C. C., Z. Chen, et al. (2002). Effects of pharmacologic antagonists of epidermal growth factor receptor, PI3K and MEK signal kinases on NF-κB and AP-1 activation and IL-8 and VEGF expression in human head and neck squamous cell carcinoma lines. *Int. J. Cancer* 99: 538–548.

Barber, T.D., B. Vogelstein, et al. (2004). Somatic mutations of EGFR in colorectal cancers and glioblastomas. *N. Engl. J. Med.* 351: 2883.

Baselga, J. and C. L. Arteage (2005). Critical update and emerging trends in epidermal growth factor receptor targeting in cancer. *J. Clin. Oncol.* 23: 2445–2459.

Baselga, J., L. Norton, et al. (1993). Antitumor effects of doxorubicin in combination with anti-epidermal growth factor receptor monoclonal antibodies. *J. Natl. Cancer Inst.* 85: 1327–1333.

Baselga, J., D. Pfister, et al. (2000). Phase I studies of anti-epidermal growth factor receptor chimeric antibody C225 alone and in combination with cisplatin. *J. Clin. Oncol.* 18: 904–914.

Baselga, J., J. M. Trigo, et al. (2005). Phase II multicenter study of the antiepidermal growth factor receptor monoclonal antibody cetuximab in combination with platinum-based chemotherapy in patients with platinum-refractory metastatic and/or recurrent squamous cell carcinoma of the head and neck. *J. Clin. Oncol.* 23: 5568–5577.

Bianco, R., G. Daniele, et al. (2005). Monoclonal antibodies targeting the epidermal growth factor receptor. *Curr. Drug Targets* 6: 275–287.

Biswas, D.K., A. P. Cruz, et al. (2000). Epidermal growth factor-induced nuclear factor κB activation: A major pathway of cell-cycle progression in estrogen-receptor negative breast cancer cells. *Proc. Natl. Acad. Sci. U. S. A.* 97: 8542–8547.

Bonner, J.A., J. Giralt, et al. (2004). Cetuximab prolongs survival in patients with locoregionally advanced squamous cell carcinoma of head and neck: A phase III study of high dose radiation therapy with or without cetuximab. *Proc. Am. Soc. Clin. Oncol.* 23: abstract #5507.

Bottero, V., V. Busuttil, et al. (2001). Activation of nuclear factor κB through the IKK complex by the topoisomerase poisons SN38 and doxorubicin: A brake to apoptosis in HeLa human carcinoma cells. *Cancer Res.* 61: 7785–7791.

Brandt, R., R. Eisenbrandt, et al. (2000). Mammary gland specific hEGF receptor transgene expression induces neoplasia and inhibits differentiation. *Oncogene.* 19: 2129–2137.

Bruns, C. J., M. T. Harbison, et al. (2000). Epidermal growth factor receptor blockade with C225 plus gemcitabine results in regression of human pancreatic carcinoma growing orthotopically in nude mice by antiangiogenic mechanisms. *Clin. Cancer Res.* 6: 1936–1948.

Cappuzzo, F., E. Magrini, et al. (2004). Akt phosphorylation and gefitinib efficacy in patients with advanced non-small-cell lung cancer. *J. Natl. Cancer Inst.* 96: 1133–1141.

Carpenter, G. (2000). The EGF receptor: A nexus for trafficking and signaling. *Bioessays* 22: 697–707.

Chadd, H. E. and S. M. Chamow (2001). Therapeutic antibody expression technology. *Curr. Opin. Biotechnol.* 12: 188–194.

Chiao, P. J., F. Z. Bischoff, et al. (1990). The current state of oncogenes and cancer: Experimental approaches for analyzing oncogenetic events in human cancer. *Cancer Metastasis Rev.* 9: 63–80.

Chung, K.Y., J. Shia, et al. (2005). Cetuximab shows activity in colorectal cancer patients with tumors that do not express the epidermal growth factor receptor by immunohistochemistry. *J. Clin. Oncol.* 23: 1803–1810.

Ciardiello, F., R. Bianco, et al. (1999). Antitumor activity of sequential treatment with topotecan and anti-epidermal growth factor receptor monoclonal antibody C225. *Clin. Cancer Res.* 5: 909–916.

Ciardiello, F., R. Bianco, et al. (2000). Antiangiogenic and antitumor activity of anti-epidermal growth factor receptor C225 monoclonal antibody in combination with vascular endothelial growth factor antisense oligonucleotide in human GEO colon cancer cells. *Clin. Cancer Res.* 6: 3739–3747.

Ciardiello, F., R. Caputo, et al. (1998). Cooperative inhibition of renal cancer growth by anti-epidermal growth factor receptor antibody and protein kinase A antisense oligonucleotide. *J. Natl. Cancer Inst.* 90: 1087–1094.

Clynes, R. A., T. L. Towers, et al. (2000). Inhibitory Fc receptors modulate in vivo cytoxicity against tumor targets. *Nat. Med.* 6: 443–446.

Cunningham, D., Y. Humblet, et al. (2004). Cetuximab monotherapy and cetuximab plus irinotecan in irinotecan-refractory metastatic colorectal cancer. *N. Engl. J. Med.* 351: 337–345.

Cusack, J. C. Jr., R. Liu, and A. S. Baldwin Jr. (2000). Inducible chemoresistance to 7-ethyl-10-[4-(1-piperidino)-1-piperidino]-carbonyloxycamptothecin (CPT-11) in colorectal cancer cells and a xenograft model is overcome by inhibition of nuclear factor-κB activation. *Cancer Res.* 60: 2323–2330.

Dassonville, O., J. L. Formento, et al. (1993). Expression of epidermal growth factor receptor and survival in upper aerodigestive tract cancer. *J. Clin. Oncol.* 11: 1873–1878.

Dent, P., D. B. Reardon, et al. (1999). Radiation-induced release of transforming growth factor α activates the epidermal growth factor receptor and mitogen-activated protein kinase pathway in carcinoma cells, leading to increased proliferation and protection from radiation-induced cell death. *Mol. Biol. Cell.* 10: 2493–2506.

Dent, P., A. Yacoub, et al. (2003). Stress and radiation-induced activation of multiple intracellular signaling pathways. *Radiat. Res.* 159: 283–300.

Di Fiore, P. P., J. H. Pierce, et al. (1987). Overexpression of the human EGF receptor confers an EGF-dependent transformed phenotype to NIH 3T3 cells. *Cell.* 51: 1063–1070.

Divgi, C. R., S. Welt, et al. (1991). Phase I and imaging trial of indium 111-labeled anti-epidermal growth factor receptor monoclonal antibody 225 in patients with squamous cell lung carcinoma. *J. Natl. Cancer Inst.* 83: 97–104.

Downward, J., Y. Yarden, et al. (1984). Close similarity of epidermal growth factor receptor and v-erb-B oncogene protein sequences. *Nature.* 307: 521–527.

Ekstrand, A. J., C. D. James, et al. (1991). Genes for epidermal growth factor receptor, transforming growth factor α, and epidermal growth factor and their expression in human gliomas in vivo. *Cancer Res.* 51: 2164–2172.

Fabian, M. A., W. H. Biggs, et al. (2005). A small molecule-kinase interaction map for clinical kinase inhibitors. *Nat. Biotechnol.* 23: 329–336.

Fan, Z., J. Baselga, et al. (1993a). Antitumor effect of anti-epidermal growth factor receptor monoclonal antibodies plus cis-diamminedichloroplatinum on well established A431 cell xenografts. *Cancer Res.* 53: 4637–4642.

Fan, Z., H. Masui, et al. (1993b). Blockade of epidermal growth factor receptor function by bivalent and monovalent fragments of 225 anti-epidermal growth factor receptor monoclonal antibodies. *Cancer Res.* 53: 4322–4328.

Fan, Z., B. Y. Shang, et al. (1997). Reciprocal changes in p27(Kip1) and p21(Cip1) in growth inhibition mediated by blockade or overstimulation of epidermal growth factor receptors. *Clin. Cancer Res.* 3: 1943–1948.

Fischel, J. L., P. Formento, and G. Milano (2005). Epidermal growth factor receptor double targeting by a tyrosine kinase inhibitor (Iressa) and a monoclonal antibody (Cetuximab): Impact on cell growth and molecular factors. *Br. J. Cancer* 92: 1063–1068.

Glennie, M. J. and J. G. van de Winkel (2003). Renaissance of cancer therapeutic antibodies. *Drug Discov. Today* 8: 503–510.

Goldstein, N. I., M. Prewett, et al. (1995). Biological efficacy of a chimeric antibody to the epidermal growth factor receptor in a human tumor xenograft model. *Clin. Cancer Res.* 1: 1311–1318.

Goldstein, N. S. and M. Armin (2001). Epidermal growth factor receptor immunohistochemical reactivity in patients with American Joint Committee on Cancer Stage IV colon adenocarcinoma: Implications for a standardized scoring system. *Cancer* 92: 1331–1346.

Grandis, J. R., M. F. Melhem, et al. (1998). Levels of TGF-α and EGFR protein in head and neck squamous cell carcinoma and patient survival. *J. Natl. Cancer Inst.* 90: 824-832.

Grant, S., L. Qiao, and P. Dent (2002). Roles of ERBB family receptor tyrosine kinases, and downstream signaling pathways, in the control of cell growth and survival. *Front. Biosci.* 7: d376–d389.

Gullick, W. J. (1998). Type I growth factor receptors: Current status and future work. *Biochem. Soc. Symp.* 63: 193–198.

Habib, A. A., S. Chatterjee, et al. (2001). The epidermal growth factor receptor engages receptor interacting protein and nuclear factor-κB (NF-κB)-inducing kinase to activate NF-κB: Identification of a novel receptor-tyrosine kinase signalosome. *J. Biol. Chem.* 276: 8865–8874.

Hagan, M., L. Wang, et al. (2000). Ionizing radiation-induced mitogen-activated protein (MAP) kinase activation in DU145 prostate carcinoma cells: MAP kinase inhibition enhances radiation-induced cell killing and G2/M-phase arrest. *Radiat. Res.* 153: 371–383.

Han, S. W., T. Y. Kim, et al. (2005). Predictive and prognostic impact of epidermal growth factor receptor mutation in non-small-cell lung cancer patients treated with gefitinib. *J. Clin. Oncol.* 23: 2493–2501.

Hancock, M. C., B. C. Langton, et al. (1991). A monoclonal antibody against the c-erbB-2 protein enhances the cytotoxicity of *cis*-diamminedichloroplatinum against human breast and ovarian tumor cell lines. *Cancer Res.* 51: 4575–4580.

Herbst, R. S., M. Arquette, M., et al. (2005). Epidermal growth factor receptor antibody cetuximab and cisplatin for recurrent and refractory squamous cell carcinoma of the head and neck: A phase II, multicenter study. *J. Clin. Oncol.* 23: 5578–5587.

Huang, S., E. A. Armstrong, et al. (2004). Dual-agent molecular targeting of the epidermal growth factor receptor (EGFR): Combining anti-EGFR antibody with tyrosine kinase inhibitor. *Cancer Res.* 64: 5355–5362.

Huang, S. M., J. Li, and P. M. Harari (2002). Molecular inhibition of angiogenesis and metastatic potential in human squamous cell carcinomas after epidermal growth factor receptor blockade. *Mol. Cancer Ther.* 1: 507–514.

Huang, S. M., J. M. Bock, and P. M. Harari (1999). Epidermal growth factor receptor blockade with C225 modulates proliferation, apoptosis, and radiosensitivity in squamous cell carcinomas of the head and neck. *Cancer Res.* 59: 1935–1940.

Huang, S. M. and P. M. Harari (2000). Modulation of radiation response after epidermal growth factor receptor blockade in squamous cell carcinomas: Inhibition of damage repair, cell cycle kinetics, and tumor angiogenesis. *Clin. Cancer Res.* 6: 2166–2174.

Huang, T. T., S. M. Wuerzberger-Davis, et al. (2000). NF-κB activation by camptothecin: A linkage between nuclear DNA damage and cytoplasmic signaling events. *J. Biol. Chem.* 275: 9501–9509.

Humphreys, R. C. and L. Hennighausen (2000). Transforming growth factor α and mouse models of human breast cancer. *Oncogene*. 19: 1085–1091.

Hunter, T. (1991). Cooperation between oncogenes. *Cell.* 64: 249–270.

Hynes, N. E. and H. A. Lane (2005). ERBB receptors and cancer: The complexity of targeted inhibitors. *Nat. Rev. Cancer*. 5: 341–354.

Inoue, K., J. W. Slaton, et al. (2000). Paclitaxel enhances the effects of the anti-epidermal growth factor receptor monoclonal antibody ImClone C225 in mice with metastatic human bladder transitional cell carcinoma. *Clin. Cancer Res.* 6: 4874–4884.

Jones, P. T., P. H. Dear, et al. (1986). Replacing the complementarity-determining regions in a human antibody with those from a mouse. *Nature*. 321: 522–555.

Jung, Y. D., P. F. Mansfield, et al. (2002). Effects of combination anti-vascular endothelial growth factor receptor and anti-epidermal growth factor receptor therapies on the growth of gastric cancer in a nude mouse model. *Eur. J. Cancer*. 38: 1133–1140.

Karashima, T., P. Sweeney, et al. (2002). Inhibition of angiogenesis by the antiepidermal growth factor receptor antibody ImClone C225 in androgen-independent prostate cancer growing orthotopically in nude mice. *Clin. Cancer Res.* 8: 1253–1264.

Kari, C., T. O. Chan, et al. (2003). Targeting the epidermal growth factor receptor in cancer: Apoptosis takes center stage. *Cancer Res.* 63: 1–5.
Khazaeli, M. B., A. F. LoBuglio, A.F., et al. (2000). *Proc. Am. Soc. Clin. Oncol.* 19: abstract #808.
Kim, E. S., E. E. Vokes, and M. S. Kies (2004). Cetuximab in cancers of the lung and head & neck. *Semin. Oncol.* Suppl. 1: 61–67.
Kiyota, A., S. Shintani, et al. (2002). Anti-epidermal growth factor receptor monoclonal antibody 225 upregulates p27(KIP1) and p15(INK4B) and induces G1 arrest in oral squamous carcinoma cell lines. *Oncology*. 63: 92–98.
Klapper, L. N., M. H. Kirschbaum, et al. (2000). Biochemical and clinical implications of the ErbB/HER signaling network of growth factor receptors. *Adv. Cancer Res.* 77: 25–79.
Klijn, J. G., P. M. Berns, et al. (1992). The clinical significance of epidermal growth factor receptor (EGF-R) in human breast cancer: A review on 5232 patients. *Endocr. Rev.* 13: 3–17.
Krause, D.S. and R. A. Van Etten (2005). Tyrosine kinases as targets for cancer therapy. *N. Engl. J. Med.* 353: 172–187.
Lee, J. W., Y. H. Soung, et al. (2005a). Somatic mutations of EGFR gene in squamous cell carcinoma of the head and neck. *Clin. Cancer Res.* 11: 2879–2882.
Lee, J. W., Y. H. Soung, et al. (2005b). Absence of EGFR mutation in the kinase domain in common human cancers besides non-small cell lung cancer. *Int. J. Cancer*. 113: 510–511.
Lenz, H. J., R. J. Mayer, et al. (2004). Activity of cetuximab in patients with colorectal cancer refractory to both irinotecan and oxaliplatin. *Proc. Am. Soc. Clin. Oncol.* 23: abstract #3510.
Lenz, H. J., R. J. Mayer, et al. (2005). Consistent response to treatment with cetuximab monotherapy in patients with metastatic colorectal cancer. *Proc. Am. Soc. Clin. Oncol.* 24: abstract #3536.
Li, N. and M. Karin (1998). Ionizing radiation and short wavelength UV activate NF-κB through two distinct mechanisms. *Proc. Natl. Acad. Sci. U. S. A.* 95: 13012–13017.
Lilenbaum, R., P. Bonomi, et al. (2005). A phase II trial of cetuximab as therapy for recurrent non-small cell lung cancer (NSCLC): Final results. *Proc. Am. Soc. Clin. Oncol.* 24: abstract #7036.
Liu, B., M. Fang, et al. (2001). Fibroblast growth factor and insulin-like growth factor differentially modulate the apoptosis and G1 arrest induced by anti-epidermal growth factor receptor monoclonal antibody. *Oncogene*. 20: 1913–1922.
Liu, B., M. Fang, et al. (2000). Induction of apoptosis and activation of the caspase cascade by anti-EGF receptor monoclonal antibodies in DiFi human colon cancer cells do not involve the c-jun N-terminal kinase activity. *Br. J. Cancer* 82: 1991–1999.
Lonberg, N. and D. Huszar (1995). Human antibodies from transgenic mice. *Int. Rev. Immunol.* 13: 65–93.
Lu, D., H. Zhang, et al. (2005). A fully human recombinant IgG-like bispecific antibody to both the epidermal growth factor receptor and the insulin-like growth factor receptor for enhanced antitumor activity. *J. Biol. Chem.* 280, 19665–19672.
Ludwig, D.L., L. Witte, et al. (2004). Conservation of receptor antagonist anti-tumor activity by epidermal growth factor receptor antibody expressed in transgenic corn seed. *Hum. Antibodies*. 13: 81–90.
Lynch, T. J., D. W. Bell, et al. (2004). Activating mutations in the epidermal growth factor receptor underlying responsiveness of non-small-cell lung cancer to gefitinib. *N. Engl. J. Med.* 350: 2129–2139.
Mandal, M., L. Adam, et al. (1998). Nuclear targeting of Bax during apoptosis in human colorectal cancer cells. *Oncogene*. 17: 999–1007.
Masui, H., T. Kawamoto, et al. (1984). Growth inhibition of human tumor cells in athymic mice by anti-epidermal growth factor receptor monoclonal antibodies. *Cancer Res.* 44: 1002–1007.
Matar, P., F. Rojo, et al. (2004). Combined epidermal growth factor receptor targeting with the tyrosine kinase inhibitor gefitinib (ZD1839) and the monoclonal antibody cetuximab (IMC-C225): Superiority over single-agent receptor targeting. *Clin. Cancer Res.* 10: 6487–6501.
Mendelsohn, J. (1997). Epidermal growth factor receptor inhibition by a monoclonal antibody as anticancer therapy. *Clin. Cancer Res.* 3: 2703–2707.
Mendelsohn, J. (2000). Jeremiah Metzger Lecture. Targeted cancer therapy. *Trans. Am. Clin. Climatol. Assoc.* 111: 95–110.
Mendelsohn, J. (2001). The epidermal growth factor receptor as a target for cancer therapy. *Endocr. Relat. Cancer* 8: 3–9.
Milas, L., K. Mason, et al. (2000). *In vivo* enhancement of tumor radioresponse by C225 antiepidermal growth factor receptor antibody. *Clin. Cancer Res.* 6: 701–708.

Mitsudomi, T., T. Kosaka, et al. (2005). Mutations of the epidermal growth factor receptor gene predict prolonged survival after gefitinib treatment in patients with non-small-cell lung cancer with postoperative recurrence. *J. Clin. Oncol.* 23: 2513–2520.

Mochlinski, K., D. Cunningham, et al. (2005). Metastatic pancreatic adenocarcinoma in complete remission after three and a half years of combination therapy with cetuximab. *Proc. Am. Soc. Clin. Oncol.* 24: abstract #4278.

Moroni, M., S. Veronese, et al. (2005). Gene copy number for epidermal growth factor receptor (EGFR) and clinical response to antiEGFR treatment in colorectal cancer: A cohort study. *Lancet Oncol.* 6: 279–286.

Morrison, S. L., M. J. Johnson, et al. (1984). Chimeric human antibody molecules: Mouse antigen-binding domains with human constant region domains. *Proc. Natl. Acad. Sci. U. S. A.* 81: 6851–6855.

Nagahara, H., K. Mimori, et al. (2005). Somatic mutations of epidermal growth factor receptor in colorectal carcinoma. *Clin. Cancer Res.* 11: 1368–1371.

Naramura, M., S. D. Gillies, et al. (1993). Therapeutic potential of chimeric and murine anti-(epidermal growth factor receptor) antibodies in a metastasis model for human melanoma. *Cancer Immunol. Immunother.* 37: 343–349.

Needle, M. N. (2002). Safety experience with IMC-C225, an anti-epidermal growth factor receptor antibody. *Semin. Oncol.* Suppl. 14: 55–60.

Ng, M. and D. Cunningham (2004). Cetuximab (Erbitux)—an emerging targeted therapy for epidermal growth factor receptor-expressing tumours. *Int. J. Clin. Pract.* 58: 970–976.

Nicholson, R. I., J. M. Gee, and M. E. Harper (2001). EGFR and cancer prognosis. *Eur. J. Cancer.* 37: S9–S15.

Nicholson, S., J. R. Sainsbury, et al. (1989). Expression of epidermal growth factor receptors associated with lack of response to endocrine therapy in recurrent breast cancer. *Lancet.* 1: 182–185.

Nishikawa, R., X. D. Ji, et al. (1994). A mutant epidermal growth factor receptor common in human glioma confers enhanced tumorigenicity. *Proc. Natl. Acad. Sci. U. S. A.* 91: 7727–7731.

Noonberg, S. B. and C. C. Benz (2000). Tyrosine kinase inhibitors targeted to the epidermal growth factor receptor subfamily: Role as anticancer agents. *Drugs.* 59: 753–767.

Normanno, N., C. Bianco, et al. (2005). The ErbB receptors and their ligands in cancer: An overview. *Curr. Drug Targets.* 6: 243–257.

Normanno, N., G. Tortora, et al. (1999). Synergistic growth inhibition and induction of apoptosis by a novel mixed backbone antisense oligonucleotide targeting CRIPTO in combination with C225 anti-EGFR monoclonal antibody and 8-Cl-cAMP in human GEO colon cancer cells. *Oncol. Rep.* 6: 1105–1109.

Paez, J. G., P. A. Janne, et al. (2004). EGFR mutations in lung cancer: Correlation with clinical response to gefitinib therapy. *Science* 304: 1497–1500.

Pao, W. and V. Miller (2005). Epidermal growth factor receptor mutations, small-molecule kinase inhibitors, and non-small-cell lung cancer: Current knowledge and future directions. *J. Clin. Oncol.* 23: 2556–2568.

Pao, W., V. Miller, V., et al. (2004a). EGF receptor gene mutations are common in lung cancers from "never smokers" and are associated with sensitivity of tumors to gefitinib and erlotinib. *Proc. Natl. Acad. Sci. U. S. A.* 101: 13306–13311.

Pao, W., V. A. Miller, et al. (2004b). Predicting sensitivity of non-small-cell lung cancer to gefitinib: Is there a role for P-Akt? *J. Natl. Cancer Inst.* 96: 1117–1119.

Paz, K. and Z. Zhu (2005). Development of angiogenesis inhibitors to vascular endothelial growth factor receptor 2: Current status and future perspective. *Front. Biosci.* 10: 1415–1439.

Pedersen, M.W., M. Meltorn, et al. (2001). The type III epidermal growth factor receptor mutation: Biological significance and potential target for anti-cancer therapy. *Ann. Oncol.* 12: 745–760.

Pegram, M., S. Hsu, et al. (1999). Inhibitory effects of combinations of HER-2/neu antibody and chemotherapeutic agents used for treatment of human breast cancers. *Oncogene.* 18: 2241–2251.

Peng, D., Z. Fan, et al. (1996). Anti-epidermal growth factor receptor monoclonal antibody 225 up-regulates p27KIP1 and induces G1 arrest in prostatic cancer cell line DU145. *Cancer Res.* 56: 3666–3669.

Perez-Soler, R. (2003). Can rash associated with HER1/EGFR inhibition be used as a marker of treatment outcome? *Oncology.* Suppl. 12: 23–28.

Perrotte, P., T. Matsumoto, T., et al. (1999). Anti-epidermal growth factor receptor antibody C225 inhibits angiogenesis in human transitional cell carcinoma growing orthotopically in nude mice. *Clin. Cancer Res.* 5: 257–265.

Petit, A. M., J. Rak, et al. (1997). Neutralizing antibodies against epidermal growth factor and ErbB-2/neu receptor tyrosine kinases down-regulate vascular endothelial growth factor production by tumor cells in vitro and in vivo: Angiogenic implications for signal transduction therapy of solid tumors. *Am. J. Pathol.* 151: 1523–1530.

Pietras, R. J., B. M. Fendly, et al. (1994). Antibody to HER-2/neu receptor blocks DNA repair after cisplatin in human breast and ovarian cancer cells. *Oncogene.* 9: 1829–1838.

Pietras, R. J., M. D. Pegram, et al. (1998). Remission of human breast cancer xenografts on therapy with humanized monoclonal antibody to HER-2 receptor and DNA-reactive drugs. *Oncogene.* 17, 2235–2249.

Pietras, R. J., J. C. Poen, et al. (1999). Monoclonal antibody to HER-2/neureceptor modulates repair of radiation-induced DNA damage and enhances radiosensitivity of human breast cancer cells overexpressing this oncogene. *Cancer Res.* 59: 1347–1355.

Prewett, M., P. Rockwell, et al. (1996a). Anti-tumor and cell cycle responses in KB cells treated with a chimeric anti-EGFR monoclonal antibody in combination with cisplatin. *Int. J. Oncol.* 9: 217–224.

Prewett, M., P. Rockwell, et al. (1996b). Altered cell cycle distribution and cyclin-CDK protein expression in A431 epidermoid carcinoma cells treated with doxorubicin and a chimeric monoclonal antibody to the epidermal growth factor receptor. *Mol. Cell. Differ.* 4: 167–186.

Prewett, M., M. Rothman, M., et al. (1998). Mouse-human chimeric anti-epidermal growth factor receptor antibody C225 inhibits the growth of human renal cell carcinoma xenografts in nude mice. *Clin. Cancer Res.* 4: 2957–2966.

Prewett, M. C., A. T. Hooper, et al. (2002). Enhanced antitumor activity of anti-epidermal growth factor receptor monoclonal antibody IMC-C225 in combination with irinotecan (CPT-11) against human colorectal tumor xenografts. *Clin. Cancer Res.* 8: 994–1003.

Raben, D., B. Helfrich, et al. (2005). The effects of cetuximab alone and in combination with radiation and/or chemotherapy in lung cancer. *Clin. Cancer Res.* 11: 795–805.

Rosell, R., C. Daniel, et al. (2004). Randomized phase II study of cetuximab in combination with cisplatin (C) and vinorelbine (V) vs. CV alone in the first-line treatment of patients (pts) with epidermal growth factor receptor (EGFR)-expressing advanced non-small cell lung cancer (NSCLC). *Proc. Am. Soc. Clin. Oncol.* 23: abstract #7012.

Rubio, E. D., J. Tabernero, J., et al. (2005). Cetuximab in combination with oxaliplatin/5-fluorouracil (5-FU)/folinic acid (FA) (FOLFOX-4) in the first-line treatment of patients with epidermal growth factor receptor (EGFR)-expressing metastatic colorectal cancer: An international phase II study. *Proc. Am. Soc. Clin. Oncol.* 24: abstract #3535.

Saleh, M. N., K. P. Raisch, et al. (1999). Combined modality therapy of A431 human epidermoid cancer using anti-EGFr antibody C225 and radiation. *Cancer Biother. Radiopharm.* 14: 451–463.

Salomon, D. S., R. Brandt, et al. (1995). Epidermal growth factor-related peptides and their receptors in human malignancies. *Crit. Rev. Oncol. Hematol.* 19: 183–232.

Saltz, L., M. Rubin, et al. (2001). Cetuximab (IMC-C225) plus irinotecan (CPT-11) is active in CPT-11 refractory colorectal cancer (CRC) that expresses epidermal growth factor receptor (EGFR). *Proc. Am. Clin. Oncol.* 20: abstract #7.

Saltz, L. B., H. Lenz, H., et al. (2005). Randomized phase II trial of cetuximab/bevacizumab/irinotecan (CBI) versus cetuximab/bevacizumab (CB) in irinotecan-refractory colorectal cancer. *Proc. Am. Clin. Oncol.* 24: abstract #3508.

Saltz, L. B., N. J. Meropol, et al. (2004). Phase II trial of cetuximab in patients with refractory colorectal cancer that expresses the epidermal growth factor receptor. *J. Clin. Oncol.* 22: 1201–1208.

Sato, J. D., T. Kawamoto, et al. (1983). Biological effects in vitro of monoclonal antibodies to human epidermal growth factor receptors. *Mol. Biol. Med.* 1: 511–529.

Scartozzi, M., I. Bearzi, et al. (2004). Epidermal growth factor receptor (EGFR) status in primary colorectal tumors does not correlate with EGFR expression in related metastatic sites: Implications for treatment with EGFR-targeted monoclonal antibodies. *J. Clin. Oncol.* 22: 4772–4778.

Schlessinger, J. (2000). Cell signaling by receptor tyrosine kinases. *Cell.* 103: 211–225.

Sclabas, G. M., S. Fujioka, et al. (2003). Restoring apoptosis in pancreatic cancer cells by targeting the nuclear factor-κB signaling pathway with the anti-epidermal growth factor antibody IMC-C225. *J. Gastrointest. Surg.* 7: 37–43.

Shepard, H. M., G. D. Lewis, et al. (1991). Monoclonal antibody therapy of human cancer: Taking the HER2 protooncogene to the clinic. *J. Clin. Immunol.* 11: 117–127.

Sheridan, M. T., T. O'Dwyer, et al. (1997). Potential indicators of radiosensitivity in squamous cell carcinoma of the head and neck. *Radiat. Oncol. Investig.* 5: 180–186.

Shia, J., D. S. Klimstra, et al. (2005). Epidermal growth factor receptor expression and gene amplification in colorectal carcinoma: An immunohistochemical and chromogenic in situ hybridization study. *Mod. Pathol.* Apr 15, Epub ahead of print.

Shigematsu, H., L. Lin, et al. (2005). Clinical and biological features associated with epidermal growth factor receptor gene mutations in lung cancers. *J. Natl. Cancer Inst.* 97: 339–346.

Shin, D. M., N. J. Donato, et al. (2001). Epidermal growth factor receptor-targeted therapy with C225 and cisplatin in patients with head and neck cancer. *Clin. Cancer Res.* 7: 1204–1213.

Skvortsov, S., B. Sarg, et al. (2004). Different proteome pattern of epidermal growth factor receptor-positive colorectal cancer cell lines that are responsive and nonresponsive to C225 antibody treatment. *Mol. Cancer Ther.* 3: 1551–1558.

Sordella, R., D. W. Bell, et al. (2004). Gefitinib-sensitizing EGFR mutations in lung cancer activate anti-apoptotic pathways. *Science.* 305: 1163–1167.

Starling, N. and D. Cunningham (2004). Monoclonal antibodies against vascular endothelial growth factor and epidermal growth factor receptor in advanced colorectal cancers: Present and future directions. *Curr. Opin. Oncol.* 16: 385–390.

Stern, M. and R. Herrmann (2005). Overview of monoclonal antibodies in cancer therapy: Present and promise. *Crit. Rev. Oncol. Hematol.* 54: 11–29.

Sun, L. and G. Carpenter (1998). Epidermal growth factor activation of NF-κB is mediated through IκBα degradation and intracellular free calcium. *Oncogene.* 16: 2095–2102.

Sunada, H., B. E. Magun, et al. (1986). Monoclonal antibody against epidermal growth factor receptor is internalized without stimulating receptor phosphorylation. *Proc. Natl. Acad. Sci. U. S. A.* 83: 3825–3829.

Taron, W., Y. Ichinose, et al. (2005). Activating mutations in the tyrosine kinase domain of the epidermal growth factor receptor are associated with improved survival in gefitinib-treated chemorefractory lung adenocarcinomas. *Clin. Cancer Res.* 11: 5878–5885.

Thomas, M. (2005). Cetuximab: Adverse event profile and recommendations for toxicity management. *Clin. J. Oncol. Nurs.* 9: 332–338.

Tibes, R., J. Trent, and R. Kurzrock (2005). Tyrosine kinase inhibitors and the dawn of molecular cancer therapeutics. *Annu. Rev. Pharmacol. Toxicol.* 45: 357–384.

Tortora, G., R. Caputo, et al. (1999). Cooperative inhibitory effect of novel mixed backbone oligonucleotide targeting protein kinase A in combination with docetaxel and anti-epidermal growth factor-receptor antibody on human breast cancer cell growth. *Clin. Cancer Res.* 5: 875–881.

Tsuchihashi, Z., S. Khambata-Ford, et al. (2005). Responsiveness to cetuximab without mutations in EGFR. *N. Engl. J. Med.* 353: 208–209.

Ullrich, A. and J. Schlessinger (1990). Signal transduction by receptors with tyrosine kinase activity. *Cell.* 61: 203–212.

Ullrich, A., L. Coussens, et al. (1984). Human epidermal growth factor receptor cDNA sequence and aberrant expression of the amplified gene in A431 epidermoid carcinoma cells. *Nature* 309: 418–425.

Vallbohmer, D., W. Zhang, et al. (2005). Molecular determinants of cetuximab efficacy. *J. Clin. Oncol.* 23: 3536–3544.

Volm, M., T. Efferth, and J. Mattern (1992). Oncoprotein (c-myc, c-erbB1, c-erbB2, c-fos) and suppressor gene product (p53) expression in squamous cell carcinomas of the lung: Clinical and biological correlations. *Anticancer Res.* 12: 11–20.

von Mehren, M., G. P. Adams, and L. M. Weiner (2003). Monoclonal antibody therapy for cancer. *Annu. Rev. Med.* 54: 343–369.

Wang, C. Y., J. C. Cusack, et al. (1999). Control of inducible chemoresistance: Enhanced antitumor therapy through increased apoptosis by inhibition of NF-κB. *Nat. Med.* 5: 412–417.

Wang, X., K. D. McCullough, et al. (2000). Epidermal growth factor receptor-dependent Akt activation by oxidative stress enhances cell survival. *J. Biol. Chem.* 275: 14624–14631.

Waskal, H. W. (1999). Role of an anti-epidermal growth factor receptor in treating cancer. *Cancer Metastasis Rev.* 18: 427–436.

Wells, A. (1999). EGF receptor. *Int. J. Biochem. Cell Biol.* 31: 637–643.

Winter, G., A. D. Griffiths, et al. (1994). Making antibodies by phage display technology. *Annu. Rev. Immunol.* 12: 433–455.

Wu, X., Z. Fan, et al. (1995). Apoptosis induced by an anti-epidermal growth factor receptor monoclonal antibody in a human colorectal carcinoma cell line and its delay by insulin. *J. Clin. Invest.* 95: 1897–1905.

Wu, X., M. Rubin, et al. (1996). Involvement p27KIP1 in G1 arrest mediated by an anti-epidermal growth factor receptor monoclonal antibody. *Oncogene.* 12: 1397–1403.

Xu, Y. and M. A. Villalona-Calero (2002). Irinotecan: Mechanisms of tumor resistance and novel strategies for modulating its activity. *Ann. Oncol.* 13: 1841–1851.

Yamanaka, Y., H. Friess, et al. (1993). Coexpression of epidermal growth factor receptor and ligands in human pancreatic cancer is associated with enhanced tumor aggressiveness. *Anticancer Res.* 13: 565–569.

Yarden, Y. (2001). The EGFR family and its ligands in human cancer signalling mechanisms and therapeutic opportunities. *Eur. J. Cancer.* 37: S3–S8.

Yarden, Y. and M. X. Sliwkowski, M. X. (2001). Untangling the ErbB signalling network. *Nat. Rev. Mol. Cell. Biol.* 2: 127–137.

Ye, D., J. Mendelsohn, and Z. Fan (1999). Augmentation of a humanized anti-HER2 mAb 4D5 induced growth inhibition by a human-mouse chimeric anti-EGF receptor mAb C225. *Oncogene.* 18: 731–738.

Yonemura, Y., H. Takamura, et al. (1992). Interrelationship between transforming growth factor-α and epidermal growth factor receptor in advanced gastric cancer. *Oncology.* 49: 157–161.

4

MONOCLONAL ANTIBODY TO HER-2 IN BREAST CANCER

FRANCISCO J. ESTEVA* and GABRIEL N. HORTOBAGYI†

* *M.D., Ph.D., University of Texas M. D. Anderson Cancer Center, Houston, Texas*
† *M.D., University of Texas M. D. Anderson Cancer Center, Houston, Texas*

The human epidermal growth factor receptor 2 (HER-2) is a tyrosine kinase that is overexpressed in 20% to 25% of invasive breast cancers. Amplification of the *her-2* gene is associated with an aggressive tumor phenotype and a reduced patient survival rate. The HER-2 status of a tumor is the critical determinant of response to the HER-2–targeted monoclonal antibody trastuzumab (Herceptin; Genentech, South San Francisco, CA). Thus, accurate assessment of HER-2 expression levels is essential for identifying breast cancer patients who will benefit from trastuzumab. Trastuzumab combined with chemotherapy increases response rates, time to disease progression, and survival duration in patients with metastatic breast cancer and those with early-stage breast cancer. This chapter reviews the development of trastuzumab from a molecular breakthrough to clinical application.

I. INTRODUCTION

The *her-2* gene encodes a tyrosine kinase that is over-expressed in 20 to 25% of invasive breast cancers (Slamon et al., 1987). In human breast cancer cells, over-expression of HER-2 occurs primarily through amplification of the wild-type *her-2* gene (King et al., 1985) and is associated with poor disease-free survival (Press et al., 1997; Slamon et al., 1987). HER-2 has become an important therapeutic target in breast cancer for several reasons: 1) high HER-2 levels correlate strongly with the pathogenesis and prognosis of breast cancer; 2) the expression level of HER-2 in human cancer cells is much higher than in normal tissues; 3) in tumors that are driven by *her-2* gene amplification, the HER-2 protein is present in the majority of cancer cells; and 4) HER-2 over-expression is found in both primary tumors and metastases, indicating that anti–HER-2 therapy may be effective in all disease sites.

Soon after the discovery of the *her-2* gene (*c-erb*B-2, also known as *neu* in rats) (Hung et al., 1986), investigators showed that murine monoclonal antibodies could be used to downregulate its expression *in vitro* (Drebin et al., 1985). A major limitation of using murine antibodies directed against the HER-2 protein in cancer patients, however, was the development of human anti-mouse antibodies, precluding repeat dosing. In the early 1990s, investigators at Genentech produced "humanized" monoclonal antibodies using recombinant DNA technology. Trastuzumab, the first such construct, was engineered by inserting the complementarity-determining regions of the anti–HER-2 murine antibody 4D5 into the framework of the consensus human immunoglobulin G1 (Carter et al., 1992). Clinical trials showed that trastuzumab therapy was safe and effective, and it became the first HER-2–targeted therapy approved by the U.S. Food and Drug Administration (FDA) for the treatment of breast cancer. In this chapter, we will discuss the preclinical and clinical development of trastuzumab as targeted therapy for HER-2 over-expressing breast cancer.

II. MECHANISM OF ACTION OF TRASTUZUMAB

Although the mechanisms by which trastuzumab induces the regression of HER-2 over-expressing tumors are incompletely defined, several molecular and cellular effects have been observed in *in vitro* and *in vivo* models (Table 4.1).

A. Diminished Receptor Signaling

HER-2 activates multiple cellular signaling pathways, including the phosphatidylinositol-3 kinase (PI3K) and mitogen-activated protein kinase cascades. Trastuzumab reduces signaling in these pathways and thus promotes cell-cycle arrest and apoptosis. Diminished receptor signaling may result from trastuzumab-mediated internalization and degradation of the HER-2 receptor (Baselga et al., 2001). It is unclear, however, whether trastuzumab actually downregulates HER-2; some groups have shown that receptor levels are unchanged in response to trastuzumab treatment. An alternative mechanism by which trastuzumab may block PI3K signaling is disruption of the interaction between HER-2 and the Src tyrosine kinase, which may lead to the inactivation of Src, subsequent activation of the PI3K inhibitor phosphatase

TABLE 4.1 Proposed Mechanisms of Trastuzumab's Action

Disruption of receptor dimerization and downstream signaling pathways
G1-phase cell-cycle arrest and reduced proliferation: induces $p27^{kip1}$-cdk2 complex formation; induces $p27^{kip1}$ expression
Apoptosis: inhibits Akt activity
Suppression of angiogenesis: reduces tumor vasculature *in vivo;* reduces expression of pro-angiogenic VEGF, TGF-α, Ang-1, and PAI-1; induces anti-angiogenic TSP-1
Immune-mediated responses: antibody-dependent cellular cytotoxicity; stimulates natural killer cells
Inhibits proteolysis of HER-2 extracellular domain
Inhibits DNA repair

cdk, Cyclin-dependent kinase; *TGF-α*, transforming growth factor alpha; *Ang-1*, angiopoietin 1; *PAI-1*, type I plasminogen activator inhibitor.

and tensin homologue (PTEN), rapid Akt dephosphorylation, and inhibition of cell proliferation (Nagata et al., 2004).

B. G1-Phase Cell-Cycle Arrest: Modulation of $p27^{kip1}$

$P27^{kip1}$ is a member of the cyclin-dependent kinase inhibitory proteins that play an important role in counteracting the activity of cyclins D_1 and E on Rb phosphorylation and subsequent G_1-S transition. Therefore, $p27^{kip1}$ coordinates the activation of cyclin E-cdk2 with accumulation of cyclin D-cdk4 and initiates the timely exit of cells from the cell-cycle in response to antimitogenic signals. Cells treated with trastuzumab undergo arrest during the G1 phase of the cell-cycle, with a concomitant reduction in cell proliferation and reduced expression of proteins involved in sequestering $p27^{kip1}$, including cyclin D1. This results in the release of $p27^{kip1}$, allowing it to bind and inhibit cyclin E-cdk2 complexes. Trastuzumab also produces an accumulation of $p27^{kip1}$, although this seems to occur secondary to the formation of $p27^{kip1}$-cdk2 complexes, perhaps as a result of the decreased targeting of $p27^{kip1}$ for degradation by the ubiquitin pathway (Lane et al., 2001; Le et al., 2003).

C. Inhibition of Cleavage of the HER-2 Extracellular Domain

The full-length HER-2 receptor undergoes slow proteolytic cleavage in HER-2 over-expressing cells, yielding a 110-kDa extra-cellular domain (ECD), which can be detected in cell culture medium, and a 95-kDa membrane-associated fragment with increased kinase activity. The HER-2 ECD can also be detected in the serum of breast cancer patients, and HER-2 p95 has been found in some breast tumors, indicating that HER-2 ECD shedding occurs *in vivo*. Although the function of HER-2 p95 is not well defined, HER-2 ECD shedding may be of clinical importance because high serum levels of HER-2 ECD correlate with poor prognosis, increased metastasis, and decreased responsiveness to endocrine therapy and chemotherapy in patients with advanced breast cancer (Carney et al., 2003). Trastuzumab has been shown to block HER-2 ECD proteolytic cleavage, and shedding *in vitro*. Furthermore, the response to trastuzumab may depend in part on ECD levels before treatment (Esteva et al., 2002, 2005).

D. Inhibition of DNA Repair

In vitro studies have shown that trastuzumab is synergistic with a variety of chemotherapeutic agents (Pegram et al., 2004). Although not all the mechanisms of synergy are known, Pietras et al. (1999) proposed that synergy with DNA-damaging drugs may be due to trastuzumab's inhibition of DNA repair. These investigators demonstrated that trastuzumab partially inhibits the repair of DNA adducts *in vitro* after treatment with cisplatin and blocks unscheduled DNA synthesis (a measure of DNA repair) after irradiation (Pietras et al., 1999). The molecular mechanism by which trastuzumab blocks DNA repair may involve the modulation of $p21^{WAF1}$, such that trastuzumab blocks the cisplatin-associated induction of p21 and inhibits the tyrosine phosphorylation of p21 after irradiation. In addition, Pietras et al. (1999) showed that trastuzumab promotes an increase in DNA strand breaks specifically in HER-2 over-expressing BT474 and SKBR3 cells and increases the transcription of genes whose products are involved in DNA repair.

E. Induction of Apoptosis

Single-agent trastuzumab can dramatically reduce tumor size in patients with HER-2 over-expressing metastatic breast cancer. This clinical response can be explained by the agent's induction of apoptosis. In a clinical trial of preoperative trastuzumab in patients with locally advanced breast cancer, Mohsin et al. (2005) obtained primary breast cancer tissue before and after treatment and found that single-agent trastuzumab therapy significantly increased apoptotic cell death and inhibited PI3K/Akt pathways. In their study however, trastuzumab therapy did not affect Ki-67 expression, which is a marker of tumor cell proliferation (Mohsin et al., 2005). In a trial by Gennari et al. (2004), preoperative trastuzumab therapy did not change the expression level of HER-2 or Ki-67, but the study showed a correlation between antibody-dependent cellular cytotoxicity (ADCC) and response to therapy. Whether the observed cytotoxic effect of trastuzumab is a direct effect of the agent on breast cancer cells or is mediated by indirect mechanisms (i.e., anti-angiogenic activity or the immune system) remains to be determined.

F. Immune Mechanisms

The immune system effects of trastuzumab are difficult to study in humans. *In vitro*, trastuzumab has been shown to activate an ADCC response in multiple breast cancer cell lines. Natural killer (NK) cells, a principal immune cell type involved in ADCC, express the Fc gamma receptor, to which the Fc domain of the trastuzumab immunoglobulin G1 binds, activating NK-mediated cell lysis. In one study, mice bearing BT474 HER-2 over-expressing tumor xenografts had a tumor regression rate of 96% when treated with trastuzumab. In contrast, in mice lacking the Fc receptor (FcR−/−), the tumor regression rate was only 29% (Clynes et al., 2000). Thus, NK cells and ADCC are important contributors to the cytotoxic activity of trastuzumab but are not solely responsible, because partial tumor regression was still obtained in FcR−/− mice.

This immune function of trastuzumab was the focus of two recent clinical trials. In the study by Gennari et al. (2004), a strong lymphoid infiltration was noted in all breast cancer patients treated with preoperative trastuzumab, and ADCC correlated with the response to therapy. Repka et al. (2003) used low-dose interleukin-2 in combination with trastuzumab in patients with HER-2 over-expressing breast cancer that had progressed during multiple prior systemic therapies. Interleukin-2 increases the number of NK cells and enhances ADCC. In their small cohort of patients, no correlation was observed between clinical response and NK cell expansion or degree of ADCC activity (Repka et al., 2003). Patients with advanced metastatic breast cancer, however, are immunosuppressed and may not be the optimal population in which to study immune response modifiers. Additional studies are needed to better understand the importance of ADCC in mediating the response to trastuzumab.

G. Inhibition of Angiogenesis

The over-expression of HER-2 in human tumor cells is closely associated with increased angiogenesis and the expression of vascular endothelial growth factor. Treating HER-2 over-expressing breast cancers with trastuzumab reduces tumor volume and decreases microvessel density *in vivo* and reduces endothelial cell migration *in vitro* (Izumi et al., 2002). Furthermore, the expression of multiple pro-angiogenic factors is reduced and the expression of anti-angiogenic factors increased in trastuzumab-treated tumors compared with control-treated tumors *in vivo*. The combination of trastuzumab with paclitaxel inhibits angiogenesis-associated events to an even greater degree than does trastuzumab alone, an observation that may reflect more efficient drug delivery due to the normalization of tumor vasculature after trastuzumab treatment (Izumi et al., 2002).

III. MOLECULAR MECHANISMS OF TRASTUZUMAB RESISTANCE

Multiple mechanisms contributing to trastuzumab resistance in breast cancer cells have been proposed (Nahta and Esteva, 2003). These include: 1) increased cell signaling through other receptors from the HER family or the insulin-like growth factor-I receptor (IGF-IR), 2) PTEN loss and Akt signaling, and 3) the downregulation of $p27^{kip1}$.

The EGFR type I growth factor receptor tyrosine kinase family consists of EGFR, HER-2, HER-3, and HER-4. All except HER-3 contain a cytoplasmic tyrosine kinase region, and all except HER-2 bind to specific ligands at their ECD. Upon binding to ligands, the receptors dimerize, using HER-2 as the preferred binding partner. Heterodimerization induces tyrosine kinase activity and the downstream mitogen-activated protein kinase and PI3K signaling pathways. Although trastuzumab reduces HER-2–mediated signaling through these pathways, it does not reduce the signaling mediated by other HER receptors. Thus, cells with EGFR/HER-3 heterodimers or EGFR homodimers may demonstrate mitogenic PI3K and mitogen-activated protein kinase signaling even in the presence of trastuzumab. For this reason, agents that target multiple HER receptors may be effective against trastuzumab-resistant cancer cells. The dual kinase inhibitor lapatinib (GlaxoSmithKline, Research Triangle Park, NC) inhibits EGFR and HER-2, and current clinical studies are examining the efficacy of this agent against trastuzumab-refractory breast cancer (Esteva, 2004). Additionally, recently developed HER-2–targeted antibodies that inhibit dimerization, such as pertuzumab (Omnitarg; Genentech), may prove to be beneficial in treating patients with breast cancers that have progressed during trastuzumab therapy. In fact, we showed that trastuzumab and pertuzumab synergistically inhibit the survival of HER-2 over-expressing breast cancer cells, suggesting that the agents induce cytotoxicity via different mechanisms (Nahta et al., 2004a).

Constitutive Akt cell signaling was previously shown to inhibit cell-cycle arrest and apoptosis mediated by trastuzumab (Yakes et al., 2002). Another recent study suggested that increased PI3K signaling also contributes to trastuzumab resistance. Our group showed that the downregulation of PTEN blocks trastuzumab-mediated inhibition of proliferation in HER-2 over expressing breast cancer cells, with an increase in PI3K signaling (Nagata et al., 2004). Importantly, we showed that patients with PTEN-deficient HER-2 over-expressing breast tumors respond poorly to trastuzumab-based therapy. Our findings also suggested that PI3K inhibitors should be explored preclinically as potential therapies in trastuzumab-resistant tumors possessing low PTEN levels.

The growth inhibitory properties of trastuzumab depend in part upon its effects on the cdk-inhibiting protein $p27^{kip1}$. Trastuzumab increases the half-life of $p27^{kip1}$ by decreasing the cyclin E-cdk2 mediated phosphorylation of $p27^{kip1}$ and blocking subsequent ubiquitin-dependent degradation (Le et al., 2003). Trastuzumab also increases the association between $p27^{kip1}$ and cdk2 complexes, resulting in G1-phase cell-cycle arrest. Importantly, antisense oligonucleotides and small interfering RNA that reduce $p27^{kip1}$ expression levels also block trastuzumab-mediated growth arrest in SKBR3 HER-2 over-expressing breast cancer cells (Le et al., 2003). We recently showed that trastuzumab-resistant cell pools derived from the SKBR3 line after continuous drug exposure express reduced $p27^{kip1}$ levels with elevated cdk2 activity (Nahta et al., 2004b). The transfection of $p27^{kip1}$ or pharmacological induction of $p27^{kip1}$ restores the sensitivity of these cells to trastuzumab, confirming that $p27^{kip1}$ is a critical mediator of response to trastuzumab. Because trastuzumab-resistant cell pools are sensitive to the proteasome inhibitor MG132, which restores $p27^{kip1}$ levels, downregulation of $p27^{kip1}$ is likely due to increased protein degradation (Nahta et al., 2004b). Cellular localization of $p27^{kip1}$ may also be important for trastuzumab response, because trastuzumab-resistant BT474 HER-2 over-expressing cells have demonstrated loss of nuclear $p27^{kip1}$ expression. Thus, $p27^{kip1}$ may serve both as a marker of trastuzumab response and as a therapeutic target in the subset of breast cancers that progress during therapy.

The IGF mitogenic-signaling pathway is an attractive therapeutic target in breast cancer because its ligands and receptors are frequently over-expressed and implicated in cellular proliferation, transformation, and metastasis. High levels of IGFs prevent apoptosis in response to chemotherapeutic agents and radiation. Interestingly, overactive IGF-IR signaling is associated with resistance to trastuzumab in HER-2 over-expressing breast cancer cells (Lu et al., 2001). Lu et al. (2001) showed that trastuzumab-mediated growth arrest is lost in SKBR3 cells engineered to over-express IGF-IR but that this arrest is regained when the IGF-IR-inhibiting IGF binding protein 3 is added to the cell culture medium. Growth factor receptors of the type I class (HER family) and the IGF-IR signal through common pathways, including mitogen-activated protein kinase and PI3K. Cross-talk between various growth factor receptors occurs in cancer cells and may occur between IGF-IR and HER-2. Such cross-talk would activate mitogenic signaling cascades despite the blockade of HER-2 by trastuzumab, resulting in tumor progression in the presence of trastuzumab.

IV. ASSESSMENT OF HER-2 STATUS

The American Society of Clinical Oncology recommends that all primary breast tumors be evaluated for HER-2 expression, either at the time of diagnosis or upon recurrence (Bast et al., 2001). The HER-2 status of a tumor provides prognostic information and is the critical determinant of response

TABLE 4.2 Methods for Assessing HER-2 Status

Method	Advantages	Disadvantages	Clinical Use
Western blot	Widely available; relatively inexpensive	Semiquantitative; antibody variability; tumor extract is required	Not in clinical use
Polymerase chain reaction	Rapid; specific; sensitive; small amount of starting material	Semiquantitative	Not in clinical use
IHC	Widely available; relatively inexpensive	Semiquantitative; antibody variability; subjective interpretation	FDA approved; most frequently used clinically
FISH	Specific; quantitative; strong correlation with response to trastuzumab	Expensive; requires specialized equipment not widely available	FDA approved; valuable for confirmation of IHC 2+ HER-2 status
ECD ELISA	Serum easily obtained	ECD levels do not always correlate with tumor load	FDA approved to monitor response to chemotherapy; multicenter prospective study ongoing in patients receiving trastuzumab

CISH, Chromogenic *in situ* hybridization; *ECD*, extracellular domain; *ELISA*, enzyme-linked immunosorbent assay; *FDA*, U.S. Food and Drug Administration; *FISH*, fluorescence *in situ* hybridization; *IHC*, immunohistochemistry.

to trastuzumab (Slamon et al., 2001). Thus, accurate assessment of HER-2 expression levels is essential for identifying which breast cancer patients are most likely to benefit from trastuzumab. Several methods for assessing the HER-2 status of tumors are listed in Table 4.2. Currently, the two most common methods of measuring HER-2 levels in the clinical setting are IHC and fluorescence *in situ* hybridization (FISH). Both methods have been approved by the FDA for selecting breast cancer patients for trastuzumab-based therapy.

The predictive role of HER-2 expression measured using IHC and FISH was evaluated in two large randomized clinical trials of adjuvant trastuzumab conducted by the National Surgical Adjuvant Breast and Bowel Project (B31 trial) and the North Central Treatment Group (9840 trial). Both studies showed discrepancies in IHC and FISH results between large-volume and reference laboratories and low-volume laboratories (Paik et al., 2002; Roche et al., 2002). Although HER-2 amplification is considered a fairly stable abnormality, some studies suggest that changes from HER-2 positivity to negativity and from HER-2 negativity to positivity occur over time (Meng et al., 2004).

A. Immunohistochemistry

IHC, the most widely used method for measuring HER-2 levels in the clinical setting, entails the staining of paraffin-embedded tissue with a HER-2-specific antibody. When commercially available kits such as HercepTest (Dako, Carpinteria, CA) and Pathway HER2 (Ventana, Tucson, AZ) are used, staining is graded semiquantitatively on a scale from 0 (no detectable HER-2) to 3+ (high HER-2 expression) by comparison with control slides of cell lines with known HER-2 receptor density. (Standardization against control slides is essential to ensuring accurate assessment of HER-2 status.) Tumors with a staining score of 3+ are the most responsive to trastuzumab (Esteva et al., 2002; Seidman et al., 2001; Vogel et al., 2002).

B. Fluorescence *in situ* Hybridization

FISH detects *her-2* gene amplification and is more specific and sensitive than IHC (Press et al., 2002). In general, IHC and FISH have a concordance rate of approximately 80%. Importantly, FISH offers quantitative results, possibly eliminating subjectivity and variability among different laboratories. Furthermore, FISH predicts prognosis and response to trastuzumab more accurately than does IHC because the subset of patients whose tumors overexpress HER-2 in the absence of gene amplification is less likely to respond to trastuzumab-based therapy. Therefore, FISH should be performed on tumors with an IHC score of 2+ (HercepTest scoring system) (Fornier et al., 2002).

C. HER-2 Extracellular Domain

Another method under investigation for predicting response to trastuzumab is the quantification of serum levels of the HER-2 ECD. The HER-2 ECD is shed into the blood and is readily measured with an enzyme-linked immunosorbent assay as a circulating tumor antigen in the serum of 70% of patients with HER-2 overexpressing metastatic breast cancer (Carney et al., 2003). The advantage of this method is that blood is relatively easy to collect, allowing real-time monitoring of changes in HER-2 status in response to HER-2–targeted therapies. Our group previously showed that the rate of response to docetaxel and trastuzumab therapy was higher for patients whose levels of HER-2 ECD were high at baseline than for patients who had low HER-2 ECD levels before treatment (Esteva et al., 2002). Monitoring HER-2 ECD levels may also have clinical utility because a significant reduction in serum levels predicts improved response rates and time to progression (Esteva et al., 2005; Kostler et al., 2004). The available data, however, are based on retrospective analysis, and the use of this assay is not recommended outside of a clinical trial.

V. CLINICAL TRIALS WITH TRASTUZUMAB

Phase I trials of trastuzumab for patients with HER-2-overexpressing metastatic breast cancer showed that the antibody is safe. Initially, pharmacokinetic analysis estimated that trastuzumab's plasma half-life was 7 to 10 days, and in the subsequent pivotal Phase II and Phase III trials, trastuzumab was administered weekly. A recent study, however, showed that trastuzumab's plasma half-life is at least 21 days (Leyland-Jones et al., 2003), which would allow less frequent treatment administration.

Response rates to trastuzumab given as a single agent have ranged from 12% to 34%, depending in part on the method used to determine HER-2 status and the prior treatments received by the patients (Cobleigh et al., 1999; Vogel et al., 2002). In a pivotal randomized clinical trial, Slamon et al. (2001) showed that combining trastuzumab with doxorubicin and cyclophosphamide or with paclitaxel produced longer times to tumor progression, higher response rates, and improved survival rates than did chemotherapy alone. These findings were largely confirmed in a French randomized trial in which patients with HER-2-positive metastatic breast cancer were treated with docetaxel with or without trastuzumab (Marty et al., 2005).

Of concern, the administration of doxorubicin, cyclophosphamide, and trastuzumab in Slamon's randomized trial caused severe cardiac dysfunction (Slamon et al., 2001), and this regimen should not be used outside a clinical trial. Although HER-2 is not over-expressed in adult cardiomyocytes, HER-2, together with its co-receptor HER-4 and the ligand heregulin, is essential for normal development of the heart ventricle. Conditional knockout mice lacking *her-2* gene expression in ventricular cardiomyocytes develop severe dilated cardiomyopathy (Crone et al., 2002).

Clinical trials are also under way to evaluate the safety of epirubicin and liposomal anthracyclines in combination with trastuzumab (Buzdar et al., 2005). In addition, non–anthracycline-containing trastuzumab-based regimens have shown clinical activity, including those containing taxanes, platinum salts, vinorelbine, and gemcitabine. Of these, triple combinations that incorporate trastuzumab, a platinum salt, and a taxane are highly synergistic *in vitro* (Pegram et al., 2004). Data from Phase II and Phase III studies of regimens combining taxanes, platinum salts, and trastuzumab have shown a high response rate and an extended time to tumor progression (Robert et al., 2004).

The most promising application of trastuzumab therapy will be in the adjuvant therapy setting, either before or after surgery for early-stage breast cancer. Three studies presented at the 2005 Annual Meeting of the American Society of Clinical Oncology showed that the addition of trastuzumab to standard chemotherapy improves disease-free survival significantly compared with chemotherapy alone after surgery in patients with node-positive and, in some cases, node-negative early-stage breast cancer. Two of these studies, conducted by the National Surgical Adjuvant Breast and Bowel Project (B31 trial) and the North Central Treatment Group (9840 trial) in the United States, evaluated sequential doxorubicin-plus-cyclophosphamide followed by paclitaxel or by paclitaxel in combination with trastuzumab. The European Herceptin Adjuvant study (known as the HERA trial), conducted in more than 50 countries outside the United States, randomized patients to receive single-agent trastuzumab for 1 to 2 years versus observation after completion of surgery, adjuvant chemotherapy (multiple regimens were allowed), and adjuvant radiotherapy. The available data from these three trials indicate that the best outcome is achieved when trastuzumab is given simultaneously with paclitaxel following doxorubicin-plus-cyclophosphamide chemotherapy. Longer follow-up is needed to determine whether the combined approach is superior to sequential chemotherapy-trastuzumab therapy. The Breast Cancer International Research Group (protocol 006) is evaluating the role of docetaxel with and without trastuzumab following doxorubicin-plus-cyclophosphamide chemotherapy. A third experimental arm incorporates a regimen combining taxanes, platinum salts, and trastuzumab. This protocol includes node-positive and high-risk node-negative patients; HER-2 status is determined using FISH at a central laboratory.

Several studies have evaluated the role of trastuzumab as a preoperative therapy in patients with early-stage breast cancer. Burstein et al. (2003) reported on a Phase II study of preoperative trastuzumab in combination with paclitaxel (total of four cycles) in which the pathological complete response rate was 18%. Buzdar et al. (2005) conducted a randomized clinical trial of preoperative paclitaxel (four cycles) followed by fluorouracil, epirubicin, and cyclophosphamide (four cycles) versus the same preoperative chemotherapy regimen in combination with trastuzumab. The pathological complete response rate was 25% for patients receiving eight cycles of chemotherapy and 65% for the chemotherapy-trastuzumab group. The study was stopped early because of these dramatically different results (Buzdar et al., 2005). If the pathological complete response rates translate into longer disease-free and overall survival, this approach may be preferable to the postoperative adjuvant trastuzumab-based therapy approach (Burstein et al., 2003). Data on the long-term safety and efficacy of trastuzumab are needed before it can be applied widely to patients with early-stage breast cancer.

CONCLUSION

In 2005, trastuzumab was the only FDA-approved monoclonal antibody therapy for patients with metastatic breast cancer whose tumors over-express HER-2. It has taken 20 years from the discovery of the *her-2* gene for cure rates to improve in patients with HER-2 over-expressing early-stage breast cancer. We expect that the lessons learned from this fascinating story will help accelerate the development of novel signal transduction inhibitors for breast cancer and other solid tumors in the near future.

Trastuzumab monoclonal antibody therapy will be used, in a future application, to deliver toxic payloads to tumors (Nahta et al., 2003b). These antibody-drug conjugates, which have proven effective in preclinical models, will be attractive for clinical use if they can be delivered safely, with limited toxicity to healthy tissues.

One of the challenges remaining is how to manage patients who develop progressive metastatic breast cancer while receiving trastuzumab therapy, because the mechanisms of trastuzumab resistance in humans are still largely unknown. Elucidating the molecular changes that occur as tumors progress during trastuzumab therapy will allow for the design of targeted therapies to be used in combination with or after trastuzumab.

Many new agents (e.g., monoclonal antibodies and small molecules) for treating breast cancer are currently in preclinical or early clinical stages of development. Targets under investigation include growth factor receptors, farnesyl transferase enzymes, PI3K/Akt, and mammalian target of rapamycin, among others (Friedman et al., 2005; Hidalgo and Rowinsky, 2000; Nahta et al., 2003a, 2004a). Combining trastuzumab with these other targeted agents on a hypothesis-driven rational basis is an attractive avenue for further therapeutic development.

ACKNOWLEDGMENTS

The authors thank Sandy Young for editorial assistance.

USEFUL WEBSITES

http://www.mdanderson.org
http://www.cancer.gov
http://www.cancer.org
http://web.wi.mit.edu/weinberg/pub/

REFERENCES

Baselga, J., J. Albanell, et al. (2001). Mechanism of action of trastuzumab and scientific update. *Semin. Oncol.* 28: 4–11.

Bast, R. C., P. Ravdin, et al. (2001). 2000 update of recommendations for the use of tumor markers in breast and colorectal cancer: Clinical practice guidelines of the American Society of Clinical Oncology. *J. Clin. Oncol.* 19: 1865–1878.

Burstein, H. J., L. N. Harris, et al. (2003). Preoperative therapy with trastuzumab and paclitaxel followed by sequential adjuvant doxorubicin/cyclophosphamide for HER2 overexpressing stage II or III breast cancer: A pilot study. *J. Clin. Oncol.* 21: 46–53.

Buzdar, A. U., N. K. Ibrahim, et al. (2005). Significantly higher pathologic complete remission rate after neoadjuvant therapy with trastuzumab, paclitaxel, and epirubicin chemotherapy: Results of a randomized trial in human epidermal growth factor receptor 2-positive operable breast cancer. *J. Clin. Oncol.* 23: 3676–3685.

Carney, W. P., R. Neumann, et al. (2003). Potential clinical utility of serum HER-2/neu oncoprotein concentrations in patients with breast cancer. *Clin. Chem.* 49: 1579–1598.

Carter, P., L. Presta, et al. (1992). Humanization of an anti-p185her2 antibody for human cancer therapy. *Proc. Natl. Acad. Sci. U. S. A.* 89: 4285–4289.

Clynes, R. A., T. L. Towers, et al. (2000). Inhibitory Fc receptors modulate in vivo cytoxicity against tumor targets. *Nat. Med.* 6: 443–446.

Cobleigh, M. A., C. L. Vogel, et al. (1999). Multinational study of the efficacy and safety of humanized anti-HER2 monoclonal antibody in women who have HER-2 overexpressing metastatic breast cancer that has progressed after chemotherapy for metastatic disease. *J. Clin. Oncol.* 17: 2639–2648.

Crone, S. A., Y. Y. Zhao, et al. (2002). ErbB2 is essential in the prevention of dilated cardiomyopathy. *Nat. Med.* 8: 459–465.

Drebin, J. A., V. C. Link, et al. (1985). Down-modulation of an oncogene protein product and reversion of the transformed phenotype by monoclonal antibodies. *Cell* 41: 697–706.

Esteva, F. J. (2004). Monoclonal antibodies, small molecules, and vaccines in the treatment of breast cancer. *Oncologist* 9 Suppl 3: 4–9.

Esteva, F. J., C. D. Cheli, et al. (2005). Clinical utility of serum HER2/neu in monitoring and prediction of progression-free survival in metastatic breast cancer patients treated with trastuzumab-based therapies. *Breast Cancer Res.* 7: R436–R443.

Esteva, F. J., V. Valero, et al. (2002). Phase II study of weekly docetaxel and trastuzumab for patients with HER-2-overexpressing metastatic breast cancer. *J. Clin. Oncol.* 20: 1800–1808.

Fornier, M., M. Risio, et al. (2002). HER2 testing and correlation with efficacy of trastuzumab therapy. *Oncology (Huntingt)* 16: 1340–1342.

Friedman, L. M., A. Rinon, et al. (2005). Synergistic down-regulation of receptor tyrosine kinases by combinations of mAbs: Implications for cancer immunotherapy. *Proc. Natl. Acad. Sci. U. S. A* 102: 1915–1920.

Gennari, R., S. Menard, et al. (2004). Pilot study of the mechanism of action of preoperative trastuzumab in patients with primary operable breast tumors overexpressing HER2. *Clin. Cancer. Res.* 10: 5650–5655.

Hidalgo, M. and E. K. Rowinsky (2000). The rapamycin-sensitive signal transduction pathway as a target for cancer therapy. *Oncogene* 19: 6680–6686.

Hung, M. C., A. L. Schechter, et al. (1986). Molecular cloning of the neu gene: Absence of gross structural alteration in oncogenic alleles. *Proc. Natl. Acad. Sci. U. S. A.* 83: 261–264.

Izumi, Y., L. Xu, et al. (2002). Tumour biology: Herceptin acts as an anti-angiogenic cocktail. *Nature* 416: 279–280.

King, C. R., M. H. Kraus, and S. A. Aaronson (1985). Amplification of a novel v-erbB-related gene in a human mammary carcinoma. *Science* 229: 974–976.

Kostler, W. J., B. Schwab, et al. (2004). Monitoring of serum Her-2/neu predicts response and progression-free survival to trastuzumab-based treatment in patients with metastatic breast cancer. *Clin. Cancer Res.* 10: 1618–1624.

Lane, H. A., A. B. Motoyama, et al. (2001). Modulation of p27/Cdk2 complex formation through 4D5-mediated inhibition of HER2 receptor signaling. *Ann. Oncol.* 12: 21–22.

Le, X. F., F. X. Claret, F.X., et al. (2003). The role of cyclin-dependent kinase inhibitor p27Kip1 in anti-HER2 antibody-induced G1 cell cycle arrest and tumor growth inhibition. *J. Biol. Chem.* 278: 23441–23450.

Leyland-Jones, B., K. Gelmon, et al. (2003). Pharmacokinetics, safety, and efficacy of trastuzumab administered every three weeks in combination with paclitaxel. *J. Clin. Oncol.* 21: 3965–3971.

Lu, Y., X. Zi, et al. (2001). Insulin-like growth factor-I receptor signaling and resistance to trastuzumab (Herceptin). *J. Natl. Cancer Inst.* 93: 1852–1857.

Marty, M., F. Cognetti, et al. (2005). Randomized phase II trial of the efficacy and safety of trastuzumab combined with docetaxel in patients with human epidermal growth factor receptor 2-positive metastatic breast cancer administered as first-line treatment: The M77001 study group. *J. Clin. Oncol.* 23: 4265–4274.

Meng, S., D. Tripathy, D., et al. (2004). HER-2 gene amplification can be acquired as breast cancer progresses. *Proc. Natl. Acad. Sci. U. S. A* 101: 9393–9398.

Mohsin, S. K., H. L. Weiss, et al. (2005). Neoadjuvant trastuzumab induces apoptosis in primary breast cancers. *J. Clin. Oncol.* 23: 2460–2468.

Nagata, Y., K. H. Lan, et al. (2004). PTEN activation contributes to tumor inhibition by trastuzumab, and loss of PTEN predicts trastuzumab resistance in patients. *Cancer Cell* 6: 117–127.

Nahta, R. and F. J. Esteva (2003). HER-2-targeted therapy: Lessons learned and future directions. *Clin Cancer Res* 9: 5078–5084.

Nahta, R., G. N. Hortobagyi, and F. J. Esteva (2003a). Growth factor receptors in breast cancer: Potential for therapeutic intervention. *Oncologist* 8: 5–17.

Nahta, R., G. N. Hortobagyi, and F. J. Esteva (2003b). Novel pharmacological approaches in the treatment of breast cancer. *Expert. Opin. Investig. Drugs* 12: 909–921.

Nahta, R., M. C. Hung, and F. J. Esteva (2004a). The HER-2-targeting antibodies trastuzumab and pertuzumab synergistically inhibit the survival of breast cancer cells. *Cancer Res.* 64: 2343–2346.

Nahta, R., T. Takahashi, et al. (2004b). P27(kip1) down-regulation is associated with trastuzumab resistance in breast cancer cells. *Cancer Res.* 64: 3981–3986.

Paik, S., J. Bryant, et al. (2002). Real-world performance of HER2 testing: National surgical adjuvant breast and bowel project experience. *J. Natl. Cancer Inst.* 94: 852–854.

Pegram, M. D., G. E. Konecny, et al. (2004). Rational combinations of trastuzumab with chemotherapeutic drugs used in the treatment of breast cancer. *J. Natl. Cancer Inst.* 96: 739–749.

Pietras, R. J., J. C. Poen, et al. (1999). Monoclonal antibody to HER-2/neureceptor modulates repair of radiation-induced DNA damage and enhances radiosensitivity of human breast cancer cells overexpressing this oncogene. *Cancer Res.* 59: 1347–1355.

Press, M. F., L. Bernstein, et al. (1997). HER-2/neu gene amplification characterized by fluorescence *in situ* hybridization: Poor prognosis in node-negative breast carcinomas. *J. Clin. Oncol.* 15: 2894–2904.

Press, M. F., D. J. Slamon, et al. (2002). Evaluation of HER-2/neu gene amplification and overexpression: Comparison of frequently used assay methods in a molecularly characterized cohort of breast cancer specimens. *J. Clin. Oncol.* 20: 3095–3105.

Repka, T., E. G. Chiorean, et al. (2003). Trastuzumab and interleukin-2 in HER2-positive metastatic breast cancer: A pilot study. *Clin. Cancer Res.* 9: 2440–2446.

Robert, N. J., B. Leyland-Jones, et al. (2004). Randomized phase III study of trastuzumab, paclitaxel, and carboplatin versus trastuzumab and paclitaxel in women with HER-2 overexpressing metastatic breast cancer: An update including survival. *J. Clin. Oncol.* 22: 573.

Roche, P. C., V. J. Suman, et al. (2002). Concordance between local and central laboratory HER2 testing in the breast intergroup trial N9831. *J. Natl. Cancer Inst.* 94: 855–857.

Seidman, A. D., M. Fornier, et al. (2001). Weekly trastuzumab and paclitaxel therapy for metastatic breast cancer with analysis of efficacy by HER2 immunophenotype and gene amplification. *J. Clin. Oncol.* 19: 2587–2595.

Slamon, D. J., G. M. Clark, et al. (1987). Human breast cancer: Correlation of relapse and survival with amplification of the HER-2/neu oncogene. *Science* 235: 177–182.

Slamon, D. J., B. Leyland-Jones, et al. (2001). Use of chemotherapy plus a monoclonal antibody against HER2 for metastatic breast cancer that overexpresses HER2. *N. Engl. J. Med.* 344: 783–792.

Vogel, C. L., M. A. Cobleigh, et al. (2002). Efficacy and safety of trastuzumab as a single agent in first-line treatment of HER2-overexpressing metastatic breast cancer. *J. Clin. Oncol.* 20: 719–726.

Yakes, F. M., W. Chinratanalab, et al. (2002). Herceptin-induced inhibition of phosphatidylinositol-3 kinase and Akt is required for antibody-mediated effects on p27, cyclin D1, and antitumor action. *Cancer Res.* 62: 4132–4141.

5 VALIDATION OF TNF AS A DRUG TARGET IN INFLAMMATORY BOWEL DISEASE

MICHAEL J. ELLIOTT

M.D., Ph.D., Centocor Research & Development, Inc., Malvern, Pennsylvania

Tumor necrosis factor alpha (TNFα; in this chapter shortened to TNF) is a member of the TNF superfamily of cytokines and their receptors and shows many biological functions of potential importance in inflammation, including effects on vascular biology, immune and hematopoietic cell function. Although often categorized as a T helper (Th)1 cell product, the major cell of origin for TNF in many inflammatory diseases is the macrophage, so TNF may also be regarded as part of a final common pathway in inflammation, common to diseases of Th1, Th2, or mixed etiology. Inflammatory bowel disease (IBD) comprises two disease entities: Crohn's disease (CD) and ulcerative colitis (UC). Despite clinical and histopathological differences, the two disorders may have a common origin, namely an exaggerated or inappropriate mucosal immune response to enteric commensal microorganisms, driven in part by genetic predisposition. The immunopathology of CD and UC appears to differ, with CD bearing the hallmarks of Th1-driven inflammation, whereas UC appears to fit neither Th1 nor Th2 patterns of cytokine expression and may instead be driven by atypical IL-13–secreting natural killer (NK) T cells. Whatever the underlying immune drive, both conditions are associated with local production of TNF in the gut wall, largely by macrophages. Animal studies support the notion that TNF can drive experimental colitis. Clinical studies with specific TNF blocking agents have been conducted in both CD and UC, with most experience being with the use of infliximab, a chimeric monoclonal antibody that binds and neutralizes soluble and membrane-bound TNF. Large-scale, controlled clinical trials have provided convincing evidence for the efficacy of infliximab in both CD and UC, and the drug has now received United States Food and Drug Administration (FDA) approval for the treatment of these disorders. The success of infliximab therapy in IBD may lie not only in its ability to neutralize TNF bioactivity, but also in its ability to engage Fc effector mechanisms to cause lysis of enteric TNF-expressing cells. The development of infliximab in IBD represents a major medical advance and demonstrates the power of targeted therapies such as monoclonal antibodies to dissect disease mechanisms and provide clear target validation in humans.

I. INTRODUCTION

In July 1993, a case report was published in the *The Lancet*, describing the successful use of a novel therapeutic agent in a young patient with Crohn's disease (Derkx et al., 1993). The agent, known at the time as cA2 but now more usually known as infliximab, is a chimeric monoclonal antibody targeting TNF. This was the first published manuscript describing the use of a TNF blocking agent in any chronic inflammatory disease in humans. Five years later, in August 1998, the FDA approved infliximab for the short-term treatment of Crohn's disease, representing the first approval for an anti-TNF agent in any indication by a regulatory authority worldwide. In September 2005, FDA approval was also granted for the treatment of ulcerative colitis.

Infliximab and other TNF blocking agents have now been developed in many other inflammatory diseases, including rheumatoid arthritis and related rheumatic disorders, and inflammatory skin diseases such as psoriasis. The focus of this review, however, is on the pioneering drug development that was conducted in patients with CD and its sister condition UC, collectively known as IBD. The clinical characteristics and distinguishing features of each of these two forms of IBD will be described in this review and the putative role of TNF in the pathogenesis of these conditions, arising from early *in vitro* and preclinical work, will be described. For both conditions, but in particular for UC, large-scale, carefully conducted, controlled clinical trials with infliximab have been key in providing final proof of the importance of TNF in disease pathogenesis. The results of these clinical studies will also be described.

The development of infliximab in IBD has provided an important new therapeutic option for patients suffering from these painful and debilitating conditions. In addition, the story illustrates of the power of a highly targeted therapeutic agent, such as a monoclonal antibody, to provide target validation in settings of scientific controversy.

II. TUMOR NECROSIS FACTOR

TNF was first described as an endotoxin-induced serum factor that could lyse transplanted tumor cells in mice, and this biological activity provided its name (Carswell et al., 1975). Once the gene for TNF was cloned and recombinant protein was obtained, numerous additional biological functions of TNF become apparent (Pennica et al., 1984). Eventually, TNF and its two receptors, TNFR1 (also known as p55 or CD120a) and TNFR2 (p75 or CD120b), were recognized as the prototypes for the TNF superfamily of cytokines and their receptors (Aggarwal, 2003), which were capable of controlling cell differentiation, proliferation, and apoptosis necessary to control many aspects of mammalian development, immune function, and hematopoiesis.

TNF is produced predominantly by macrophages upon lipopolysaccharide interaction with CD14, the toll-like receptor 4, as part of the adaptive immune response (Tapping et al., 2000), as well as by T cells, B cells, NK cells, and granulocytes. Expressed on the cell surface as a 26-kDa type II membrane protein, three TNF monomers self-associate to form a homotrimer that is eventually released as a soluble TNF homotrimer following proteolytic cleavage by TNF-alpha converting enzyme (Black et al., 1997;

Moss et al., 1997; TACE). The homotrimer is the bioactive form of both soluble and transmembrane TNF (Perez et al., 1990; Smith and Baglion, 1987).

While TNFRI is constitutively expressed on virtually all nucleated cells, expression of TNFR2 is limited to immune and endothelial cells (Aggarwal, 2003). Early studies suggested that most of the activities attributed to TNF were mediated by TNFR1, as it was shown to have a significantly higher affinity for soluble TNF than TNFR2 (Grell et al., 1998).

III. INFLAMMATORY BOWEL DISEASE

A. Crohn's Disease

Crohn's disease (CD) is a chronic inflammatory disorder of the gastrointestinal (GI) tract, with clinical features including abdominal pain, diarrhea, weight loss, fever, and fistula formation. The disease most commonly afflicts subjects in their 20s and 30s, and in the United States and Western Europe, the estimated incidence of the disorder is 2 cases per 100,000 population, with a higher prevalence (20 to 40 cases per 100,000) that reflects the chronic nature of the disease (Glickman, 1998).

Although the disease can affect any portion of the intestinal tract, it has a particular predilection for the terminal ileum, and involvement of more proximal sections (e.g., the mouth, tongue, esophagus, stomach, and duodenum) is much less common (Glickman, 1998; Stenson, 2000). In about 30% of cases, gross pathological involvement is limited to the small intestine, usually the terminal ileum; 30% display only colonic involvement, and 40% manifest ileocolonic involvement. Noncaseating granulomas, a defining characteristic of CD, have been detected in up to 50% of resected bowel sections. Pathologically, the disease is characterized by inflammation that typically extends through all layers of the intestinal wall, involving the mesentery and regional lymph nodes, yet intestinal involvement is discontinuous: "skip areas" of apparently normal tissue separate severely involved segments. As the disease progresses, the bowel thickens, narrowing the lumen, resulting in stenosis. Intestinal obstruction, a frequent complication, occurs in 20% to 30% of patients. Additionally, fistula formation affects 15% to 32% of patients. The fistulas may communicate internally (e.g., bowel to bowel, bowel to bladder, or rectum to vagina) or may be enterocutaneous (extending from the bowel through to the abdominal wall or into the perineum). The presence of fistulas has a major impact on the morbidity and quality of life in CD and spontaneous closure of fistulas is rare (Glickman, 1998). Fistulizing CD has the poorest prognosis, with surgical resection usually required after 3 to 4 years.

Treatment of CD may include dietary manipulations (bowel rest or supplemental nutrition) and judicious surgery, particularly for disease complications such as intestinal stenosis or fistula formation, but the mainstay of disease management is pharmacotherapy (Hanauer and Meyers, 1997). Standard nonbiological approaches to the treatment of CD include anti-inflammatory drugs such as 5-aminosalicylate (5-ASA) compounds (e.g., mesalamine, olsalazine, and sulfasalazine), corticosteroids (e.g., prednisone, methylprednisolone, and budesonide), and immunomodulators (e.g., azathioprine [AZA], 6-mercaptopurine [6-MP], and cyclosporine), often in combination. Many of these treatments can induce disease remission in some patients, but others respond poorly, and disease relapse upon discontinuation of therapy is common. Antibiotics, particularly metronidazole can be effective in the treatment of patients with CD, particularly those who have fistulizing disease.

B. Ulcerative Colitis

Like CD, ulcerative colitis (UC) is a chronic inflammatory disorder of the GI tract of unknown etiology. Occurrence peaks between the ages of 15 and 35 years, but patients of all ages can be affected. The disorder is more common than CD, with an incidence in the United States and Western Europe of 6 to 8 cases per 100,000, and a prevalence of 70 to 150 cases per 100,000 (Glickman, 1998). Clinically, patients with UC suffer from diarrhea, rectal bleeding, weight loss, and fever. Profuse diarrhea and bleeding are seen in patients with fulminant UC. The diagnosis of UC is generally based on clinical features and proctosigmoidoscopy or colonoscopy with a rectal biopsy (Glickman, 1998; Stenson, 2000).

Unlike CD, UC is characterized by inflammation that is uniform and continuous, with no intervening areas of normal mucosa. In most cases, inflammation involves the rectum and extends proximally for a variable distance. When inflammation affects the entire colon (pancolitis), involvement may also extend to the last part of the terminal ileum. In UC, the inflammatory reaction involves the surface mucosa, the crypt epithelium, and the submucosa. The affected mucosa bleeds easily, contains ulcerated areas, or is completely denuded. In severe UC, as seen with toxic megacolon, the bowel wall may become extremely thin, the mucosa denuded, and inflammation may extend to the serosa, leading to dilatation and subsequent perforation (Glickman, 1998; Stenson, 2000).

Many UC patients may also display prominent extraintestinal manifestations, most commonly colitic arthritis and ankylosing spondylitis (Hyams, 2000; Stenson, 2000). Colitic arthritis affects predominantly the large joints and manifests as joint pain, swelling, and stiffness, symptoms that can parallel the GI disease course. Ankylosing spondylitis is a progressive and potentially crippling condition, characterized by morning stiffness, low back pain, and stooped posture. The most common hepatic manifestation, pericholangitis, is characterized by inflammation of the portal tracts with lymphocyte and eosinophil infiltrates accompanied by degenerative changes in the bile ductules. The two most common dermal complications are lesions (pyoderma gangrenosum) and nodules (erythema nodosum) of the lower extremities (Baldassano et al., 1999; Stenson, 2000). Extraintestinal manifestations unique to pediatric UC and CD patients include growth failure and delayed sexual development (Baldassano and Piccoli, 1999; Hyams, 2000).

Patients with UC, have an increased risk of carcinoma when compared with the general population. The estimated risk of colorectal carcinoma increases from 2% at 10 years to 8% at 20 years to 18% at 30 years, with the risk increasing progressively as the duration and extent of disease increase (Eaden et al., 2001).

As for CD, the treatment of UC includes a variety of supportive and symptomatic therapies and surgery, but once again, drug therapy is central to disease control. Treatment with 5-ASA compounds, corticosteroids, and immunomodulators can suppress the inflammatory process (Hyams, 2000; Kornbluth and Sachar, 2004). 5-ASA compounds are primarily effective in treating mild disease and can be used to prevent relapses. Periodic enemas with corticosteroids or mesalamine can also be used during periods of active disease. Although corticosteroids can effectively control acute episodes of colitis, the side effects of corticosteroids, which include acne, moon facies, osteoporosis, compression fractures, and sleep and mood disturbances, prohibit their long-term use (Carter et al., 2004). In addition, up to 30% of patients treated with corticosteroids become resistant, necessitating proctocolectomy or therapy with cyclosporine (Hanauer and Meyers, 2004). One year after starting corticosteroids, 22% of patients with UC will be steroid dependant, 49% will be in response, and 29% will have had surgery (Faubion et al., 2001). Immunomodulators are primarily used when 5-ASA compounds and corticosteroids have been ineffective, or as steroid sparing agents in corticosteroid-dependent patients (Carter et al., 2004). Agents such as azathioprine or 6-MP can reduce or eliminate the need for corticosteroids in some patients, and they may also help maintain remission in patients who have not responded to other medications. Immunomodulators, however, can take as long as 3 months to work, may only be partially effective, and are associated with significant toxicity, including neutropenia, pancytopenia, pancreatitis, and hepatotoxicity. Cyclosporine can also be effective in UC, though the associated side effects (e.g., nephrotoxicity, infection, seizures, and hypertension) and the need for careful monitoring has limited its use to tertiary-care centers.

Despite receiving standard therapy, approximately 12% to 58% of subjects experience some degree of relapse of their disease on medical therapy. Of those with a severe relapse, approximately 33% will require a proctocolectomy (Kumar et al., 2004). Although subjects undergoing proctocolectomy can be "cured," the procedure itself is not without risk. Proctocolectomy has a perioperative mortality rate that ranges from 1% to 3%. The procedure generally requires a temporary ileostomy, and once the ileorectal pouch has been completed, the patient generally has an increased frequency of loose stools, often has fecal incontinence, and commonly develops chronic pouchitis (Kornbluth and Sachar, 2004).

IV. PATHOPHYSIOLOGY OF IBD AND THE PUTATIVE ROLE OF TNF

A. Pathophysiology of IBD

Despite being clinically distinguishable and with distinct gross and microscopic pathological features, it is recognized that CD and UC likely share an important underlying pathophysiology. The current dominant hypothesis is that IBD results from an abnormal mucosal immune response to enteric flora, a response that is in part genetically determined (Hendrickson et al., 2002; Bouma and Strober, 2003). Multiple lines of evidence support this hypothesis, but one particularly important observation is that in general, experimental colitis does not develop when otherwise-susceptible mice are kept in a germ-free environment (Podolsky, 2002; Shanahan, 2002). Although the individual organisms responsible for disease induction or maintenance have not been identified, it is clear that some members of the normal enteric microflora are not pathogenic. Indeed, one experimental approach to the treatment of IBD involves the oral administration of certain strains of normal enteric organisms (or probiotics), an approach aimed at favorably altering the balance of gut flora.

In normal circumstances, enteric-derived antigens induce regulatory T cell responses, or T cell anergy, or deletion of antigen-specific T cells, a phenomenon known as oral tolerance. Such oral tolerance in response to mucosal microflora establishes a homeostasis that ensures that most mucosal responses are self-limited and fail to result in inflammation. Failure of such homeostatic mechanisms may underpin the chronic nature of inflammation in IBD (Shanahan, 2002). In one commonly accepted model of the pathogenesis of IBD, bacterial products or dietary antigens penetrate the mucosal barrier and interact with antigen presenting cells to promote a classic adaptive immune response. In an alternative model, bacterial products may interact with toll-like receptors and other components of the innate immune system, with production by epithelial cells of cytokines and other growth factors that can recruit and activate mucosal immune cells.

B. Genetic Contribution to IBD

As with other immune-mediated disorders, many lines of evidence support a role for genetics in the development of IBD. These include data from animal models, ethnic differences in disease frequency, familial aggregation, concordance rates in monozygotic compared with dizygotic twins, and linkage with specific chromosomal regions (Bouma and Strober, 2003; Shanahan, 2002). Different mouse strains with different genetic defects may develop spontaneous mucosal inflammation resembling IBD, showing that that a variety of genetic abnormalities may lead to a similar clinical picture. In humans, the concordance rate for CD in identical twins is as high as 58%, whereas the dizygotic twin concordance is not significantly different from that for all siblings. Concordance rates for monozygotic and dizygotic twins in UC range from 6% to 17% and 0% to 5% respectively. These data support the notion that particularly in CD, susceptibility to IBD is inherited. It also indicates that IBD is not inherited as a mendelian trait, but rather as a complex genetic basis. In contrast to other immune-mediated conditions such as rheumatoid arthritis and type 1 diabetes, no clear associations between disease susceptibility and the major histocompatibility complex (MHC) locus have been defined. A genetic locus in the peri-centromeric region of chromosome 16, now known as IBD 1, has been identified, which confers susceptibility to CD. Further analysis identified a strong association between this locus and a single gene, NOD2, also known as caspase recruitment domain protein 15 (CARD15). Persons who are homozygous for a particular variant of NOD2/CARD15 have a 20-fold increase susceptibility to CD (Podolsky, 2002).

Multiple polymorphisms of the NOD2 gene have been identified in patients with CD, but not UC, and the possible association between NOD2 variants and the enteric microflora hypothesis of CD pathogenesis is intriguing. NOD2 protein is expressed intracellularly in monocytes, granulocytes, dendritic cells, and epithelial cells and stimulation of NOD2 transfected cells with bacterial proteins results in activation of the NF-kappa B pathway. The bacterial component that is recognized by NOD2 protein is muramyl dipeptide, a component of peptidoglycan that is associated with various bacterial strains. The hypothesis that links NOD2 to CD invokes defective activation of macrophages as a result of NOD2 polymorphisms, that leads in turn to persistent infection of macrophages with intracellular organisms. Persistent intracellular macrophage infection, however, has not been detected in CD, so other possible links must also be considered (Bouma and Strober, 2003).

C. Environmental Influences in IBD

The incomplete concordance rate for CD within monozygotic twins, together with risk variations within ethnic groups living in different geographical locations shows that environmental factors also play an important role in the pathogenesis of IBD. While the major interest in environmental factors in IBD focuses on the enteric microflora, other factors may also play a role. For example, nonsteroidal anti-inflammatory drugs can lead to disease flares in IBD, possibly related to alterations in the mucosal barrier, and early appendectomy is associated with protection from the development of UC. Smoking appears to protect against UC but increases the risk of CD.

D. Th1 versus Th2 in IBD

Based on data from animal models of IBD, as well as analysis of human tissue, CD has come to be viewed as a classic T helper (Th)1 cell immune response disorder. CD is characterized by increased secretion of cytokines associated with the Th1 phenotype, namely interleukin (IL)-12, IL-18, interferon (IFN) gamma, TNF, IL-17, and IL-21 (Bouma and Strober, 2003; Gordon et al., 2005; Hendrickson et al., 2002). The data include immunohistochemical studies showing over-production of IL-12 by intestinal macrophages in CD but not in UC, increased levels of stat 4 expression, indicative of IL-12 signaling, and increased expression of the IL-12 receptor beta 2 chain, also characteristic of Th1 cells. Recent data from CD patients treated with a monoclonal antibody specific for the p40 component of IL-12 demonstrated prompt and marked clinical response in most patients, supporting the notion that CD is an IL-12 driven Th1-mediated disorder (Mannon et al., 2004).

The data in human UC differs markedly from that described in CD (Bouma and Strober, 2003; Gordon et al., 2005). Th2-type features of UC include the production of auto-antibodies such as anti-neutrophil cytoplasmic antibody (pANCA) and anti-tropomyosin, suggesting Th2 help in the activation of B cells and the induction of a humoral immune response, and the presence in UC patients of Th2-related subclasses of immunoglobulins, such as IgG1 and IgG4. In addition, although the characteristic Th2 cytokine, IL4, is not expressed in increased levels in UC, increased secretion of IL-5, another Th2 cytokine, has been described. In addition, IL-13–producing natural killer (NK) T cells, which have been shown to have an important role in oxazalone-induced colitis (Heller et al., 2002), have also been demonstrated in the lamina propria in patients with UC, but not CD (Fuss et al., 2004). IL-13 is cytotoxic to epithelial cell targets and increases epithelial permeability (Heller et al., 2005), thereby potentially contributing to abnormal access of enteric antigens or bacterial products to the intestinal wall.

These findings have led some to characterize UC as an atypical Th2 disease (Fuss et al., 1996; Inoue et al., 1999), or as an atypical NK T cell–driven, IL-13 mediated inflammatory disorder (Fuss et al., 2004; Heller et al., 2002, 2005). The findings illustrate the growing understanding that although experimental models of disease can often be clearly classified as Th1 or Th2 in nature, real human diseases are more complex and may display overlapping features (Gor et al., 2003).

E. TNF in Human IBD

Irrespective of the underlying Th1, Th2, or the mixed immunopathology of IBD, and despite some early conflict within the published dataset, it is now clear that TNF is highly expressed in patients with IBD. That TNF is expressed in both CD and UC may reflect the fact that it is not only a Th1 cytokine, but also a macrophage product, and may be considered part of the final common pathway of inflammation, irrespective of the Th1 or Th2 origins of a disease.

In one early investigation of cytokine expression using single-cell cytokine immunoassays, biopsies from children with CD and UC showed elevated frequencies of TNF-expressing cells, compared with biopsies from control subjects (Macdonald et al., 1990). In another study using the polymerized chain reaction (PCR) in normal and IBD intestinal specimens, the authors reported high-level expression of IL-1 beta and IL-6, but no difference in TNF expression levels between IBD and control specimens (Stevens et al., 1992). In a study by Reinecker et al. (1993), however, the expression of IL-1 beta, IL-6, and TNF from biopsy-derived lamina propria mononuclear cells was studied using *in vitro* culture techniques and Enzyme Linked Immunosorbant Assay (ELISA). The results showed low levels of spontaneous TNF secretion by isolated lamina propria mononuclear cells taken from normal mucosa and from noninvolved UC and CD mucosa but elevated levels of spontaneous secretion from cells obtained from involved UC or CD mucosal biopsies. This level of secretion was further enhanced by *in vitro* mitogen stimulation. In another study (Murch et al., 1993), intestinal specimens from patients with CD and UC were compared to those from control patients, and the location and tissue density of cells immunoreactive for TNF were studied. The results showed significantly increased density of TNF-expressing cells in the lamina propria in both conditions, although the distribution differed. In UC, most of the TNF was expressed in subepithelial macrophages, while in CD, the cells were distributed evenly throughout the lamina propria. Sub-mucosal immunoreactivity was found only in CD, once again in macrophages. The authors commented on the fact that most of the mucosal TNF-secreting cells were macrophages, and speculated on the potential for direct activation of subepithelial macrophages by lipopolysaccharide, or other products of the luminal flora, particularly after any breach in epithelial integrity. Another group examined mucosal biopsies from patients with both colonic and ileal CD using *in situ* hybridization for IL-1 beta, IL-6 and TNF (Woywodt et al., 1994). Tissue sections from controls showed only occasional positive cells, while in CD tissues, large numbers of cells were found producing mRNA specific for the three cytokines.

Other studies have examined the concentrations of TNF in serum, blood cells, and stool samples. In subjects with UC and CD, serum concentrations of TNF and IL-6 are elevated during active disease and decrease during disease remission (Maeda et al., 1992; Murch et al., 1991; Schürmann et al., 1992). Braegger et al. (1992) measured TNF concentrations in stool samples from normal children, from infants with diarrhea, and from children with IBD including both subjects with active and inactive disease. The results were striking, with markedly increased levels of stool TNF in subjects with either active CD or UC, but not in subjects with inactive disease or in infants with nonspecific diarrhea or disease. The levels were much higher than had previously been measured in serum. The authors speculate that the stool TNF likely arises from two sources: leakage across epithelium after synthesis in the lamina propria and from the passage of cells containing TNF into the lumen with subsequent release of cytoplasmic TNF. Elsässer-Beile et al. (1994) examined levels of cytokine expression in mitogen-stimulated whole blood-cell cultures from patients with CD and UC and healthy controls. Relative to controls, patients with CD and UC expressed higher levels of TNF, IL-1 alpha, and IL-1 beta. In another study, TACE expression was also increased in the colonic mucosa of subjects with UC (Brynskov et al., 2002).

In summary, most studies have shown elevated TNF levels in IBD, and the enzyme responsible for cleavage of membrane-bound TNF, TACE, is upregulated in the colonic mucosa. The early data showing much higher TNF concentrations in the stool than in the serum suggested local production of TNF within the bowel wall, and in support of this, TNF producing lamina propria cells are present in IBD gut tissue and are associated with disease activity. Furthermore, while the distribution of TNF immunoreactivity has been reported to be distinct in the gut wall of subjects with CD and UC, in both conditions TNF immunoreactivity was observed in association with the macrophage.

Downstream effects of TNF and other inflammatory cytokines expressed in IBD include the generation of activated matrix metalloproteinase (MMPs), effects on local microvasculature including increased vascular permeability, and upregulation of adhesion molecules, with resulting influx of effector cells such as neutrophils and monocytes (Shanahan, 2002).

F. Manipulating TNF in Experimental Models of IBD

The role of TNF in IBD is also illustrated by the development of intestinal inflammation in mouse strains in which TNF is over-produced, either because of a deletion in the regulatory elements of the TNF gene (TNF^{ARE} mice) or in STAT-3 knockout mice, which show exaggerated production of IFN gamma and TNF (Hendrickson et al., 2002). In a study in a rodent model of chronic colitis, administration of anti-TNF antibodies downregulated production or activity of other inflammatory cytokines and reduced the severity of intestinal inflammation (Powrie et al., 1994).

V. CLINICAL EXPERIENCE WITH TNF-BLOCKING THERAPY IN IBD

A. Infliximab in CD

Infliximab (cA2, Remicade®) is a recombinant IgG1k, human-murine chimeric monoclonal antibody that specifically binds and neutralizes soluble TNF alpha and its membrane-bound precursor. This high-affinity binding prevents the interaction of TNF with its cellular receptors (Knight et al., 1993). Infliximab does not bind lymphotoxin. Infliximab binding to membrane-bound TNF can induce complement-mediated lysis and/or apoptosis of TNF-expressing cells. Infliximab is administered by intravenous (IV) infusion, most commonly with an initial induction regimen at weeks zero, two, and six of the treatment course, followed by less frequent dosing during the maintenance phase of treatment. Details of dose and regimen for different approved indications can be found in the Remicade® product information.

The first description of the use of infliximab in CD was published by a group at the Academic Medical Center, Amsterdam (Derkx et al., 1993). The reports described a girl who presented at age 12 with clinical, colonoscopic, and histological findings typical of colonic CD. The patient was treated over a period of 16 months with standard therapies, including prednisone at doses of up to 30 mg per day, mesalazine at doses of up to 500 mg three times a day, azathioprine 75 mg a day, and enemas containing aspirin and corticosteroids. In addition, she was treated with a semielemental diet and metronidazole. The patient experienced a poor response to therapy and because of her disabling disease, received 10 mg/kg of infliximab by IV infusion at weeks zero and two. Her other therapy remained unchanged. The patient showed rapid improvement following the first dose of infliximab, with improvement in clinical symptoms, resolution of elevated body temperature, and weight gain.

In 1995, the same group set out to extend their experience with a 10-patient, open-label trial of infliximab in patients with active CD (Van Dullemen et al., 1995). Each patient received a single IV infusion of infliximab at a dose of 10 mg/kg. Patients were continued on stable doses of prednisone, and other concomitant treatments such as immunosuppressive drugs. The therapy was, in general, well tolerated. Of the nine evaluable patients, eight had substantial improvement in subjective symptoms within 1 week after treatment and the mean CD Activity Index (CDAI) score dropped from a baseline value of 257 before treatment to 69 at week eight. The CDAI is a composite disease activity index, incorporating eight variables: the number of liquid or very soft stools, the severity of abdominal pain or cramping, general well-being, the presence of extra-intestinal manifestations, abdominal mass, use of antidiarrheal drugs, hematocrit, and body weight; the items yield a composite score ranging from 0 to approximately 600, scores below 150 indicate remission, whereas scores above 450 indicate severe illness. The duration of a clinical response was variable ranging from 3 to 6 months. Although this was an open-label, uncontrolled trial, the authors noted the consistency of the clinical response, the accompanying improvements in C-reactive protein (CRP) and erythrocyte sedimentation rate (ESR), and the observations of mucosal healing on video endoscopies. The authors also noted improvement in certain extra-intestinal manifestations of CD, including pyoderma gangrenosum in one patient, arthritis and/or arthralgia in another, and acne in a third.

In this, and a subsequent publication describing use of infliximab in patients with difficult to treat perineal CD following surgery (Van Dullemen et al., 1998), the authors highlighted the role of TNF in granulomatous inflammation in general.

These early, uncontrolled experiments were followed in 1997 by the publication of the first randomized, placebo-controlled study of infliximab in CD (Targan et al., 1997). In their justification for administering a TNF-blocking agent in CD, the authors focused on the Th1 nature of the disease, but highlighted the T cell as the likely source of TNF. One-hundred and eight subjects with active CD receiving stable doses of mesalamine, corticosteroids, and/or immunomodulators were randomly assigned to receive a single IV infusion of either placebo or infliximab at a dose of 5, 10, or 20 mg/kg. At week four, 65% of subjects in the active treatment groups were in response, compared to 17% in the placebo group, and 33% of the infliximab-treated patients were in remission, compared with only 4% of the placebo group. All differences were highly statistically significant. Clinical response was seen rapidly with up to 61% of infliximab-treated patients in response by week two. The changes in overall clinical activity were mirrored by decreases in CRP, with a maximal reduction occurring within the first 2 weeks. Overall, the treatments were well tolerated. The data mirrored those reported by the Amsterdam group and confirmed the short-term efficacy of infliximab in CD.

The study by Targan et al. (1997) provided conclusive proof for the efficacy of infliximab in CD, and therefore for the importance of TNF in the pathophysiology of that disease, but many other clinical questions remained. Present et al. (1999) set out to answer one of these, with a study of the use of infliximab in the management of difficult-to-treat fistulizing CD. In this study, 94 patients who had draining abdominal or perianal fistulas of at least 3 months duration were randomly assigned to receive placebo or infliximab 5 mg/kg or 10 mg/kg. Treatments were administered at weeks zero, two, and six and the primary endpoint was a 50% or greater reduction from baseline in the number of draining fistulas observed at two or more consecutive study visits. Sixty-eight percent of patients receiving 5 mg/kg and 56% of those receiving 10 mg/kg achieved the primary endpoint, as compared with 26% of patients in the placebo group. In addition, for the most stringent endpoint of closure of all fistulas, 55% of patients receiving 5 mg/kg and 38% of those receiving 10 mg/kg achieved this endpoint as compared with 13% assigned to the placebo group. Once again, the differences between active and placebo were highly statistically significant. On average, the fistulas remained closed for three months.

The early clinical experience had established that the agent was effective in the short-term treatment of luminal and fistulizing disease and in the induction of clinical response in remission in patients with disease refractory to other therapies. The question, however, remained regarding the ability of infliximab to maintain response or remission over time. The first study that provided definitive evidence for the long-term utility of infliximab therapy in CD was the ACCENT 1 study, published by Hanauer et al. (2002). In this study, 573 patients with active CD received a single infusion of infliximab 5 mg/kg at week zero. Response assessments were made at week two and patients in response at this time (defined as a reduction in CDAI of at least 70 points) were randomized to receive repeat infusions of placebo, infliximab 5 mg/kg or 10 mg/kg at regular intervals through week 46. The prespecified co-primary endpoints were the proportion of patients randomized at week 2 who were in remission in week 30 (defined as a CDAI score of less than 150) and the time to loss of response up to week 54. In this study, 58% of patients responded at week two and by week 30, 21% of placebo maintenance patients were in remission compared with 39% and 45% of patients treated with maintenance infliximab 5 mg/kg and 10 mg/kg, respectively ($p = 0.003$, $p = 0.0002$, respectively). Thus, patients receiving maintenance infliximab therapy were more likely to achieve and sustain clinical remission than patients receiving maintenance placebo. Through week 54, the median time to loss of response was also longer in those groups receiving infliximab compared to placebo. The time course of response and loss of response through week 54 can be seen in Figure 5.1, which displays the median CDAI scores over time in each of the three groups. This study allowed for the tapering of corticosteroids in patients responding to infliximab. At week 54, three times as many patients in the combined infliximab maintenance groups had discontinued corticosteroids while in clinical remission compared with patients with the placebo maintenance groups. CDAI remission was associated with reduced hospitalization and surgery rates, increased employment, and normalized quality of life (Lichtenstein et al., 2003).

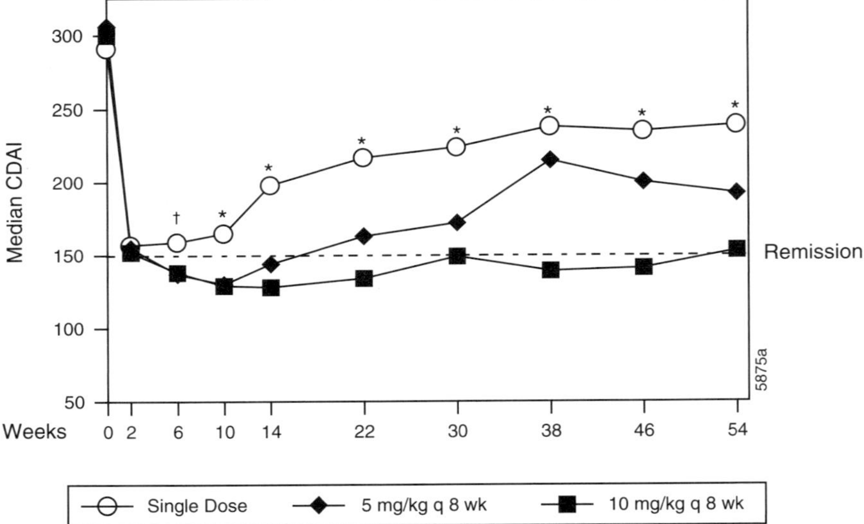

FIGURE 5.1 **Clinical response to infliximab therapy in patients with Crohn's disease.** The figure shows median Crohn's Disease Activity Index (CDAI) scores among week-two responders. Open circles: single dose infliximab 5 mg/kg at week zero, followed by placebo infusions at weeks two, six and then every 8 weeks. Closed diamonds: infliximab 5 mg/kg at weeks zero, two, six and then every 8 weeks. Closed squares: infliximab 5 mg/kg at weeks zero, two, six followed by infliximab 10 mg/kg every 8 weeks. *significant vs active treatment groups.

Sands et al. (2004a) extended the beneficial effects of maintenance infliximab therapy in CD with the publication of results from the ACCENT II study, testing the role of infliximab maintenance therapy for fistulizing CD. This multicenter, double blind, randomized, placebo-controlled trial studied the use of infliximab in 306 adult patients with CD and one or more draining abdominal or perianal fistulas of at least 3 months duration. The patients received 5 mg/kg infliximab at weeks zero, two, and six, and responding patients were then randomized to receive either placebo or 5 mg/kg infliximab maintenance therapy every 8 weeks through 1 year. The primary analysis was the time to loss of response amongst those patients who had a response at week 14. The results paralleled those of the Hanauer study, demonstrating significant benefit for infliximab maintenance therapy, compared to placebo. The time to loss of response for those who received infliximab maintenance was 40 weeks, compared to only 14 weeks for those receiving placebo ($p < 0.001$). At week 54, only 19% of patients in the placebo maintenance group had a complete absence of draining fistulas compared with 36% of patients in the infliximab maintenance group ($p = 0.009$). The study also demonstrated beneficial effects of infliximab maintenance therapy on the closure of an important subgroup of fistulas, recto-vaginal fistulas (Sands et al., 2004), and the beneficial effect of infliximab maintenance treatment on hospitalizations, surgeries, and other procedures (Lichtenstein et al., 2005).

Two additional papers based on the ACCENT 1 study data highlight the importance of the infliximab maintenance treatment schedule on long-term response (Hanauer et al., 2004; Rutgeerts et al., 2004). In particular, the use of regular scheduled maintenance treatment in the ACCENT 1 study produced a superior outcome at 1 year compared with episodic maintenance treatment, a result that may be in part related to a reduced tendency for the formation of anti-infliximab antibodies.

Although the most experience to date with the use of infliximab in CD has been in adult patients, a clinical trial examining the use of the drug in pediatric patients with CD has also been conducted recently. Responses to infliximab in children with CD seem to be at least as robust as those seen in adults (unpublished observations).

In the clinical trials described above, in general around two-thirds of CD patients showed either a partial or complete clinical response to infliximab therapy. The clinical and biological disease characteristics that might predict response (or nonresponse) are of interest, and were summarized recently by Louis et al. (2004). Certain demographic and clinical characteristics including younger age, colonic location of the disease, co-treatment with immunosuppressive drugs, and nonsmoking have shown trends to association with a response to infliximab therapy. Of more interest have been studies that have looked at possible relationships between TNF and TNF receptor gene polymorphisms, as well as NOD2/CARD15 gene mutations, and clinical response. In general, these have not disclosed strong associations with treatment response. In the study by Louis et al., however, the biological or clinical response to treatment with infliximab in 200 patients with luminal or fistulizing CD was examined in relation to polymorphisms of the gene coding for Fc gamma R3A, a macrophage and NK cell membrane receptor responsible for binding the Fc portion of Ig molecules. In particular, it is known that the Fc gamma R3A-158V (valine) allotype has a higher affinity for IgG1 than the Fc gamma R3A-158F (phenylalanine) allotype, and the authors hypothesized that patients homozygous or heterozygous for the V allotype would show higher potencies in antibody-dependent cell-mediated cytotoxicity (ADCC) of TNF-bearing immune cells and may therefore show better responses to infliximab. The results of the study showed a significant association between the biological response to infliximab (assessed by a reduction in CRP level) and the high-affinity Fc gamma R3A allotype. The most striking difference concerned the proportion of biological nonresponders; none were seen in V/V patients versus 30.2% in F carriers (V/F and F/F patient groups combined). Although intriguing, a similar genetic analysis of a subpopulation of patients from the ACCENT 1 study failed to show the response association (A. Olson, personal communication). The issue is therefore unresolved, but the Lewis data could provide a partial explanation for nonresponse or lesser response in certain patient groups, and suggest that ADCC of membrane TNF-expressing immune cells forms at least part of the mechanism of action of infliximab in CD. Indeed, the authors speculate that the failure of etanercept to show efficacy in the treatment of CD (see below) may be because of the inability of that drug to induce ADCC in TNF positive mononuclear cells.

It is also possible that this cytolytic effect of infliximab on TNF-expressing cells represents a reversal of a fundamental immune cell defect in CD. Lamina propria T cells from normal gut mucosa show high levels of activation-induced Fas-mediated apoptosis, and tissue sections from normal gut indicate a high level of *in vivo* apoptosis on a continuing basis. In both CD and UC, apoptosis appears to be reduced. A number of cytokines found in IBD tissues may deliver anti-apoptotic signals, including IL-2, IL-15, IL-17, IL-18 and in particular, IL-6 (Shanahan, 2002).

B. Other TNF Blocking Agents in CD

Although most of the published data on TNF-blocking agents in CD describe the use of infliximab, clinical studies with other biological agents targeting TNF have also been described. Etanercept (Enbrel®) is a dimeric fusion protein consisting of two identical chains of the recombinant human TNF receptor p75 monomer fused with the Fc domain of human IgG1. Etanercept binds and inactivates soluble TNF alpha and lymphotoxin. Cells expressing transmembrane TNF that bind etanercept are not lysed, either in the presence of absence of complement. Etanercept is administered as a subcutaneous injection once or twice a week at a dose of 25 to 50 mg, depending on the indication (Enbrel® US product information). In the only substantial published experience with etanercept in IBD, Sandborn et al. (2001) enrolled 43 subjects with active moderate to severe CD in a randomized, double-blind placebo-controlled trial of etanercept 25 mg or placebo, given twice weekly by subcutaneous injection. The primary outcome measure was clinical response at week four using the standard 70-point decrease in the CDAI score, or a CDAI score of less than 150. At week four, there was no statistically significant difference between etanercept treated and placebo-treated patients, and indeed the response trend was in favor of placebo. The authors concluded that etanercept at the administered dose was ineffective for the treatment of patients with CD, and speculated that higher doses or more frequent dosing might have produced a favorable outcome. In support of this 'dose intensity' hypothesis, etanercept is effective for the treatment of rheumatoid arthritis at a dose of 25 mg twice weekly, but in psoriasis, the recommended starting dose is 50 mg twice weekly (Enbrel® US product information), suggesting that some inflammatory diseases may require higher levels of TNF blockade for efficacy. Alternatively, however, the results may reflect the failure of etanercept to bind membrane-bound TNF and thus to induce ADCC of immune cell populations within the inflamed gut.

In other studies of TNF-blocking agents in CD, the use of certolizumab pegol (a PEGylated, anti-TNF Fab′ fragment) was evaluated in 292 patients with moderate to severe active CD (Schreiber et al., 2005). Patients were treated with subcutaneous certolizumab at a dose of 100, 200 or 400 mg or placebo at weeks zero, four, and eight, and the primary endpoint was clinical response at week 12 (a fall in CDAI of at least 100 points). The study demonstrated a trend favoring certolizumab over placebo at the primary endpoint, with significantly better responses for certolizumab at other timepoints and in certain subgroups, namely those with high CRP at baseline. Overall, the data suggest likely benefit for certolizumab in CD. Most recently, the effects of the human anti-TNF monoclonal antibody (adalimumab) were examined in a similar patient population (Sanborn et al., 2005) and demonstrated clear evidence of efficacy extending out through 6 months.

C. Infliximab in UC

The early clinical experience with infliximab in UC is noteworthy for inconsistent outcomes. Sands et al. (2001) enrolled 11 patients with severe, active, steroid-refractory UC in a double-blind, placebo-controlled trial. The patients received a single IV infusion of infliximab at a dose of 5, 10, or 20 mg/kg and treatment response was assessed 2 weeks later. Four of eight patients who received infliximab were considered treatment successes at 2 weeks, while none of three patients who received placebo achieved success. In a study by Chey (2001), subjects with treatment-refractory, severely active UC were evaluated by clinical response (lack of symptoms) and endoscopic and histological outcomes after receiving one or two infusions of 5 mg/kg infliximab. Dramatic clinical, endoscopic and histological improvement was demonstrated in 14 out of 16 subjects. In contrast were the results from the study conducted by Probert et al. (2003), the largest early experience with infliximab in UC. This study also employed a randomized, placebo-controlled design and tested the use of infliximab 5 mg/kg vs. placebo in the treatment of 43 glucocorticoid resistant UC patients. Infusions were given at weeks zero and two and treatment response was assessed over 8 weeks of follow-up. After 2 weeks, there was no statistically significant difference between infliximab and placebo groups. After 6 weeks, there was a trend in favor of infliximab for treatment remission, but this also failed to reach statistical significance. The authors concluded that the results did not support the use of infliximab in the management of glucocorticoid resistant UC.

A fourth study published by Ochsenkűhn et al. (2004) compared three intravenous infusions of infliximab 5 mg/kg with a high-dose prednisolone treatment regimen in patients with non-steroid–refractory acute UC. In this small study, the majority of patients in both treatment groups showed therapy success after 3 and 13 weeks, and the authors concluded that infliximab could be effective in the treatment of acute moderate to severe UC.

The mixed experience with infliximab in these early, small, clinical trials led to doubts as to the validity of TNF as a therapeutic target in UC. In this setting, the results of the two large, company-sponsored trials of infliximab in UC, ACT I and ACT II, were of particular interest (Rutgeerts et al., 2005). These two multi-center, randomized, double-blind, placebo-controlled trials enrolled 364 patients each. Patients had active UC according to the composite clinical and endoscopic Mayo score and moderate to severe disease activity on sigmoidoscopy despite concurrent treatment with corticosteroids and/or immunomodulators or with a history of failure of response to such therapies. Patients were randomized to receive infusions of placebo or infliximab 5 mg or 10 mg/kg at weeks zero, two, and six, and then every 8 weeks through 6 or 12 months. The results were nearly identical in the two studies. Approximately two-thirds of patients receiving 5 or 10 mg/kg infliximab had a clinical response at week eight compared with about one-third receiving placebo ($p > 0.001$ for both comparisons). Responses and remissions were sustained through weeks 30 and 54 (Figure 5.2).

The robust outcomes in the ACT trials were important for two reasons. Firstly, they provided convincing clinical evidence for the utility of infliximab in the treatment of patients with moderate to severe UC, and led the way for the approval of infliximab in this indication in the US in early 2005. Additionally, from the viewpoint of target validation, the results provided unequivocal evidence for the importance of TNF in disease pathogenesis in UC.

A recent publication reported results from a study that examined the benefits of infliximab in hospitalized patients with fulminant UC, unresponsive to IV corticosteroids, who were considered likely to need colectomy (Järnerot et al., 2005). The proportion of subjects who eventually underwent colectomy was compared between the group randomly assigned to receive infliximab 5 mg/kg and the group who received placebo. At 90 days, the proportion of subjects who required colectomy was significantly smaller in the infliximab

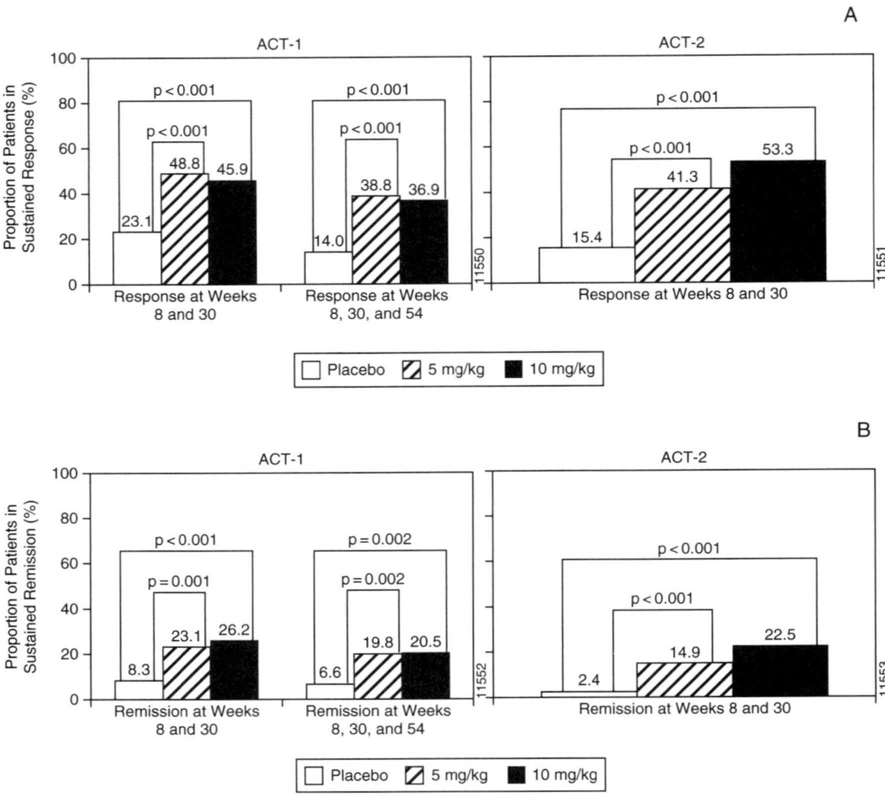

FIGURE 5.2 Clinical response to infliximab therapy in patients with ulcerative colitis. The figure shows proportion of patients with a sustained clinical response (Panel A) and in sustained clinical remission (Panel B) in ACT 1 and ACT 2.

treatment group vs. placebo (29%, versus 67%, $p < 0.017$). This study is important because it extends the benefit of TNF blockade in UC to patients with the most severe form of the disease, and demonstrated an ability for infliximab to prevent the need for major surgery, with all its consequences.

There are currently only limited data on the utility of infliximab in pediatric UC (Mamula et al., 2004) but these data also support the utility of infliximab in childhood UC. A clinical study of infliximab use in pediatric UC is planned.

D. Safety of TNF-Blocking Agents

TNF-blocking agents represent a powerful new class of drugs for the treatment of inflammatory disease, but as with most drugs, they are associated with safety issues. Some of these can be related directly to the mechanism of action of the drugs. For example, inhibition of TNF-driven granulomatous inflammation likely underlies the reactivation of latent tuberculosis and other granulomatous diseases, such as histoplasmosis, reported with TNF-blocking drugs. Other safety issues can be clearly related to the development of antibody responses to the injected biological agent, with consequent infusion reactions or local injection site reactions on repeat dosing. Other safety signals with TNF-blocking drugs are of less certain origin, for example the possible worsening of heart failure in patients with preexisting disease, and the possible association with neurological disease, lymphoma, and other malignancies. A full description of the safety issues associated with TNF-blocking drugs is beyond the scope of this review, and the reader is referred to the product labels for these agents for a review of the safety data.

CONCLUSION

TNF has many biological activities that made it a strong contender as a drug target in inflammatory disorders such as CD and UC, and the early, if somewhat inconsistent, demonstrations of disease association for TNF in IBD strengthened that claim. TNF, however, was one of many potential drug targets that might make such a claim in IBD. Indeed, the recent early success in the clinic of a monoclonal antibody targeting IL-12 in patients with CD demonstrates that at least one other of the many potential molecular targets in IBD is likely to be clinically important. On this background, the convincing demonstrations of efficacy for infliximab in both CD and UC in large, well-controlled clinical trials are particularly important. These outcomes have provided a powerful new treatment for patients with IBD, and at the same time represent a major success story for the pharmaceutical industry and the imprecise science of drug development.

ACKNOWLEDGMENTS

The author thanks Diane Hilferty for assistance with typing the manuscript, Dr. David Shealy for assistance with the section dealing with TNF biology, and Dr. Marion Blank for her expert review of the manuscript.

REFERENCES

Aggarwal, B. (2003). Signaling pathways of the TNF superfamily: A double-edged sword. *Nature Rev.* 3: 745–756.

Baldassano, R. and D. Piccoli (1999). Inflammatory bowel disease in pediatric and adolescent patients. *Gastroenterol. Clin. North Am.* 28: 445–458.

Black, R., C. Rauch, et al. (1997). A metalloproteinase distintegrin that releases tumour-necrosis factor-α from cells. *Nature* 365: 729–732.

Bouma, G. and W. Strober (2003). The immunological and genetic basis of inflammatory bowel disease. *Immunology* 3: 521–533.

Braegger, C., S. Nicholls, et al. (1992). Tumour necrosis factor alpha in stool as a marker of intestinal inflammation. *Lancet* 339: 89–91.

Brynskov, J., P. Foegh, et al. (2002). Tumour necrosis factor α converting enzyme (TACE) activity in the colonic mucosa of patients with inflammatory bowel disease. *Gut* 51: 37–43.

Carswell, E., L. Old, et al. (1975). An endotoxin-induced serum factor that causes necrosis of tumors. *Proc. Nat. Acad. Sci. U. S. A.* 72: 3666–3670.

Carter, M., A. Lobo, and S. Travis (2004). Guidelines for the management of inflammatory bowel disease in adults. *Gut* 53: 1–16.

Chey, W. (2001). Infliximab for patients with refractory ulcerative colitis. *Inflamm. Bowel Dis.* 7: S30–S33.

Derkx, B., J. Taminlau, et al. (1993). Tumour-necrosis-factor antibody treatment in Crohn's disease. *Lancet* 342: 173–174.

Eaden, J., K. Abrams, and J. Mayberry (2001). The risk of colorectal cancer in ulcerative colitis: A meta-analysis. *Gut* 48: 526–535.

Elsässer-Beile, U., S. von Kleist, et al. (1994). Cytokine production in whole blood cell cultures of patients with Crohn's disease and ulcerative colitis. *J. Clin. Lab. Anal.* 8: 447–451.

ENBREL® (package insert). Thousand Oaks, CA: Immunex Corp., 2003.

Faubion, W., E. Loftus, et al. (2001). The natural history of corticosteroid therapy for inflammatory bowel disease: A population-based study. *Gastroenterology* 121: 255–260.

Fuss, I., F. Heller, et al. (2004). Non-classical CD1d-restricted NK T cells that produce IL-13 characterize an atypical Th2 response in ulcerative colitis. *J. Clin. Invest.* 113: 1490–1497.

Fuss, I., M. Neurath, et al. (1996). Disparate $CD4^+$ lamina propria (LP) lymphokine secretion profiles in inflammatory bowel disease. *J. Immunol.* 157: 1261–1270.

Glickman, R. (1998). Inflammatory bowel disease: Ulcerative colitis and Crohn's disease. *In* "Harrison's Principles of Internal Medicine, 14th ed." (A. Faluce and E. Braunwald, eds.), pp. 1633–1645. McGraw Hill, New York, NY.

Gor, D., N. Rose, and N. Greenspan (2003). $T_h1 - T_h2$: A Procrustean paradigm. *Nature Immunol.* 4: 503–505.

Gordon, J., A. Di Sabatino, and T. MacDonald (2005). The pathophysiologic rationale for biological therapies in inflammatory bowel disease. *Curr. Opin. Gastroenterol.* 21: 431–437.

Grell, M., H. Wajant, et al. (1998). The type 1 receptor (CD120a) is the high-affinity receptor for soluble tumor necrosis factor. *Proc. Natl. Acad. Sci. U. S. A.* 95: 570–575.

Hanauer, S., B. Feegan, et al. (2002). Maintenance infliximab for Crohn's disease: The ACCENT 1 randomised trial. *Lancet* 359: 1541–1549.

Hanauer, S. and S. Meyers (1997). Management of Crohn's disease in adults. *Am. J. Gastroenterol.* 92: 559–566.

Hanauer, S. and M. Sparrow (2004). Therapy of ulcerative colitis. *Curr. Opin. Gastroenterol.* 20: 345–350.

Hanauer, S., C. Wagner, et al. (2004). Incidence and importance of antibody responses to infliximab after maintenance or episodic treatment in Crohn's disease. *Clin. Gastroenterol. Hepatal.* 2: 542–553.

Heller, F., P. Florian, et al. (2005). Interleukin-13 is the key effector Th2 cytokine in ulcerative colitis that affects epithelial tight junctions, apoptosis, and cell restitution. *Gastroenterol.* 129: 550–564.

Heller, F., I. Fuss, et al. (2002). Oxazolone colitis: A Th2 colitis model resembling ulcerative colitis, is mediated by IL-13-producing NK-T cells. *Immunity* 17: 629–638.

Hendrickson, B., R. Gokhale, and J. Cho (2002). Clinical aspects and pathophysiology of inflammatory bowel disease. *Clin. Microbiol. Reviews* 15: 79–94.

Hyams, J. (2000). Inflammatory bowel disease. *Pediatrics in Review* 21: 291–295.

Inoue, S., T. Matsumoto, et al. (1999). Characterization of cytokine expression in the rectal mucosa of ulcerative colitis: Correlation with disease activity. *Am. J. Gastroenterol.* 94: 2441–2446.

Järnerot, G., E. Hertervig, et al. (2005). Infliximab as rescue therapy in severe to moderately severe ulcerative colitis: A randomized placebo-controlled study. *Gastroenterol.* 128: 1805–1811.

Knight, D., H. Trinh, et al. (1993). Construction and initial characterization of a mouse-human chimeric anti-TNF antibody. *Molecular Immunology* 30: 1443–1453.

Kornbluth, A. and D. Sachar (2004). Ulcerative colitis practice guidelines in adults (update): American College of Gastroenterology, Practice Parameters Committee. *Am. J. Gastroenterol.* 99: 1371–1385.

Kumar, S., U. Ghoshal, et al. (2004). Severe ulcerative colitis: Prospective study of parameters determining outcome. *J. Gastroenterol. & Hepatol.* 19: 1247–1252.

Lichtenstein, G. R., S. Yan, et al. (2005). Infliximab maintenance treatment reduces hospitalizations, surgeries, and procedures in fistulizing Crohn's disease. *Gastroenterology* 128: 862–869.

Lichtenstein, G., S. Yan, et al. (2003). Remission in patients with Crohn's disease is associated with improvement in employment and quality of life and a decrease in hospitalizations and surgeries. *Am. J. Gastro.* 99: 91–96.

Louis, E., Z. El Ghoul, et al. (2004). Association between polymorphism in IgG Fc receptor IIIa coding gene and biological response to infliximab in Crohn's disease. *Aliment Pharmacol Ther.* 19: 511–519.

MacDonald, T., P. Hutchings, et al. (1990). Tumour necrosis factor-alpha and interferon-gamma production measured at the single cell level in normal and inflamed human intestine. *Clin. Exp. Immunol.* 81: 301–305.

Maeda, M., N. Watanabe, et al. (1992). Serum tumor necrosis factor activity in inflammatory bowel disease. *Immunopharmacol. and Immunotoxicol.* 14: 451–461.

Mamula, P., J. Markowitz, et al. (2004). Infliximab in pediatric ulcerative colitis: Two-year follow up. *J. Ped. Gastroenterol. Nut.* 38: 298–301.

Mannon, P., I. Fuss, et al. (2004). Anti-interleukin-12 antibody for active Crohn's disease. *N. Engl. J. Med.* 351: 2069–2079.

Moss, M., S. Jin, et al. (1997). Cloning of a disintegrin metalloproteinase that processes precursor tumour-necrosis factor-α. (1997). *Nature* 385: 733–736.

Murch, S., C. Braegger, et al. (1993). Location of tumour necrosis factor α by immunohistochemistry in chronic inflammatory bowel disease. *Gut* 34: 1705–1709.

Murch, S., V. Lamkin, et al. (1991). Serum concentrations of tumour necrosis factor α in childhood chronic inflammatory bowel disease. *Gut.* 32, 913–917.

Ochsenkűhn, T., M. Sackmann, and B. Gőke (2004). Infliximab for acute, not steroid-refractory ulcerative colitis: A randomized pilot study. *European J. Gastroenterol. & Hepatol.* 16: 1167–1171.

Pennica, D., G. Nedwin, et al. (1984). Human tumour necrosis factor: Precursor structure, expression and homology to lymphotoxin. *Nature* 312: 724–729.

Perez, C., I. Albert, et al. (1990). A nonsecretable cell surface mutant of tumor necrosis factor (TNF) kills by cell-to-cell contact. *Cell* 63: 251–258.

Pololsky, D. K. (2002). Inflammatory bowel disease. *N. Engl. J. Med.* 347(6): 417–429.

Powrie, F., M. Leach, et al. (1994). Inhibition of Th1 responses prevents inflammatory bowel disease in *scid* mice reconstituted with CD45RB^{hi}CD4^{+} T cells. *Immunity.* 1: 553–562.

Present, D., P. Rutgeerts, et al. (1999). Infliximab for the treatment of fistulas in patients with Crohn's disease. *N. Engl. J. Med.* 340, 1398–1405.

Probert, C., S. Hearing, et al. (2003). Infliximab in moderately severe glucocorticoid resistant ulcerative colitis: A randomized controlled trial. *Gut* 52: 998–1002.

Reinecker, H., M. Steffen, et al. (1993). Enhanced secretion of tumour necrosis factor-alpha, IL-6, and IL-1β by isolated lamina propria mononuclear cells from patients with ulcerative colitis and Crohn's disease. *Clin. Exp. Immunol.* 94: 174–181.

Rutgeerts, P., B. Feagan, et al. (2004). Comparison of scheduled and episodic treatment strategies of infliximab in Crohn's disease. *Gastroenterol.* 126: 402–413.

Rutgeerts, P., W. Sandborn, et al. (2005). Infliximab for induction and maintenance therapy for ulcerative colitis. *N. Engl. J. Med.* 353: 2462–2476.

Sandborn, W., S. Hanauer, et al. (2001). Etanercept for active Crohn's disease: A randomized, double-blind, placebo-controlled trial. *Gastroenterol.* 121: 1088–1094.

Sandborn, W., S. Hanauer, et al. (2005). Clinical remission and response in patients with Crohn's disease treated open-label for six months with fully human anti-TNF monoclonal antibody adalimumab. *Gut* 54: A18.

Sands, B., F. Anderson, et al. (2004a). Infliximab maintenance therapy for fistulizing Crohn's disease. *N. Engl. J. Med.* 350: 876–885.

Sands, B., M. Blank, et al. (2004b). Long-term treatment of rectovaginal fistulas in Crohn's disease: Response to infliximab in the ACCENT 11 study. *Clin. Gastro. Hepatol.* 2: 912–920.

Sands, B., W. Tremaine, et al. (2001). Infliximab in the treatment of severe, steroid-refractory ulcerative colitis: A pilot study. *Inflammatory Bowel Dis.* 7: 83–88.

Schreiber, S., M. Khaliq-Kareemi, et al. (2005). Certolizumab pegol, a humanised anti-TNF pegylated Fab' fragment, is safe and effective in the maintenance of response and remission following induction in active Crohn's disease: A phase III study (PRECISE). *Gut* 54: A82.

Schreiber, S., P. Rutgeerts, et al. (2005). A randomized, placebo-controlled trial of certolizumab pegol (CDP870) for treatment of Crohn's disease. *Gastroenterol.* 129: 807–818.

Schürmann, G., M. Betzler, et al. (1992). Soluble interleukin-2 receptor, interleukin-6 and interleukin-1 in patients with Crohn's disease and ulcerative colitis: Preoperative levels and postoperative changes of serum concentrations. *Digestion* 51: 51–59.

Shanahan, F. (2002). Crohn's disease. *Lancet* 359: 62–69.

Smith, R. and C. Baglioni (1987). The active form of tumor necrosis factor is a trimer. *J. Biological Chemistry* 262: 6951–6954.

Stenson, W. (2000). Inflammatory bowel disease. *In* "Cecil Textbook of Medicine, 21st ed." (I. Goldman and J. Bennett, ed.), pp. 722–729. W. B. Saunders Co., Philadelphia, PA.

Stevens, C., G. Walz, et al. (1992). Tumor necrosis factor-α, interleukin-1β, and interleukin-6 expression in inflammatory bowel disease. *Digestive Diseases and Sciences* 37: 818–826.

Targan, S., S. Hanauer, et al. (1997). A short-term study of chimeric monoclonal antibody cA2 to tumor necrosis factor α for Crohn's disease. *N. Engl. J. Med.* 337: 1029–1034.

Tapping, R., S. Akashi, et al. (2000). Toll-like receptor 4, but not toll-like receptor 2, is a signaling receptor for *Escherichia* and *Salmonella* lipopolysaccharides. (2000). *J. Immunol.* 165: 5780–5787.

van Dullemen, H., E. de Jong, et al. (1998). Treatment of therapy-resistant perineal metastatic Crohn's disease after proctectomy using anti-tumor necrosis factor chimeric monoclonal antibody, cA2. *Dis. Colon Rectum* 41: 98–102.

Van Dullemen, H., S. Van Deventer, et al. (1995). Treatment of Crohn's disease with anti-tumor necrosis factor chimeric monoclonal antibody (cA2). *Gastroenterol.* 109: 129–135.

Woywodt, A., P. Neustock, et al. (1994). Cytokine expression in intestinal mucosal biopsies. *In situ* hybridisation of the mRNA for interleukin-1β, interleukin-6 and tumour necrosis factor-α in inflammatory bowel disease. *Eur. Cytokine Netw.* 5: 387–395.

6 ANTI–CCL-2/MCP-1: DIRECTED BIOLOGICALS FOR INFLAMMATORY AND MALIGNANT DISEASES

ANUK DAS* AND LI YAN†

*Ph.D., *Centocor Research and Development, Inc., Radnor, Pennsylvania*
†M.D., Ph.D., *Centocor Research and Development, Inc., Malvern, Pennsylvania; Peking University, Beijing, China*

The involvement of inflammation in various diseases has been gaining increasing attention recently. Besides immune-mediated inflammatory diseases, inflammatory processes have been recognized to play a key role in the cause and progression of cardiovascular, metabolic, and malignant diseases. An underlying characteristic shared by many of these diseases is tissue infiltration of monocytes, a process driven primarily by CC chemokine ligand-2 (CCL-2), also known as monocyte chemoattractant protein 1 (MCP-1). We will review the biology of CCL-2 using renal diseases and cancer models as examples for target validation. These CCL-2–dependent bioactivities are analyzed *in vitro* using various methodologies including assays such as chemotaxis, intracellular calcium flux, and signaling event analysis. These assays are also essential in screening and characterizing pharmaceutical antagonists to CCL-2–dependent signaling events and associated biological activities. *In vivo* assays like angiogenesis, renal disease, and different tumor models are described for target validation studies. Modulating CCL-2 activity via gene regulation or neutralizing antibodies in these models has generated informative and promising results, encouraging the research and development of biologicals targeting CCL-2 pathways in order to manipulate the local chemokine network and have therapeutic benefits in various diseases.

I. INTRODUCTION

CC chemokine ligand-2 (CCL-2), also known as monocyte chemoattractant protein 1 (MCP-1), belongs to a large family of *chemo*attractant cyto*kines*, or, chemokines. Chemokines are small proteins (molecular weights range from 8 to 10 kDa) and share approximately 20% to 70% homology (Luster, 1998). Chemokines were first characterized as proinflammatory proteins because of their ability to attract specific subsets of leukocytes to sites of inflammation. There are four chemokine sub-families organized according to structural homology and the positioning of conserved cysteine residues present in the structure of chemokines (Zlotnik and Yoshie, 2000). The sub-families are termed CC, CXC, CX3C, and C, where X denotes amino acid residues between the first two conserved cysteines. The majority of the chemokines reside within the CC and CXC chemokine sub-families. Chemokine receptors are G protein-coupled seven-transmembrane-spanning receptors and are also separated into sub-families based on the chemokines that bind to them; for example, CC chemokine receptors (CCRs) and CXC chemokine receptors (CXCRs) (Murphy et al., 2000). In general, multiple chemokines signal through one receptor and a chemokine may signal through multiple receptors.

CCL-2 belongs to the CC chemokine sub-family and is currently believed to mediate its actions through one receptor, CC chemokine receptor 2 (CCR2). CCL-2 shares CCR2 with three other chemokines, CCL-7, -8 and -13 (MCP-3, -2 and -4, respectively). Unlike CCL-2, CCL-7 and -13 can also bind and signal through other chemokine receptors; namely CCR1 and CCR3, respectively (Zlotnik and Yoshie, 2000) (Table 6.1). Two CCR2 receptors, CCR2A and CCR2B, have been cloned. Both receptors signal, through Gαi, in response to nanomolar (nM) concentrations of CCL-2. CCR2A and CCR2B differ in their carboxyl tails as a result of alternative splicing (Charo et al., 1994). Evidence of alternative signaling pathways for CCL-2 through receptors other than CCR2 has also been speculated (Schecter et al., 2004).The induction of CCL-2 was first described in murine fibroblasts. Stimulation of 3T3 fibroblasts with platelet-derived growth factor induced the early expression of the JE gene (Cochran et al., 1983). Homology studies with a protein released from human fibroblasts and gliomas revealed that CCL-2 was the human orthologue of the mouse JE gene product (Rollins et al., 1989; Yoshimura et al., 1989b). More recently, it has been demonstrated that murine MCP-5 (CCL-12) bears higher amino-acid identity to human CCL-2 than JE (64% vs. 55%, respectively) (Sarafi et al., 1997; Van Coillie et al., 1999) Therefore, in mice, two ortholgoues exist for human CCL-2. The initial biological activity attributed to CCL-2 was mononuclear cell chemoattraction (Yoshimura et al., 1989a). Indeed, organ-specific over-expression of JE in mice induces tissue accumulation of monocytes (Fuentes et al., 1995; Grewal et al., 1997;

TABLE 6.1 Affinities of Human CCL-2, -7, -8, -13 for CCR Receptors; Affinities of CCL-2 and Its Murine Orthologues JE and MCP-5 for Human and Murine CCR2.

	Receptor			
Ligand	**CCR1**	**CCR2**	**CCR3**	**CCR5**
CCL2/MCP1	+ (Neote et al., 1993)	++		
CCL8/MCP2	+ (Gong et al., 1997)	++	+	++
CCL7/MCP3	+ (Ben-Baruch et al., 1995)	++	+	+
CCL13/MCP4	++	++	++	+
Receptor	*Ligand Affinity*			
Human CCR2	Human CCL2=muMCP5>JE (Sarafi et al., 1997)			
Murine CCR2	JE>muMCP5 (Sarafi et al., 1997)			

+ = low affinity; ++ = high affinity

Kolattukudy et al., 1998). Studies have also demonstrated that CCL-2 is a potent chemoattractant for memory lymphocytes (Carr et al., 1994), natural killer cells (Allavena et al., 1994; Maghazachi et al., 1994), and dendritic cells (Xu et al., 1996) and is an activator of basophils and mast cells (Alam et al., 1994; Bischoff et al., 1992; Dahinden et al., 1994). Stimulated-secretion of CCL-2 has been demonstrated in a variety of cell types, including type II pneumocytes (Standiford et al., 1991), endothelial cells (Strieter et al., 1989), monocytes (Yoshimura et al., 1989b), and fibroblasts (Koyama et al., 1998; Witowski et al., 2001).

Inflammation is a characteristic feature of a majority of chronic diseases. As such, it is postulated that CCL-2 may play an important role in multiple features of disease pathology through its pro-inflammatory activities. Emerging evidence suggests that, together with cell chemoattraction described above, CCL-2 also mediates other biological activities, including cell activation (Alam et al., 1992), angiogenesis (Salcedo et al., 2000), and fibrosis (Gharaee-Kermani et al., 1996; Yamamoto and Nishioka, 2003). Because of these diverse activities of CCL-2 and the association of inflammation with chronic diseases, a number of different diseases may be amenable through targeting CCL-2, including artherosclerosis, renal diseases, and cancer. This chapter will review the biology of CCL-2 as it relates to the bioactivities outlined above. Further, the *in vitro* and *in vivo* studies that may be performed to validate the importance of CCL-2 in inflammatory processes will also be discussed. Two disease areas have been highlighted to illustrate the biology of CCL-2 *in vivo*: renal diseases and cancer.

II. *IN VITRO* ASSAYS TO ESTABLISH THE PRO-INFLAMMATORY ACTIVITIES OF CCL-2

The pro-inflammatory activities attributed to CCL-2 are chemotaxis, cell activation, angiogenesis, and fibrosis. A number of *in vitro* assays may be established to demonstrate these biological activities of CCL-2 and the assays can be used to validate the biology of CCL-2.

A. Chemotaxis

The first functional bioactivity attributed to CCL-2 was chemotaxis of monocytes. Chemotaxis is described as the directional movement of cells toward increasing concentrations of a chemoattractant. For further details on the biology associated with chemotaxis, the reader is referred to excellent reviews on the subject (Chung et al., 2001; Wu, 2005).

Inhibition of monocyte recruitment through neutralization of CCL-2 bioactivity serves as an important *in vitro* assay for validating the importance of CCL-2 in inflammatory diseases. Chemotaxis may be performed to recombinant or synthetic CCL-2 (Kruszynski et al., 2005), to native CCL-2 contained in supernatants from *in vitro* cell-culture studies, or in tissue homogenates from disease organs. For the purposes of evaluating the contribution of CCL-2 in mediating cell migration, CCL-2–specific inhibitors can be included in the chemotaxis assay. A number of different cell types may be used for chemotaxis assays. The cell lines THP-1 and MonoMac 6 (available from www.atcc.org) express CCR2 (Wong et al., 1997) and can be used on a routine basis. Primary peripheral blood cells, such as monocytes, express CCR2, but the inconvenience of isolating the cells for each assay limits the utility of primary cells to confirmatory assays.

The majority of chemotaxis assays rely on the basic principle of cell migration from wells at the top of a chamber, through a filter of defined pore size (8 μm for monocyte-like cells) toward a chemoattractant in lower wells. Historically, the most widely used methodology for performing chemotaxis assays relied on a 48-well filter-plate Boyden chamber where the leading edge of cells migrating to the bottom of the filter were quantitated histologically (Frow et al., 2004; Wilkinson, 1998). More recent modifications to the Boyden chamber have resulted in a 96-well disposable plate format (available for www.neuroprobe.com) that enables the quantitation of fluorescent-labeled cell migration to the bottom chamber and facilitates a higher throughput rate for the assay (e.g. Frevert et al., 1998; Kruszynski et al., 2005). The step of fluorescent labeling of cells may be by-passed by incorporation of a bioluminescent reporter into the cells (Vishwanath et al., 2005).

B. Intracellular Calcium Flux

In addition to inducing chemotaxis, binding of CCL-2 to CCR-2 (Charo et al., 1994) also elicits other monocyte responses, including intracellular calcium mobilization flux and respiratory burst (Rollins et al., 1991)—characteristics of chemotactic agents on phagocytes (Richmond et al., 1988). Calcium flux is monitored using CCR-2–expressing cells loaded with the fluorescent probe Fura-2, and occurs in a rapid, transient, and dose-dependent fashion (Sozzani et al., 1993). Increases in intracellular calcium induced by CCL-2 are entirely due to influx from extracellular sources and not from the release of calcium from intracellular stores (Sozzani et al., 1993). In addition, this CCL-2–induced calcium flux represents a downstream event of protein kinase C (PKC) activation triggered by CCR-2 and coupled $G\alpha_i$, and is partially blocked by pertussis toxin (Myers et al., 1995; Sozzani et al., 1991, 1993). The measurement of intracellular calcium fluxes in real time is widely applied for pharmacological characterization or to screen for new therapeutic candidates. The development of new calcium-sensitive dyes and assays has provided sensitive, homogeneous assays that can be readily applied to high-throughput screening (HTS) using a fluorometric imaging plate reader (FLIPR) (Chambers et al., 2003).

C. Alternative Assays

There are inherent disadvantages associated with monocyte chemotaxis or calcium flux assays. Low signal-to-noise ratio and variability in the chemotaxis assays limit their usefulness in high-throughput screening or in long-term comparison studies. In addition, the bell-shaped dose-response curve of chemotaxis to CCL-2 stimulation also complicates data interpretation. Calcium flux assays provide a sigmoid dose-response curve and an improvement over the chemotaxis assays in sensitivity. These assays do require special automation and imaging equipment given the extremely transient nature of this event, reaching the maximal concentration in approximately 20 seconds (Chambers et al., 2003; Sozzani et al., 1993). Recently, alternative assaying systems to measure activities of anti–CCl-2 antibodies or

CCR-2 antagonists have been described. Instead of directly measuring biological responses to CCL-2 stimulation, such as chemotaxis or calcium flux, these assays monitor other signaling events after receptor activation. One of such assays uses p44/42 mitogen-activated protein kinase (MAPK) phosphorylation as a readout (Hirata Terra et al., 2004). In this cell-based ELISA format assay, CCL-2 induces a dose-dependent increase in the level of p44/42 MAPK phosphorylation that is quantifiable using the phosphor-p44/42 MAPK-specific antibodies, and can be used for screening and characterizing anti–CCL-2 antibodies or small-molecule CCR2 inhibitors (Hirata Terra et al., 2004).

D. Angiogenesis

Angiogenesis, the formation of new blood vessels from an existing vasculature, is essential for tumor growth beyond 1 to 2 mm in diameter. This process occurs in all types of tumors and the intensity of the angiogenic process, as assessed by microvessel quantitation, correlates with primary tumor growth, invasiveness, and metastatic spread of disease (Folkman, 1995; O'Byrne et al., 2000). Tumor angiogenesis provides a means to supply oxygen and nutrients and to remove waste products from tissues that are hypoxic or hypovascularized (Hyder and Stancel, 1999). The link between CCL-2 and tumor angiogenesis was first described in mouse colon cancer models, where transfection of CCL-2 (also known as monocyte chemotactic and activating factor, MCAF) enhanced primary tumor angiogenesis and increased the incidence rates of spontaneous metastasis (Nakashima et al., 1995). There is accumulating clinical evidence supporting the role of CCL-2 in promoting tumor angiogenesis. CCL-2 expression in primary tumors has been associated with tumor macrophage infiltration, microvessel density, and production of pro-angiogenic growth factors in hemangioma (Isik et al., 1996), breast cancer (Goede et al., 1999; Saji et al., 2001; Ueno et al., 2000), esophageal squamous cell carcinomas (Koide et al., 2004; Ohta et al., 2002), and gastric carcinomas (Ohta et al., 2003). It was initially believed that CCL-2 induced angiogenesis solely via recruitment of macrophages, and angiogenic growth factors—such as vascular endothelial growth factor (VEGF), tumor necrosis factor alpha (TNFα), interleukin 6 (IL-6), and IL-8—expressed by these tumor infiltrating macrophages (Goede et al., 1999; Liss et al., 2001; Ueno et al., 2000). More recently, it became clear that CCL-2 also exerts direct stimulatory effects on endothelial cells to promote angiogenesis even in the absence of inflammatory cell recruitment (Salcedo et al., 2000). In these CCR2-positive endothelial cells, CCL-2 induces VEGF production (Hong et al., 2005) and modulates cell surface expression, clustering, activity, and function of membrane type 1 matrix metalloproteinase (MT1-MMP) (Galvez et al., 2005). These combined direct and indirect activities enable CCL-2 to induce angiogenesis at a similar potency as VEGF (Goede et al., 1999).

CCL-2–mediated angiogenesis provides an array of *in vitro* and *in vivo* assays for screening and characterizing CCL-2-neutralizing antibodies or CCR2 antagonists. Like monocytes, endothelial cells also migrate to chemotactic stimulation of CCL-2 (Salcedo et al., 2000), providing an additional cell type for chemotactic assays. The aortic-ring-sprouting assay is used to monitor angiogenic responses to CCL-2 in the absence of leukocytic infiltrates (Salcedo et al., 2000). *In vivo* angiogenic activity of CCL-2 could be measured by chick chorioallantoic membrane (CAM), corneal angiogenesis, or the Matrigel plug assays (Goede et al., 1999; Salcedo et al., 2000; Yan et al., 2003). In the corneal angiogenesis assay, CCL-2 elicits angiogenic responses with similar potency to the well-known angiogenic mediator VEGF-A (Goede et al., 1999). CCL-2 stimulates CAM angiogenesis in a bell-shaped curve mimicking the dose-dependence of chemotaxis (Rollins et al., 1991). When tumor cells expressing different levels of CCL-2 were compared for their *in vivo* angiogenic activity in the Matrigel plug angiogenesis assay, a correlative relationship was observed despite the expression of other angiogenic growth factors including VEGF and basic fibroblast growth factor (bFGF) (Yan et al., 2003). Furthermore, we have used a monoclonal antibody against human CCL-2 to explore the causal effects of CCL-2 during tumor angiogenesis. When tumor cell–derived CCL-2 was neutralized, angiogenesis was significantly inhibited in a dose-dependent fashion when either Panc-1 or MDA-MB-435S tumor cells were used to stimulate angiogenesis (Yan et al., 2003). Given the complexity and redundancy of angiogenic growth factor expression profile in these tumor cells, CCL-2 appears to play a critical role in stimulating tumor angiogenesis in these models. These observations, however, may not be entirely applicable to clinical situations since tumor angiogenesis is a complex and multi-step process regulated by numerous pro-angiogenic growth factors as well as affected by angiostatic factors. The involvement of CCL-2 may likely vary in different cancer types and at different stages of malignant disease process.

E. Fibroblast Assays

Fibrosis is a result of extracellular cellular matrix (ECM) deposition (e.g., collagen) by activated effectors cells such as fibrocytes and fibroblasts. CCL-2 has been demonstrated to have a direct role in fibrosis by activating these effector cells of fibrosis (summarized below). *In vitro* assays with fibrocytes and fibroblasts may be established to validate the biology of CCL-2 in fibrosis. No fibrocyte cell lines exist making working with these cells low throughput. In contrast, a number of organ-specific and disease-specific primary human fibroblasts are available commercially (www.atcc.org, www.cambrex.org). Further, mouse fibroblast cell lines (e.g., NIH 3T3 embryonic fibroblasts) are also commercially available (www.atcc.org).

Fibrocytes are circulating, bone-marrow–derived, collagen-producing cells (Abe et al., 2001; Bucala et al., 1994; Phillips et al., 2004) that are characterized by the presence of distinct cell surface and intracellular markers (CD34+, CD45+ and collagen I+). Fibrocytes accumulate at sites of injury and the profile of chemokine receptors on the cell surface (CCR2, CCR7, and CXCR4) (Abe et al., 2001; Hashimoto et al., 2004; Moore et al., 2005; Phillips et al., 2004) suggests an important role for chemokines in the recruitment of these cells. Very recent data have shown that mouse fibrocytes migrate to JE *in vitro* and JE stimulates collagen secretion from these cells (Moore et al., 2005).

Fibroblasts are mesenchymal cells. When activated by different stimuli, these cells proliferate, secrete ECM proteins, and differentiate into myofibroblasts. CCL-2 has been shown to induce procollagen α(I) mRNA expression in normal human dermal fibroblasts (Yamamoto and Nishioka, 2003) and rat-lung fibroblasts through TGFβ1 (Gharaee-Kermani et al., 1996).

For validation studies to establish the role of CCL-2 or JE/MCP-5 in fibrocyte accumulation, human or mouse circulating fibrocytes may be isolated from the peripheral blood (Abe et al., 2001; Bucala et al., 1994; Chesney et al., 1998). Chemotaxis assays (described above) can be performed to validate the role of CCL-2 or JE/MCP-5 in the recruitment of these cells (Abe et al., 2001; Moore et al., 2005; Phillips et al., 2004). Stimulation of collagen secretion by CCL-2 or JE/MCP-5 can also be utilized as a validation assay in fibrocytes and fibroblasts. Collagen protein synthesis/secretion may be measured by a number of different methodologies: ^{3}H-proline incorporation (e.g., Fenwick et al., 2001; Micera et al., 2005), Sircol™ (www.biocolor.co.uk; e.g., El Chaar et al., 2005), and hydroxyproline assay (e.g., Campa et al., 1990; Gu et al., 2004). Finally, proliferation assays may also be performed by utilizing any of the following techniques: ^{3}H-thymidine incorporation may be used to measure increased DNA synthesis (e.g., Blanc-Brude et al., 2001). A number of different biochemical assays, such as MTT (3-(4,5-dimethylthiazolyl-2)-2,5-diphenyltetrazolium bromide) and BrdU (5-bromo-2′-deoxyuridine), may also be used to measure proliferation (www.probes.invitrogen.com; e.g., Zhang et al., 2004).

III. *IN VIVO* VALIDATION STUDIES

The establishment of a scientific rationale for a specific target in the pathogenesis of a disease needs to address the following questions:

- Is the target associated with human-disease pathology?
- Can the appropriate cells associated with the disease express the target *in vitro*?
- Do the *in vivo* models of the disease express the target?
- Are target inhibition strategies efficacious in the *in vivo* models?

A. The Rationale for Targeting CCL-2 in Renal Diseases

There is a marked association of tissue macrophages with the progression of tubulointerstitial renal diseases (reviewed by: Sean Eardley and Cockwell, 2005). An association of CCL-2 with a number of different renal diseases has been demonstrated. Urinary levels of CCL-2 correlates with proteininuria in patients with glomerulopathies and diabetic nephropathy (Banba et al., 2000; Morii et al., 2003; Rovin et al., 1996; Tashiro et al., 2002). Further, the elevated urine levels of CCL-2 correlated with disease severity (Morii et al., 2003; Rovin et al., 1996; Tashiro et al., 2002; Wada et al., 2000) and extent of renal tubular damage (Morii et al., 2003; Tashiro et al., 2002; Wada et al., 2000). Histological studies have demonstrated a correlation between urinary CCL-2 levels and increased infiltrating mononuclear cells in the glomeruli of patients with glomerulopathies, suggesting one potential mechanism through which CCL-2 may contribute to disease pathogenesis (Rovin et al., 1996) and, in the interstitium of patients with IgA nephropathy, crescentic glomerulonephritis and diabetic nephropathy (Grandaliano et al., 1996; Wada et al., 1999, 2000; Yokoyama et al., 1998). In crescentic glomerulonephritis, urinary CCL-2 levels also correlated with the total number of crescents and the number of fibrotic crescents (Wada et al., 1999). Further, expression of CCL-2 was localized to tubular epithelial cells, interstitial infiltrating cells, and cells in the glomerular tufts and crescents (Liu et al., 2003; Segerer et al., 2000). CCL-2 expression has also been described in vascular endothelial cells, tubular epithelial cells and infiltrating mononuclear cells in patients with immune-mediated renal diseases, such as lupus nephritis, Wegener's granulamotosis, and IgA nephropathy (Cockwell et al., 1998; Dai et al., 2001; Grandaliano et al., 1996; Wada et al., 1996b; Yokoyama et al., 1998). Finally, CCL-2 expression is found in the tubular epithelium of patients with diabetic nephropathy (Mezzano et al., 2003) and infiltrating cells expressing the chemokine have been found in the interstium (Wada et al., 2000).

Many stimuli can induce the expression of CCL-2 in resident renal cells *in vitro*. Physiological stimuli associated with renal disease progression, such as elevated levels of very low density lipoprotein (associated with focal glomerulosclerosis) (Lynn et al., 2000), high glucose (associated with the development and progression of diabetic nephropathy) (Ihm et al., 1998), and mechanical stretch (induced by glomerular hypertension) (Gruden et al., 2005; Suda et al., 2001), induce CCL-2 expression and production in mesangial cells. It is conceivable that this increased production of CCL-2 by renal cells elicits the infiltration of circulating mononuclear cells. Direct contact of monocytes with renal fibroblasts induces expression of CCL-2 in renal fibroblasts (Hao et al., 2003), providing another mechanism that contributes to the exacerbation and perpetuation of the inflammatory milleu. Interestingly, the presence of myofibroblasts (which are differentiated, collagen-producing fibroblasts) has been shown to correlate with renal disease pathogenesis and progression (Goumenos et al., 1998; Roberts et al., 1997; Tamimi et al., 1998; Wu et al., 2001). Stimulation of mesangial or proximal tubular epithelial cells with pro-inflammatory cytokines such as TNFα and IL-1 also elicits CCL-2 expression (Gong et al., 2004; Pai et al., 1996; Prodjosudjadi et al., 1995).

A number of *in vivo* rodent models are utilized to model human renal diseases. All the models utilize injury to induce kidney disease. For the purposes of this chapter, the relevant *in vivo* models of kidney diseases will be reviewed regarding their appropriateness for modeling CCL-2 biology. Studies demonstrating efficacy and validating the role for rodent CCL-2 orthologues, by antibody-directed chemokine neutralization or gene deletion strategies, are also highlighted. It is important to note that all *in vivo* studies to date have evaluated the role of mouse JE, but not mouse MCP-5.

Human crescentic glomerulonephritis is a rapidly progressive glomerular injury disease with poor prognosis. This disease is modeled in rodents by the administration of nephrotoxic serum (NTS), an anti-sera raised against the glomerular basement membrane. The *in vivo* model exhibits an accelerated development of glomerular nephritis characterized by an inflammatory cell infiltrate, glomerular crescent formation and fibrosis. The expression of rodent orthologues for CCL-2 has been demonstrated in the NTS model in mice (Lloyd et al., 1997; Neugarten et al., 1995; Tesch et al., 1999b) and rats (Fujinaka et al., 1997; Panzer et al., 2001; Rovin et al., 1994; Tang et al., 1995; Wada et al., 1996a; Wu et al., 1997). The development of the disease pathology (crescentic glomeruli) and loss of renal function (as measured by urinary protein) have been shown to correlate with the glomerular infiltration of monocytes and macrophages in Wistar-Kyoto rats (Fujinaka et al., 1997; Isome et al., 2004). Depletion of renal macrophages with liposome-encapsulated dichloromethylene diphosphonate protected against the development of crescents and urine protein secretion (Isome et al., 2004). The role of rodent CCL-2 in mediating the disease in this model has been demonstrated through utilization of JE gene-deletion and antibody neutralization strategies. JE gene–deleted mice were protected against tubular damage and was associated with the presence of fewer interstitial macrophages following NTS administration (Tesch et al., 1999b). The gene-deleted mice, however, were not protected against glomerular damage, glomerular macrophage influx and urinary protein secretion. Interestingly, following NTS administration, no increase in mRNA expression of the other murine orthologue for CCL-2, MCP-5, mRNA was observed in wild-type or JE gene–deleted mice. This suggests that for the CCL-2 system, rodent and human biology may differ and caution must be taken when interpreting *in vivo* rodent data. Antibody neutralization studies have generated contrasting data to that shown with JE gene–deleted mice. In rats, antibody neutralization of rat CCL-2 reduced glomerular macrophage numbers (Fujinaka et al., 1997; Panzer et al., 2001; Wada et al., 1996a), improved renal dysfunction (as measured by decreased blood urea nitrogen levels, urinary protein secretion or creatinine clearance) (Fujinaka et al., 1997; Panzer et al., 2001; Wada et al., 1996a), and prevented the development of crescents at the early time points (Fujinaka et al., 1997; Wada et al., 1996a) and the development of glomerulosclerosis at later time points (Wada et al., 1996a). Similarly, in mice, neutralization of mouse JE resulted in similar efficacy: reduction of proteinuria, glomerular and interstitial inflammation, crescent formation, and interstitial fibrosis (as measured by collagen I deposition) (Lloyd et al., 1997).

The anti-Thy 1.1 model of mesangial proliferative glomerulonephritis is a widely used acute *in vivo* model that mimics the events following mesangial cell-death–induced human-glomerular diseases, such as IgA nephropathy and diabetic nephropathy. In this model, injection of an antibody to the Thy-1 antigen, which is present on the surface of mesangial cells, induces complement-dependent mesangial cell death. The subsequent proliferation of mesangial cells, inflammatory cell influx, and fibrosis result in renal dysfunction (reviewed by: Jefferson and Johnson, 1999). Expression of rat CCL-2 in glomeruli has been demonstrated at early time points following anti–Thy-1 antibody administration (Stahl et al., 1993). Treatment of rats with anti-rat CCL2 antibody decreased the ensuing glomerular mononuclear cell infiltration following anti–Thy-1 administration (Wenzel et al., 1997) and collagen IV protein deposition (Schneider et al., 1999). The anti-fibrotic effects observed with anti-rat CCL2 antibody treatment appear to be mediated through inhibition of the pro-fibrotic cytokine transforming growth factor beta (Schneider et al., 1999).

Two *in vivo* murine models of systemic lupus erythematosus may be utilized to validate the role of JE orthologues in renal diseases:, MRL-Fas^{lpr} and New Zealand black × New Zealand white (NZB/W) mice. In both models, immune complex-mediated renal damage occurs over time. In these chronic models, JE expression has been demonstrated to be upregulated in the glomerular and interstitial areas of the kidney prior to the infiltration of mononuclear cells, renal damage, and dysfunction (Perez de Lema et al., 2001; Tesch et al., 1999a; Zoja et al., 1997). To date, no antibody-neutralization studies have been performed. Utilization of JE-deficient MRL-Fas^{lpr} mice has demonstrated the importance of this mouse orthologue in mediating the renal inflammatory cell infiltrate, proteinuria and survival (Tesch et al., 1999a).

The final two models of renal diseases that are highlighted here have not been utilized to date to validate the role of the rodent orthologues of CCL-2 in mediating disease. Sub-total (5/6) nephrectomy, a model of chronic renal disease, induces glomerular hypertension leading to nephron loss and progressive glomerulosclerosis and tubulointerstitial fibrosis. The model is characterized by an influx of mononuclear cells into the glomeruli and interstitium of the remnant kidney (Floege et al., 1992; Ota et al., 2002; Schiller and Moran, 1997; Taal et al., 2000), and JE expression in tubular cells and glomerular tufts (Gong et al., 2004; Ota et al., 2002; Schiller and Moran, 1997; Taal et al., 2000). The unilateral ureteral obstruction model (UUO) is characterized by the development of interstitial fibrosis, which is preceded by the influx of mononuclear cells and expression of rodent CCL-2 expression (Crisman et al., 2001; Diamond et al., 1994; Morrissey and Klahr, 1998).

B. The Rationale for Targeting CCL-2 in Malignant Diseases

The involvement of CCL-2 in cancer was first appreciated when its expression was assessed in a number of cancer types, including human carcinomas of breast (Goede et al., 1999; Saji et al., 2001; Ueno et al., 2000), bladder (Amann et al., 1998), lung (Saji et al., 2003), as well as gastric carcinoma (Ohta et al., 2003), esophageal squamous cell carcinoma (Ohta et al., 2002), gliomas (Leung et al., 1997; Takeshima et al., 1994), and hemangioma (Isik et al., 1996). CCL-2 was found to be expressed mainly by tumor cells, and—in some tumors—by surrounding stromal cells and tumor-infiltrating macrophages. Furthermore, CCL-2 expression in these tumors was associated with increased angiogenesis, tumor cell proliferation, macrophage infiltration, and matrix metalloproteinase production. In some of these cancer types (including breast, ovarian, and bladder), CCL-2 expression in the primary cancer or CCL-2 levels in serum or urine correlated with disease progression; that is, a higher percentage of CCL-2-positive tumors or higher serum/urinary CCL-2 levels were observed in patients with late stage diseases or with distant metastasis (Amann et al., 1998; Hefler et al., 1999; Ueno et al., 2000). CCL-2 also serves as a prognostic marker of natural disease course and outcome (Amann et al., 1998; Parekattil et al., 2003). CCL-2 polymorphisms at the –2518 position in the promoter region influence CCL-2 production level. Monocytes from individuals carrying A/G or G/G produced more CCL-2 than those from A/A homozygous individuals (Rovin et al., 1999). In a prospective study, the G allele was further linked with increased risk of developing metastasis independently of the initial stage in breast cancer patients (Ghilardi et al., 2005). In these cancers, CCL-2 presumably functions as a monocyte chemokine to attract and recruit monocytes to tumor sites that subsequently differentiate into macrophages. These tumor-associated macrophages are meant to wage a battle against cancer cells by mediating phagocytic functions. In contrast, they are "hijacked" by tumor cells and may actually facilitate tumor progression instead of promoting antitumor immunity (Pollard, 2004).

There are, however, also reports of negative correlation between tumor CCL-2 expression and disease progression in pancreatic (Monti et al., 2003) or gastric cancers (Tonouchi et al., 2002). High-circulating CCL-2 levels, either in portal vein or peripheral blood, were associated with a good prognosis (Monti et al., 2003; Tonouchi et al., 2002). It is worth pointing out that in these two reports, CCL-2 levels were measured in serum rather than in tumor tissue. Therefore, this negative correlation could be attributed to the differences in measuring methods, differences in biological significance between tumor tissue CCL-2 and that in the circulation, and the dual functionality of tumor-associated macrophages. Alternatively, the negative correlation reported here could also be attributed to the intrinsic differences in cancer types or stages.

The effort to validate CCL-2 as a potential target for anticancer therapeutics has also encountered similar contradictory and ambiguous situations regarding the relationship between CCL-2 expression and clinical disease progression. Inhibiting tumor CCL-2 production with recombinant DNA engineering resulted in significant tumor inhibition. Transfection of CCL-2 into a mouse colon cancer cell line resulted in enhanced primary tumor angiogenesis and increased the incidence rates of spontaneous metastasis (Nakashima et al., 1995). Similarly, CCL-2-transfected rat CNS-1 gliomas were massively infiltrated by microglial cells, with increased intracerebral tumor growth (Platten et al., 2003). Divergent effects of CCL-2 gene transduction were observed depending on tumor models, expression level, number of tumor cells implanted, and species-specificity of CCL-2. When low CCL-2-producing melanoma cells from a biologically early, nontumorigenic stage were transduced to over-express the CCL-2 gene, low-level CCL-2 secretion with modest monocyte infiltration resulted in tumor formation (Nesbit et al., 2001). High CCL-2 secretion was associated with marked monocyte/macrophage infiltration into the tumor mass, leading to its destruction within a few days after injection into mice (Nesbit et al., 2001). Inhibition of tumorigenicity or retardation of tumor growth were also seen in CHO cells (Rollins and Sunday, 1991), melanoma (Bottazzi et al., 1992), and renal cell cancer (Huang et al., 1995) after CCL-2 gene transduction. This finding in CHO cells was not reproduced in a subsequent study (Hirose et al., 1995), or the anti-cancer effect of CCL-2 required additional immunostimulation, such as IL-2 treatment (Bottazzi et al., 1992).

To better understand the role of CCL-2 in cancer, investigators resorted to other approaches, including neutralizing antibodies and gene knockout. Treatment of immunodeficient mice bearing human breast carcinoma cells with a neutralizing polyclonal antibody to CCL-2 resulted in significant increases in survival and inhibition of the growth of lung micrometastases (Salcedo et al., 2000). Using a monoclonal antibody, we have also demonstrated tumor inhibitory activity in multiple cancer types including breast, pancreatic, and colon cancers (Kesavan et al., 2005). In JE gene–deleted mice, lack of JE resulted in reduced tumor-macrophage infiltration and suppressed tumor angiogenesis, which in turn led to tumor inhibition of IL-1β-expressing tumor cells (Ono et al., 2005). These lines of evidence suggest a potential use of blocking or modulating CCL-2 biological activity in treating various cancers.

IV. SUMMARY

Research into chemokine networks has revealed marked significance of inflammation in multiple diseases. With encouraging preclinical target validation data, it is plausible to target chemokine gradients such as CCL-2 in these diseases to control tissue inflammatory reactions. Given the complexity of inflammatory processes in disease progression, however, the success of therapeutic interventions targeting chemokines will be critically dependent on choosing the right patient population with the appropriate underlying pathogenic changes at the correct time during the disease course. Therefore, the identification and application of surrogate biomarkers based on a comprehensive understanding of the role of inflammatory responses and the functionality of various inflammatory factors will be the path-forward to meet such challenges.

ACKNOWLEDGMENTS

The authors would like to thank Dr. Lynne Murray for help with Table 6.1.

REFERENCES

Abe, R., S. C. Donnelly, et al. (2001). Peripheral blood fibrocytes: Differentiation pathway and migration to wound sites. *J. Immunol.* 166: 7556–7562.

Alam, R., D. Kumar, et al. (1994). Macrophage inflammatory protein-1 alpha and monocyte chemoattractant peptide-1 elicit immediate and late cutaneous reactions and activate murine mast cells in vivo. *J. Immunol.* 152: 1298–1303.

Alam, R., M. A. Lett-Brown, et al. (1992). Monocyte chemotactic and activating factor is a potent histamine-releasing factor for basophils. *J. Clin. Invest.* 89: 723–728.

Allavena, P., G. Bianchi, et al. (1994). Induction of natural killer cell migration by monocyte chemotactic protein-1, -2 and -3. *Eur. J. Immunol.* 24: 3233–3236.

Amann, B., F. G. Perabo, et al. (1998). Urinary levels of monocyte chemo-attractant protein-1 correlate with tumour stage and grade in patients with bladder cancer. *Br. J. Urol.* 82: 118–121.

Banba, N., T. Nakamura, et al. (2000). Possible relationship of monocyte chemoattractant protein-1 with diabetic nephropathy. *Kidney Int.* 58: 684–690.

Ben-Baruch, A. and L. Xu, et al. (1995). Monocyte chemotactic protein-3 (MCP3) interacts with multiple leukocyte receptors: C-C CKR1, a receptor for macrophage inflammatory protein-1 alpha/Rantes, is also a functional receptor for MCP3. *J. Biol. Chem.* 270: 22123–22128.

Bischoff, S. C., M. Krieger, et al. (1992). Monocyte chemotactic protein 1 is a potent activator of human basophils. *J. Exp. Med.* 175: 1271–1275.

Blanc-Brude, O. P., R. C. Chambers, et al. (2001). Factor Xa is a fibroblast mitogen via binding to effector-cell protease receptor-1 and autocrine release of PDGF. *Am. J. Physiol. Cell Physiol.* 281: C681–689.

Bottazzi, B., S. Walter, et al. (1992). Monocyte chemotactic cytokine gene transfer modulates macrophage infiltration, growth, and susceptibility to IL-2 therapy of a murine melanoma. *J. Immunol.* 148: 1280–1285.

Bucala, R., L. A. Spiegel, et al. (1994). Circulating fibrocytes define a new leukocyte subpopulation that mediates tissue repair. *Mol. Med.* 1: 71–81.

Campa, J. S., R. J. McAnulty, and G. J. Laurent (1990). Application of high-pressure liquid chromatography to studies of collagen production by isolated cells in culture. *Anal. Biochem.* 186: 257–263.

Carr, M. W., S. J. Roth, et al. (1994). Monocyte chemoattractant protein 1 acts as a T-lymphocyte chemoattractant. *Proc. Natl. Acad. Sci. U. S. A.* 91: 365–3656.

Chambers, C., F. Smith, et al. (2003). Measuring intracellular calcium fluxes in high throughput mode. *Comb. Chem. High Throughput Screen* 6: 355–362.

Charo, I. F., S. J. Myers, et al. (1994). Molecular cloning and functional expression of two monocyte chemoattractant protein 1 receptors reveals alternative splicing of the carboxyl-terminal tails. *Proc. Natl. Acad. Sci. U. S. A.* 91: 2752–2756.

Chesney, J., C. Metz, C., et al. (1998). Regulated production of type I collagen and inflammatory cytokines by peripheral blood fibrocytes. *J. Immunol.* 160: 419–425.

Chung, C. Y., S. Funamoto, and R. A. Firtel (2001). Signaling pathways controlling cell polarity and chemotaxis. *Trends Biochem. Sci.* 26: 557–566.

Cochran, B. H., A. C. Reffel, and C. D. Stiles (1983). Molecular cloning of gene sequences regulated by platelet-derived growth factor. *Cell* 33: 939–947.

Cockwell, P., A. J. Howie, et al. (1998). *In situ* analysis of C-C chemokine mRNA in human glomerulonephritis. *Kidney Int.* 54: 827–836.

Crisman, J. M., L. L. Richards, et al. (2001). Chemokine expression in the obstructed kidney. *Exp. Nephrol.* 9: 241–248.

Dahinden, C. A., T. Geiser, et al. (1994). Monocyte chemotactic protein 3 is a most effective basophil- and eosinophil-activating chemokine. *J. Exp. Med.* 179: 751–756.

Dai, C., Z. Liu, et al. (2001). Monocyte chemoattractant protein-1 expression in renal tissue is associated with monocyte recruitment and tubulo-interstitial lesions in patients with lupus nephritis. *Chin. Med. J. (Engl.)* 114: 864–868.

Diamond, J. R., D. Kees-Folts, et al. (1994). Macrophages, monocyte chemoattractant peptide-1, and TGF-beta 1 in experimental hydronephrosis. *Am. J. Physiol.* 266: F926–933.

El Chaar, M., E. Attia, et al. (2005). Cyclooxygenase-2 inhibitor decreases extracellular matrix synthesis in stretched renal fibroblasts. *Nephron Exp. Nephrol.* 100: e150–155.

Fenwick, S. A., V. Curry, et al. (2001). 96-well plate-based method for total collagen analysis of cell cultures. *Biotechniques* 30: 1010–1014.

Floege, J., M. W. Burns, et al. (1992). Glomerular cell proliferation and PDGF expression precede glomerulosclerosis in the remnant kidney model. *Kidney Int.* 41: 297–309.

Folkman, J. (1995). Seminars in Medicine of the Beth Israel Hospital, Boston: Clinical applications of research on angiogenesis. *N. Engl. J. Med.* 333: 1757–1763.

Frevert, C. W., V. A. Wong, et al. (1998). Rapid fluorescence-based measurement of neutrophil migration in vitro. *J. Immunol. Methods* 213: 41–52.

Frow, E. K., J. Reckless, and D. J. Grainger (2004). Tools for anti-inflammatory drug design: In vitro models of leukocyte migration. *Med. Res. Rev.* 24: 276–298.

Fuentes, M. E., S. K. Durham, et al. (1995). Controlled recruitment of monocytes and macrophages to specific organs through transgenic expression of monocyte chemoattractant protein-1. *J. Immunol.* 155: 5769–5776.

Fujinaka, H., T. Yamamoto, et al. (1997). Suppression of anti-glomerular basement membrane nephritis by administration of anti-monocyte chemoattractant protein-1 antibody in WKY rats. *J. Am. Soc. Nephrol.* 8: 1174–1178.

Galvez, B. G., L. Genis, et al. (2005). Membrane type 1-matrix metalloproteinase is regulated by chemokines monocyte-chemoattractant protein-1/ccl2 and interleukin-8/CXCL8 in endothelial cells during angiogenesis. *J. Biol. Chem.* 280: 1292–1298.

Gharaee-Kermani, M., E. M. Denholm, and S. H. Phan (1996). Costimulation of fibroblast collagen and transforming growth factor beta1 gene expression by monocyte chemoattractant protein-1 via specific receptors. *J. Biol. Chem.* 271: 17779–17784.

Ghilardi, G., M. L. Biondi, et al. (2005). Breast cancer progression and host polymorphisms in the chemokine system: Role of the macrophage chemoattractant protein-1 (MCP-1) -2518 G allele. *Clin. Chem.* 51: 452–455.

Goede, V., L. Brogelli, et al. (1999). Induction of inflammatory angiogenesis by monocyte chemoattractant protein-1. *Int. J. Cancer* 82: 765–770.

Gong, R., A. Rifai, et al. (2004). Hepatocyte growth factor ameliorates renal interstitial inflammation in rat remnant kidney by modulating tubular expression of macrophage chemoattractant protein-1 and RANTES. *J. Am. Soc. Nephrol.* 15: 2868–2881.

Gong, X., W. Gong, et al. (1997). Monocyte chemotactic protein-2 (MCP-2) uses CCR1 and CCR2B as its functional receptors. *J. Biol. Chem.* 272: 11682–11685.

Goumenos, D., K. Tsomi, et al. (1998). Myofibroblasts and the progression of crescentic glomerulonephritis. *Nephrol. Dial. Transplant.* 13: 1652–1661.

Grandaliano, G., L. Gesualdo, et al. (1996). Monocyte chemotactic peptide-1 expression in acute and chronic human nephritides: A pathogenetic role in interstitial monocytes recruitment. *J. Am. Soc. Nephrol.* 7: 906–913.

Grewal, I. S., B. J. Rutledge, et al. (1997). Transgenic monocyte chemoattractant protein-1 (MCP-1) in pancreatic islets produces monocyte-rich insulitis without diabetes: Abrogation by a second transgene expressing systemic MCP-1. *J. Immunol.* 159: 401–408.

Gruden, G., G. Setti, et al. (2005). Mechanical stretch induces monocyte chemoattractant activity via an NF-kappaB-dependent monocyte chemoattractant protein-1-mediated pathway in human mesangial cells: Inhibition by rosiglitazone. *J. Am. Soc. Nephrol.* 16: 688–696.

Gu, L., Y. J. Zhu, et al. (2004). Effect of IFN-gamma and dexamethasone on TGF-beta1-induced human fetal lung fibroblast-myofibroblast differentiation. *Acta. Pharmacol. Sin.* 25: 1479–1488.

Hao, L., H. Okada, et al. (2003). Direct contact between human peripheral blood mononuclear cells and renal fibroblasts facilitates the expression of monocyte chemoattractant protein-1. *Am. J. Nephrol.* 23: 208–213.

Hashimoto, N., H. Jin, et al. (2004). Bone marrow-derived progenitor cells in pulmonary fibrosis. *J. Clin. Invest.* 113: 243–252.

Hefler, L., C. Tempfer, et al. (1999). Monocyte chemoattractant protein-1 serum levels in ovarian cancer patients. *Br. J. Cancer* 81: 855–859.

Hirata Terra, J., I. Montano, et al. (2004). Development of a microplate bioassay for monocyte chemoattractant protein-1 based on activation of p44/42 mitogen-activated protein kinase. *Anal. Biochem.* 327: 119–125.

Hirose, K., M. Hakozaki, et al. (1995). Chemokine gene transfection into tumour cells reduced tumorigenicity in nude mice in association with neutrophilic infiltration. *Br. J. Cancer* 72: 708–714.

Hong, K. H., J. Ryu, and K. H. Han (2005). Monocyte chemoattractant protein-1-induced angiogenesis is mediated by vascular endothelial growth factor-A. *Blood* 105: 1405–1407.

Huang, S., K. Xie, et al. (1995). Suppression of tumor growth and metastasis of murine renal adenocarcinoma by syngeneic fibroblasts genetically engineered to secrete the JE/MCP-1 cytokine. *J. Interferon. Cytokine. Res.* 15: 655–665.

Hyder, S. M. and G. M. Stancel (1999). Regulation of angiogenic growth factors in the female reproductive tract by estrogens and progestins. *Mol. Endocrinol.* 13(6): 806-811.

Ihm, C. G., J. K. Park, et al. (1998). A high glucose concentration stimulates the expression of monocyte chemotactic peptide 1 in human mesangial cells. *Nephron* 79: 33–37.

Isik, F. F., R. P. Rand, et al. (1996). Monocyte chemoattractant protein-1 mRNA expression in hemangiomas and vascular malformations. *J. Surg. Res.* 61: 71–76.

Isome, M., H. Fujinaka, et al. (2004). Important role for macrophages in induction of crescentic anti-GBM glomerulonephritis in WKY rats. *Nephrol. Dial. Transplant* 19: 2997–3004.

Jefferson, J. A. and R. J. Johnson (1999). Experimental mesangial proliferative glomerulonephritis (the anti-Thy-1.1 model). *J. Nephrol.* 12: 297–307.

Kesavan, P., F. McCabe, et al. (2005). Anti-CCL-2 / MCP-1 (monocyte chemoattractant protein-1) monoclonal antibodies effectively inhibit tumor angiogenesis and growth of human breast carcinoma. In *Proceedings of 96th AACR Annual Meeting.*

Koide, N., A. Nishio, et al. (2004). Significance of macrophage chemoattractant protein-1 expression and macrophage infiltration in squamous cell carcinoma of the esophagus. *Am. J. Gastroenterol.* 99: 1667–1674.

Kolattukudy, P. E., T. Quach, et al. (1998). Myocarditis induced by targeted expression of the MCP-1 gene in murine cardiac muscle. *Am. J. Pathol.* 152: 101–111.

Koyama, S., E. Sato, et al. (1998). Human lung fibroblasts release chemokinetic activity for monocytes constitutively. *Am. J. Physiol.* 275: L223–230.

Kruszynski, M., N. Stowell, et al. (2005). Synthesis and biological characterization of human monocyte chemoattractant protein 1 (MCP-1) and its analogs. *J. Pept. Sci.* 12(1): 25-32.

Leung, S. Y., M. P. Wong, et al. (1997). Monocyte chemoattractant protein-1 expression and macrophage infiltration in gliomas. *Acta. Neuropathol. (Berl).* 93(5): 518–527.

Liss, C., M. J. Fekete, et al. (2001). Paracrine angiogenic loop between head-and-neck squamous-cell carcinomas and macrophages. *Int. J. Cancer* 93: 781–785.

Liu, Z. H., S. F. Chen, et al. (2003). Glomerular expression of C-C chemokines in different types of human crescentic glomerulonephritis. *Nephrol. Dial. Transplant* 18: 1526–1534.

Lloyd, C. M., M. E. Dorf, et al. (1997). Role of MCP-1 and RANTES in inflammation and progression to fibrosis during murine crescentic nephritis. *J. Leukoc. Biol.* 62: 676–680.

Luster, A. D. (1998). Chemokines—chemotactic cytokines that mediate inflammation. *N. Engl. J. Med.* 338: 436–445.

Lynn, E. G., Y. L. Siow, and K. O (2000). Very low-density lipoprotein stimulates the expression of monocyte chemoattractant protein-1 in mesangial cells. *Kidney Int.* 57: 1472–1483.

Maghazachi, A. A., A. al-Aoukaty, and T. J. Schall (1994). C-C chemokines induce the chemotaxis of NK and IL-2-activated NK cells: Role for G proteins. *J. Immunol.* 153: 4969–4977.

Mezzano, S., A. Droguett, et al. (2003). Renin-angiotensin system activation and interstitial inflammation in human diabetic nephropathy. *Kidney Int. Suppl.* S64–70.

Micera, A., I. Puxeddu, et al. (2005). The pro-fibrogenic effect of nerve growth factor on conjunctival fibroblasts is mediated by transforming growth factor-beta. *Clin. Exp. Allergy* 35: 650–656.

Monti, P., B. E. Leone, et al. (2003). The CC chemokine MCP-1/CCL2 in pancreatic cancer progression: Regulation of expression and potential mechanisms of antimalignant activity. *Cancer Res.* 63: 7451–7461.

Moore, B. B., J. E. Kolodsick, et al. (2005). CCR2-mediated recruitment of fibrocytes to the alveolar space after fibrotic injury. *Am. J. Pathol.* 166: 675–684.

Morii, T., H. Fujita, et al. (2003). Association of monocyte chemoattractant protein-1 with renal tubular damage in diabetic nephropathy. *J. Diabetes Complications* 17: 11–15.

Morrissey, J. J. and S. Klahr (1998). Differential effects of ACE and AT1 receptor inhibition on chemoattractant and adhesion molecule synthesis. *Am. J. Physiol.* 274: F580–586.

Murphy, P. M., M. Baggiolini, et al. (2000). International union of pharmacology. XXII. Nomenclature for chemokine receptors. *Pharmacol. Rev.* 52: 145–176.

Myers, S. J., L. M. Wong, and I. F. Charo (1995). Signal transduction and ligand specificity of the human monocyte chemoattractant protein-1 receptor in transfected embryonic kidney cells. *J. Biol. Chem.* 270: 5786–5792.

Nakashima, E., N. Mukaida, et al. (1995). Human MCAF gene transfer enhances the metastatic capacity of a mouse cachectic adenocarcinoma cell line *in vivo*. *Pharm. Res.* 12: 1598–1604.

Neote, K., D. DiGregorio, et al. (1993). Molecular cloning, functional expression, and signaling characteristics of a C-C chemokine receptor. *Cell* 72: 415–425.

Nesbit, M., H. Schaider, et al. (2001). Low-level monocyte chemoattractant protein-1 stimulation of monocytes leads to tumor formation in nontumorigenic melanoma cells. *J. Immunol.* 166: 6483–6490.

Neugarten, J., G. W. Feith, et al. (1995). Role of macrophages and colony-stimulating factor-1 in murine antiglomerular basement membrane glomerulonephritis. *J. Am. Soc. Nephrol.* 5: 1903–1909.

O'Byrne, K. J., A. G. Dalgleish, et al. (2000). The relationship between angiogenesis and the immune response in carcinogenesis and the progression of malignant disease. *Eur. J. Cancer* 36: 151–169.

Ohta, M., Y. Kitadai, et al. (2002). Monocyte chemoattractant protein-1 expression correlates with macrophage infiltration and tumor vascularity in human esophageal squamous cell carcinomas. *Int. J. Cancer* 102: 220–224.

Ohta, M., Y. Kitadai, et al. (2003). Monocyte chemoattractant protein-1 expression correlates with macrophage infiltration and tumor vascularity in human gastric carcinomas. *Int. J. Oncol.* 22: 773–778.

Ono, M., S. Nakao, et al. (2005). The control of tumor growth and angiogenesis by inflammatory cytokines and infiltration of macrophages in tumor microenvironment. In *Proceedings of 96th AACR Annual Meeting*.

Ota, T., M. Tamura, et al. (2002). Expression of monocyte chemoattractant protein-1 in proximal tubular epithelial cells in a rat model of progressive kidney failure. *J. Lab. Clin. Med.* 140: 43–51.

Pai, R., H. Ha, et al. (1996). Role of tumor necrosis factor-alpha on mesangial cell MCP-1 expression and monocyte migration: Mechanisms mediated by signal transduction. *J. Am. Soc. Nephrol.* 7: 914–923.

Panzer, U., F. Thaiss, et al. (2001). Monocyte chemoattractant protein-1 and osteopontin differentially regulate monocytes recruitment in experimental glomerulonephritis. *Kidney Int.* 59: 1762–1769.

Parekattil, S. J., H. A. Fisher, and B. A. Kogan (2003). Neural network using combined urine nuclear matrix protein-22, monocyte chemoattractant protein-1 and urinary intercellular adhesion molecule-1 to detect bladder cancer. *J. Urol.* 169: 917–920.

Perez de Lema, G., H. Maier, et al. (2001). Chemokine expression precedes inflammatory cell infiltration and chemokine receptor and cytokine expression during the initiation of murine lupus nephritis. *J. Am. Soc. Nephrol.* 12: 1369–1382.

Phillips, R. J., M. D. Burdick, et al. (2004). Circulating fibrocytes traffic to the lungs in response to CXCL12 and mediate fibrosis. *J. Clin. Invest.* 114: 438–446.

Platten, M., A. Kretz, et al. (2003). Monocyte chemoattractant protein-1 increases microglial infiltration and aggressiveness of gliomas. *Ann. Neurol.* 54: 388–392.

Pollard, J. W. (2004). Tumour-educated macrophages promote tumour progression and metastasis. *Nat. Rev. Cancer* 4: 71–78.

Prodjosudjadi, W., J. S. Gerritsma, J. S., et al. (1995). Production and cytokine-mediated regulation of monocyte chemoattractant protein-1 by human proximal tubular epithelial cells. *Kidney Int.* 48: 1477–1486.

Richmond, A., E. Balentien, et al. (1988). Molecular characterization and chromosomal mapping of melanoma growth stimulatory activity, a growth factor structurally related to beta-thromboglobulin. *EMBO J.* 7: 2025–2033.

Roberts, I. S., C. Burrows, et al. (1997). Interstitial myofibroblasts: Predictors of progression in membranous nephropathy. *J. Clin. Pathol.* 50: 123–127.

Rollins, B. J., P. Stier, et al. (1989). The human homolog of the JE gene encodes a monocyte secretory protein. *Mol. Cell Biol.* 9: 4687–4695.

Rollins, B. J. and M. E. Sunday (1991). Suppression of tumor formation *in vivo* by expression of the JE gene in malignant cells. *Mol. Cell. Biol.* 11: 3125–3131.

Rollins, B. J., A. Walz, and M. Baggiolini (1991). Recombinant human MCP-1/JE induces chemotaxis, calcium flux, and the respiratory burst in human monocytes. *Blood* 78: 1112–1116.

Rovin, B. H., N. Doe, and L. C. Tan (1996). Monocyte chemoattractant protein-1 levels in patients with glomerular disease. *Am. J. Kidney Dis.* 27: 640–646.

Rovin, B. H., L. Lu, and R. Saxena (1999). A novel polymorphism in the MCP-1 gene regulatory region that influences MCP-1 expression. *Biochem. Biophys. Res. Commun.* 259: 344–348.

Rovin, B. H., M. Rumancik, et al. (1994). Glomerular expression of monocyte chemoattractant protein-1 in experimental and human glomerulonephritis. *Lab. Invest.* 71: 536–542.

Saji, H., M. Koike, et al. (2001). Significant correlation of monocyte chemoattractant protein-1 expression with neovascularization and progression of breast carcinoma. *Cancer* 92: 1085–1091.

Saji, H., H. Nakamura, et al. (2003). Significance of expression of TGF-beta in pulmonary metastasis in non-small cell lung cancer tissues. *Ann. Thorac. Cardiovasc. Surg.* 9: 295–300.

Salcedo, R., M. L. Ponce, et al. (2000). Human endothelial cells express CCR2 and respond to MCP-1: Direct role of MCP-1 in angiogenesis and tumor progression. *Blood* 96: 34–40.

Sarafi, M. N., E. A. Garcia-Zepeda, et al. (1997). Murine monocyte chemoattractant protein (MCP)-5: A novel CC chemokine that is a structural and functional homologue of human MCP-1. *J. Exp. Med.* 185: 99–109.

Schecter, A. D., A. B. Berman, et al. (2004). MCP-1-dependent signaling in CCR2(-/-) aortic smooth muscle cells. *J. Leukoc. Biol.* 75: 1079–1085.

Schiller, B. and J. Moran (1997). Focal glomerulosclerosis in the remnant kidney model: An inflammatory disease mediated by cytokines. *Nephrol. Dial. Transplant* 12: 430–437.

Schneider, A., U. Panzer, et al. (1999). Monocyte chemoattractant protein-1 mediates collagen deposition in experimental glomerulonephritis by transforming growth factor-beta. *Kidney Int.* 56: 135–144.

Sean Eardley, K. and P. Cockwell (2005). Macrophages and progressive tubulointerstitial disease. *Kidney Int.* 68: 437–455.

Segerer, S., Y. Cui, et al. (2000). Expression of the chemokine monocyte chemoattractant protein-1 and its receptor chemokine receptor 2 in human crescentic glomerulonephritis. *J. Am. Soc. Nephrol.* 11: 2231–2242.

Sozzani, S., W. Luini, et al. (1991). The signal transduction pathway involved in the migration induced by a monocyte chemotactic cytokine. *J. Immunol.* 147: 2215–2221.

Sozzani, S., M. Molino, et al. (1993). Receptor-activated calcium influx in human monocytes exposed to monocyte chemotactic protein-1 and related cytokines. *J. Immunol.* 150: 1544–1553.

Stahl, R. A., F. Thaiss, et al. (1993). Increased expression of monocyte chemoattractant protein-1 in anti-thymocyte antibody-induced glomerulonephritis. *Kidney Int.* 44: 1036–1047.

Standiford, T. J., S. L. Kunkel, et al. (1991). Alveolar macrophage-derived cytokines induce monocyte chemoattractant protein-1 expression from human pulmonary type II-like epithelial cells. *J. Biol. Chem.* 266: 9912–9918.

Strieter, R. M., R. Wiggins, et al. (1989). Monocyte chemotactic protein gene expression by cytokine-treated human fibroblasts and endothelial cells. *Biochem. Biophys. Res. Commun.* 162: 694–700.

Suda, T., A. Osajima, et al. (2001). Pressure-induced expression of monocyte chemoattractant protein-1 through activation of MAP kinase. *Kidney Int.* 60: 1705–1715.

Taal, M. W., K. Zandi-Nejad, et al. (2000). Proinflammatory gene expression and macrophage recruitment in the rat remnant kidney. *Kidney Int.* 58: 1664–1676.

Takeshima, H., J. Kuratsu, et al. (1994). Expression and localization of messenger RNA and protein for monocyte chemoattractant protein-1 in human malignant glioma. *J. Neurosurg.* 80(6): 1056–1062.

Tamimi, N. A., P. E. Stevens, et al. (1998). Expression of cytoskeletal proteins differentiates between progressors and non-progressors in treated idiopathic membranous nephropathy. *Exp. Nephrol.* 6: 217–225.

Tang, W. W., S. Yin, et al. (1995). Chemokine gene expression in anti-glomerular basement membrane antibody glomerulonephritis. *Am. J. Physiol.* 269: F323–330.

Tashiro, K., I. Koyanagi, et al. (2002). Urinary levels of monocyte chemoattractant protein-1 (MCP-1) and interleukin-8 (IL-8), and renal injuries in patients with type 2 diabetic nephropathy. *J. Clin. Lab. Anal.* 16: 1–4.

Tesch, G. H., S. Maifert, et al. (1999a). Monocyte chemoattractant protein 1-dependent leukocytic infiltrates are responsible for autoimmune disease in MRL-Fas(lpr) mice. *J. Exp. Med.* 190: 1813–1824.

Tesch, G. H., A. Schwarting, et al. (1999b). Monocyte chemoattractant protein-1 promotes macrophage-mediated tubular injury, but not glomerular injury, in nephrotoxic serum nephritis. *J. Clin. Invest.* 103: 73–80.

Tonouchi, H., C. Miki, et al. (2002). Profile of monocyte chemoattractant protein-1 circulating levels in gastric cancer patients. *Scand. J. Gastroenterol.* 37: 830–833.

Ueno, T., M. Toi, et al. (2000). Significance of macrophage chemoattractant protein-1 in macrophage recruitment, angiogenesis, and survival in human breast cancer. *Clin. Cancer Res.* 6: 3282–3289.

Van Coillie, E., J. Van Damme, and G. Opdenakker (1999). The MCP/eotaxin subfamily of CC chemokines. *Cytokine Growth Factor Rev.* 10: 61–86.

Vishwanath, R. P., C. E. Brown, et al. (2005). A quantitative high-throughput chemotaxis assay using bioluminescent reporter cells. *J. Immunol. Methods* 2(1–2): 78–89

Wada, T., K. Furuichi, et al. (2000). Up-regulation of monocyte chemoattractant protein-1 in tubulointerstitial lesions of human diabetic nephropathy. *Kidney Int.* 58: 1492–1499.

Wada, T., K. Furuichi, et al. (1999). MIP-1alpha and MCP-1 contribute to crescents and interstitial lesions in human crescentic glomerulonephritis. *Kidney Int.* 56: 995–1003.

Wada, T., H. Yokoyama, et al. (1996a). Intervention of crescentic glomerulonephritis by antibodies to monocyte chemotactic and activating factor (MCAF/MCP-1). *FASEB J.* 10: 1418–1425.

Wada, T., H. Yokoyama, et al. (1996b). Monitoring urinary levels of monocyte chemotactic and activating factor reflects disease activity of lupus nephritis. *Kidney Int.* 49: 761–767.

Wenzel, U., A. Schneider, et al. (1997). Monocyte chemoattractant protein-1 mediates monocyte/macrophage influx in anti-thymocyte antibody-induced glomerulonephritis. *Kidney Int.* 51: 770–776.

Wilkinson, P. C. (1998). Assays of leukocyte locomotion and chemotaxis. *J. Immunol. Methods* 216: 139–153.

Witowski, J., A. Thiel, et al. (2001). Synthesis of C-X-C and C-C chemokines by human peritoneal fibroblasts: Induction by macrophage-derived cytokines. *Am. J. Pathol.* 158: 1441–1450.

Wong, L. M., S. J. Myers, et al. (1997). Organization and differential expression of the human monocyte chemoattractant protein 1 receptor gene: Evidence for the role of the carboxyl-terminal tail in receptor trafficking. *J. Biol. Chem.* 272: 1038–1045.

Wu, D. (2005). Signaling mechanisms for regulation of chemotaxis. *Cell. Res.* 15: 52–56.

Wu, Q., K. Jinde, et al. (2001). Analysis of prognostic predictors in idiopathic membranous nephropathy. *Am. J. Kidney Dis.* 37: 380–387.

Wu, X., G. J. Dolecki, et al. (1997). Chemokines are expressed in a myeloid cell-dependent fashion and mediate distinct functions in immune complex glomerulonephritis in rat. *J. Immunol.* 158: 3917–3924.

Xu, L. L., M. K. Warren, et al. (1996). Human recombinant monocyte chemotactic protein and other C-C chemokines bind and induce directional migration of dendritic cells in vitro. *J. Leukoc. Biol.* 60: 365–371.

Yamamoto, T. and K. Nishioka (2003). Role of monocyte chemoattractant protein-1 and its receptor, CCR-2, in the pathogenesis of bleomycin-induced scleroderma. *J. Invest. Dermatol.* 121: 510–516.

Yan, L., P. Kesavan, et al. (2003). Contribution of inflammatory cells in tumor angiogenesis: MCP-1 and tumor angiogenesis. In *Proceedings of AACR-NCI-EORTC International Conference, Molecular Targets and Cancer Therapeutics*, pp. 132.

Yokoyama, H., T. Wada, et al. (1998). Urinary levels of chemokines (MCAF/MCP-1, IL-8) reflect distinct disease activities and phases of human IgA nephropathy. *J. Leukoc. Biol.* 63: 493–499.

Yoshimura, T., E. A. Robinson, et al. (1989a). Purification and amino acid analysis of two human glioma-derived monocyte chemoattractants. *J. Exp. Med.* 169: 1449–1459.

Yoshimura, T., N. Yuhki, et al. (1989b). Human monocyte chemoattractant protein-1 (MCP-1): Full-length cDNA cloning, expression in mitogen-stimulated blood mononuclear leukocytes, and sequence similarity to mouse competence gene JE. *FEBS Lett.* 244: 487–493.

Zhang, X. F., S. Z. Guo, et al. (2004). Different roles of PKC and PKA in effect of interferon-gamma on proliferation and collagen synthesis of fibroblasts. *Acta. Pharmacol. Sin.* 25: 1320–1326.

Zlotnik, A. and O. Yoshie (2000). Chemokines: A new classification system and their role in immunity. *Immunity* 12: 121–127.

Zoja, C., X. H. Liu, et al. (1997). Renal expression of monocyte chemoattractant protein-1 in lupus autoimmune mice. *J. Am. Soc. Nephrol.* 8: 720–729.

7

TARGETING IL-12p40 FOR IMMUNE-MEDIATED DISEASE

JACQUELINE BENSON

Ph.D., Centocor Research and Development, Inc., Radnor, Pennsylvania

The p40 protein subunit of interleukins 12 (IL-12) and 23 (IL-23) is believed to be central to CD4+ T cell responses and a key driver of cell-mediated immunity. This chapter summarizes the biology of IL-12 and IL-23 and the antagonist technologies, *in vitro* assays, and *in vivo* studies that validated the shared IL-12p40 subunit as a therapeutic target. This collection of work spanned many years and multiple investigators whose consistent findings have established much of our understanding of CD4+ T cell biology. Fortunately, drug development based on IL-12p40 research has recently translated into successful clinical trials that demonstrated amelioration of two human immune-mediated diseases. As we have learned from the evolution of tumor necrosis factor alpha (TNFα) inhibitors, immune disorders often share common pathways despite their different etiologies and tissue specificities. Therefore, there is hope that the benefits of targeting IL-12p40 will eventually address unmet medical needs for a spectrum of immune disorders.

I. INTRODUCTION

Dysregulation of cell-mediated immunity is thought to be the underlying pathology of many autoimmune diseases. In particular, over-activation of a subset of CD4+ T lymphocytes—also called helper T cells (Th)—can result in chronic debilitating diseases, including psoriasis, Crohn's disease, rheumatoid arthritis, and multiple sclerosis. Although this collection of disorders varies greatly in their tissue specificities and pathologies, a common hallmark is the increased expression of pro-inflammatory cytokines and interleukins in the affected tissues. Indeed, expression of IL-12 is often increased in these tissues when compared to non-diseased controls. These observations spurred great interest in understanding more about the biology of IL-12.

IL-12, previously called T cell differentiation factor, cytotoxic lymphocyte maturation factor, or natural killer stimulatory factor, was reported as discovered in 1989 and cloned in the early 1990s (Kobayashi et al., 1989; Stern et al., 1990). IL-12 is produced by antigen-presenting cells (APCs) in response to cellular stress or toll-like receptor signaling that is believed to occur during bacterial or viral infections. IL-12 is a secreted heterodimeric cytokine comprised of two disulfide-linked glycosylated protein subunits, designated p35 and p40 for their approximate molecular weights. IL-12 drives cell-mediated immunity by binding to a two-chain receptor complex expressed on the surface of Th or natural killer (NK) cells, resulting in intracellular signaling events and activation of the receptor-bearing cell. The IL-12 receptor beta-1 (IL-12Rβ1) chain binds to the p40 subunit of IL-12, providing the primary interaction between IL-12 and its receptor. However, it is IL-12p35 ligation of the second receptor chain, IL-12Rβ2, that confers intracellular signaling specific for IL-12 (Presky et al., 1996). For CD4+ T cells, IL-12 signaling concurrent with antigen presentation is thought to invoke differentiation toward the T helper 1 (Th1) phenotype. Th1 cells are characterized by the production of a robust proinflammatory cytokine, interferon gamma (IFNγ). Under normal conditions, Th1 cells are believed to promote immunity to some intracellular pathogens, generate complement-fixing antibody isotypes, and perhaps provide protective tumor immunosurveillance. IL-12 and Th1 cell populations are also believed to be abnormally regulated in many immune-mediated diseases (Trinchieri, 2003).

The p40 protein subunit of IL-12 (IL-12p40) can also associate with a separate protein subunit, designated p19, to form a novel cytokine, IL-23 (Oppmann et al., 2000). IL-23 is also produced by APCs, although the conditions that distinguish IL-23 from IL-12 production are still unclear. IL-23 signals through a two-chain receptor complex expressed primarily on T cell populations. Because the p40 subunit is shared between IL-12 and IL-23, it follows that the IL-12Rβ1 chain is also shared between IL-12 and IL-23. It is the IL-23p19 ligation of the second component of the IL-23 receptor complex, IL-23R, however, that confers IL-23 specific intracellular signaling (Parham et al., 2002). IL-23 was recently proposed to elicit the differentiation of a unique subset of Th cells that produce IL-17 (Th_{IL-17}) (Langrish et al., 2005). It remains to be determined if the IL-23-induced Th_{IL-17} population will be as prevalent with human T cell populations as it is in the murine system. Both IL-12 and IL-23 exist only as secreted heterodimeric cytokines since neither IL-12p35 nor IL-23p19 subunits can be secreted without covalent association with p40. In contrast to TNFα, none of the IL-12 or IL-23 subunits contain a transmembrane domain and therefore these cytokines cannot exist as cell surface proteins. Despite their molecular similarities, IL-12 and IL-23 will likely evolve to have different biologies and impact on immune functions.

The majority of studies to date have been directed toward the shared p40 subunit. Therefore, it is unclear which biological functions are attributable to IL-12 versus IL-23. While the contribution of IL-12 to pathogen immunity is well established, how IL-23 contributes to these pathways is still under investigation. There is increasing evidence, however, for the role of IL-23 in immune-mediated disease (Trinchieri, 2003). Because the p40 subunit is shared between IL-12 and IL-23, and each of these cytokines appears to contribute to immune pathologies, IL-12p40 warranted further investigation as a potential therapeutic target.

II. *IN VITRO* TARGET VALIDATION OF IL-12p40

IL-12p40 mRNA and protein expression has been associated with many immune-mediated disease tissues, which suggests that inhibition—or antagonist activity—may provide therapeutic benefit. Further *in vitro* and *in vivo* studies, however, were required to validate IL-12p40 as a therapeutic target. Therefore, the first step was to generate tools to demonstrate IL-12p40 specific inhibition. Secondly, IL-12p40 blockade was characterized through *in vitro* assays in which IL-12p40 antagonists effectively suppressed immune responses that were believed to contribute to human disease. Finally, IL-12p40 antagonists were tested *in vivo* to validate that specific inhibition would provide a desirable biological effect despite contribution from multiple inflammatory networks. Indeed, successful completion of these three stages validated IL-12p40 as an appropriate therapeutic target. Fortunately, this has recently translated to clinical benefit in human patients with psoriasis or Crohn's disease after treatment with IL-12p40 specific antagonists (Kauffman et al., 2004; Mannon et al., 2004).

A. IL-12p40 Antagonist Technologies

There are several options by which to approach IL-12 or IL-23 inhibition. For example, the IL-12 pathway can be interrupted through targeting IL-12p35, IL-12p40, IL-12Rβ1, or IL-12Rβ2. Similarly, IL-23 signaling could be blocked by targeting IL-23p19, IL-12p40, IL-12Rβ1, or IL-23R (Figure 7.1). Therefore, inhibition of both IL-12 and IL-23 is accomplished by targeting the shared cytokine and receptor subunits, namely IL-12p40 and IL-12Rβ1. Now that reagents are available to many of these IL-12 and IL-23 cytokine and receptor subunits, the mechanism and specificity of antagonists are more clearly understood. IL-12p40 target validation has utilized multiple technologies, such as monoclonal antibodies, small-molecule inhibitors, protein mimetics, and genetically altered mice. IL-12p40 target validation was remarkably successful because information was consistent across multiple technical approaches.

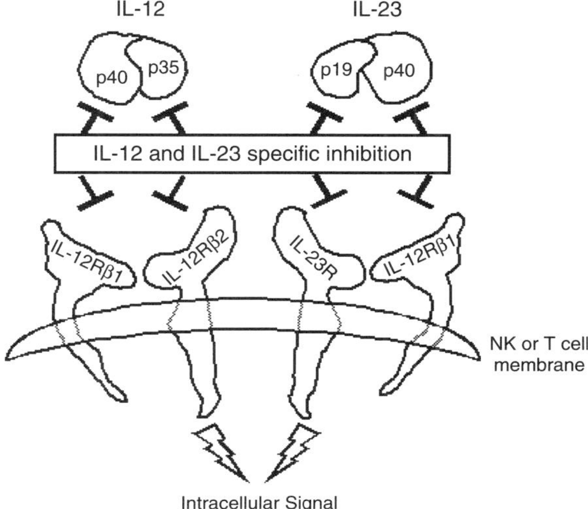

FIGURE 7.1 **IL-12 and IL-23 inhibition.** IL-12 and IL-23 are heterodimeric cytokines that share a common p40 subunit and the IL-12Rβ1 chain of their receptor complexes found on NK and T cells. Targeting IL-12p40 or IL-12Rβ1 will neutralize both IL-12 and IL-23 signaling, whereas targeting IL-12p35 or IL-12Rβ2 will offer IL-12 specific inhibition and targeting IL-23p19 or IL-23R will provide IL-23 specific suppression.

Monoclonal antibodies (mAbs) to many of the IL-12 and IL-23 cytokine and receptor subunits have been generated by multiple laboratories, and some are available commercially for use in a wide range of *in vitro* and *in vivo* studies. In general, mAb generation is straightforward and can be achieved through hybridoma or phage-display technologies. Not all mAbs inhibit functional activity of their target, however. Neutralization of IL-12 or IL-23 activity can depend upon the epitope and affinity of the mAb/target binding. In addition, specificity is an important consideration in this cascade given the shared subunits. Therefore, if an IL-12 or IL-23 specific reagent is desired, mAb generation approaches that were directed toward the intended target from the beginning were more likely to yield the specificity desired in the end. Otherwise, IL-12 versus IL-23 specificity is an important filter for *in vitro* characterization of mAb antagonists, which must occur prior to any target validation studies. An additional factor to consider with mAb technology is species reactivity. Often, mAbs will not bind to both human and murine IL-12 or IL-23 proteins, even though their homology is high. Therefore, to utilize mAb technology for target validation, a panel of both human and murine reagents was required.

Small-molecule inhibitors that have been reported to affect IL-12 or IL-23 pathways primarily function via one of three mechanisms: NF-κB inhibition, increase of intracellular cAMP, or interference with IL-12 signal transduction (reviewed by Vandenbroeck et al., 2004). In addition, molecules that interfere with nuclear histone deacetylases, cyclophilins, ion pumps, channels, transporters, or compounds that modify posttranslational processing are also thought to inhibit IL-12. With a small-molecule approach, one can consider intracellular and extracellular targets beyond IL-12p40 and IL-12Rβ1. In contrast to mAb technologies, specificity is the highest challenge for small-molecule compounds. Identifying inhibitors that are specific for IL-12 or IL-23 is still a focus of active research. Once discovered, a specific small-molecule inhibitor can have tremendous impact as a therapeutic entity.

Protein mimetics have also been utilized in target validation studies for IL-12p40. Unlike the IL-12 and IL-23 heterodimeric proteins, the p40 subunit can be secreted by APCs as a monomer or homodimer. There has been great speculation in the scientific literature about whether or not this acts as an endogenous antagonist for IL-12 activity since binding of p40 or $(p40)_2$ to IL-12β1 prevents IL-12 from docking and transmitting intracellular signals through IL-12Rβ2 (Gillessen et al., 1995). There seems to be more evidence for this phenomenon in mouse systems than in human. Application of p40 monomers or homodimers to *in vitro* or *in vivo* systems, however, has been utilized to demonstrate the effects of blockade of IL-12 and IL-23 cascades (Costa et al., 2001; Ling et al., 1995).

Genetically altered mice, either rendered deficient ("knock-out") or transgenic over-expressors, have been another important tool in IL-12 and IL-23 target validation. In fact, many of the differences between IL-12 and IL-23 biology have been suggested by comparing the effects of IL-12p35, IL-12p40, and IL-23p19 knock-out or IL-23p19 transgenic animals in mouse models of immune-mediated disease, discussed in more detail later in this chapter. While significant biological clues can be elucidated utilizing knock-out mice, there are some limitations to determining if selective antagonism would be effective. The cytokine system is notoriously redundant and often other biologies will compensate for the missing genes. In addition, because the depletion is absolute, there is no ability to determine if partial or temporal inhibition would be beneficial. In the case of IL-12 and IL-23, striking biological observations have been made utilizing knock-out and transgenic technologies, which further validates the central role these cytokines play in immune responses.

B. *In vitro* Assays for IL-12p40 Activity

Regardless of which antagonist technology was utilized, *in vitro* systems were a powerful tool to validate IL-12p40 as a target for therapeutic intervention. In addition, antagonists with a variety of mechanisms of action could be tested *in vitro*. For example, antagonists that prevent IL-12 or IL-23/receptor binding, receptor signaling, or gene transcription were included in functional *in vitro* assays to validate core principals of IL-12 and IL-23 biology.

Cell-based functional assays, or "bioassays," are perhaps the most informative for IL-12 and IL-23 because these cytokines are extremely potent in activating their receptor-bearing cells. Physiologically relevant concentrations of IL-12 or IL-23 will elicit robust effects that are easily detected under many assay conditions. In addition, these assays most represent what we understand about the *in vivo* situation. IL-12 and IL-23 are known to bind receptor complexes on human and murine T and NK cells. Therefore, cell lines and primary cultures that are likely to bind IL-12 and IL-23 can be easily identified. IL-12 or IL-23 receptor ligation can activate multiple cellular responses such as: intracellular STAT3/4 protein phosphorylation, cellular proliferation, cytotoxic cell killing, and the production of effector cytokines (e.g., IFNγ, IL-17). Protocols for the following assays are discussed in detail in "Current Protocols of Immunology." STAT3 and STAT4 phosphorylation and cytokine production can be detected by a variety of laboratory techniques, such as Western blotting, intracellular flow cytometry, RT-PCR for mRNA, and enzyme-linked immunosorbant assays (ELISA). Although IL-12 or IL-23 will not induce proliferation of resting T or NK cells, IL-12 will elicit proliferation of previously activated T or NK cell populations. Cellular proliferation can be assessed by ^{3}H-thymidine incorporation, ATP-Lite, BrdU, or CFSE analysis. A common assay set up would include culturing the receptor-bearing cells with either IL-12 or IL-23 alone or combinations of cytokine and antagonist dilutions. Cells or culture supernatants could then be analyzed by the mentioned technologies for IL-12- or IL-23-mediated cellular responses. IL-12 will also enhance the cytotoxic activity of CD8+ T cells and NK cells, which is typically measured through ^{51}Cr-release upon lyses of antigen-labeled target cells. These assays require an appropriate ratio of target to effector cells, which should be determined prior to testing an IL-12p40 antagonist. Effective inhibition of all of these events through IL-12p40 antagonists has been demonstrated by multiple laboratories (reviewed by Gately et al., 1998). These data suggest that IL-12p40 antagonists can effectively prevent multiple cellular responses that are believed to contribute to immune-mediated diseases.

Bioassays will determine if a specific antagonist can prevent a biological effect, but, the precise mechanism of action of IL-12 and IL-23 antagonists can be further defined through biochemical and cell-based binding assays. In biochemical receptor binding assays, individual receptor proteins—such as IL-12Rβ1, IL-12Rβ2, or IL-23R—are coated on beads or microtiter plates then either IL-12 or IL-23 alone or combinations of cytokine and antagonist dilutions are added. Similar to ELISAs, detection of cytokine binding can be determined by utilizing directly labeled cytokine or adding a labeled secondary detection reagent. The assay signal will reflect the level of IL-12 or IL-23 binding to each receptor chain with or without the addition of the antagonist. This will help determine if the tested antagonist prevents cytokine binding to individual receptor subunits as part of the mechanism of action. Because IL-12 and IL-23 biologically function through two-chain receptor complexes, it is often necessary to also study antagonists in the context of both receptor chains. The most relevant system in this case would be cell-based binding. These assays are also useful if the receptors for the ligand of interest have not yet been identified or are not available as reagents for biochemical based assays. Although not as quantitative as the biochemical assay, one format of cell-based binding that is particularly useful is flow cytometry. Similar to the biochemical receptor binding assays, cell cultures are incubated with either IL-12 or IL-23 alone, or combinations of cytokine and antagonist dilutions. Detection of cytokine binding can be determined by utilizing fluorescently labeled cytokine or adding a fluorescently labeled secondary detection reagent. Together, these analyses provide valuable information about which cytokine/receptor interaction is inhibited and if this neutralization is efficient when the receptors are expressed in their native heterodimeric complex.

The combination of functional bioassays and biochemical/cell-based binding assays has determined biological effects of IL-12p40 specific inhibition and where particular antagonists act in the IL-12 and IL-23 cascades. Antagonists can function by preventing cytokine/receptor interaction, by binding to the receptor but preventing receptor signaling, or by allowing receptor signaling but interrupting downstream gene transcription. While these may provide equivalent effects *in vitro* to support target validation, subsequent *in vivo* analysis will determine if these approaches can be effective, which is ultimately required for drug discovery and development.

III. *IN VIVO* PROOF-OF-CONCEPT FOR IL-12p40 INHIBITION

IL-12p40 is believed to contribute to several human immune-mediated conditions, such as psoriasis, Crohn's disease, rheumatoid arthritis, and multiple sclerosis. *In vivo* target validation for IL-12p40 derived from a variety of sources that include mRNA and protein expression in these human disease tissues, disease amelioration through IL-12p40 antagonists, analysis of genetically deficient mice in disease models, and ultimately successful disease amelioration in humans after IL-12p40 specific inhibition. *In vivo* studies with IL-12p40 antagonists continue to provide valuable insights in regards to the necessary timing and localization of treatment in relationship to disease progression. In addition, the limitations of therapy can also be elucidated as well as downstream immunological effects of IL-12p40 inhibition. As these studies continue in both animal models and human patient populations, we will undoubtedly learn important lessons about how well our *in vitro* and *in vivo* target-validation studies predicted human disease outcome.

A. Psoriasis

Psoriasis is a chronic immune-mediated skin disease characterized by painful and itchy erythematous plaques believed to be a consequence of keratinocyte hyperproliferation and dense infiltration of activated lymphocytes (Lebwohl, 2003). Although the etiology of psoriasis is unknown, substantial evidence suggests that IL-12 or IL-23 contributes to pathogenesis. CD4+ and CD8+ T cells from plaques and peripheral blood produce primarily Th1 cytokines upon *in vitro* re-stimulation (Austin et al., 1999), and elevated levels of IFNγ and IL-12p40 mRNA, and IL-12 protein can be detected within psoriatic epidermis and dermis (Yawalkar et al., 1998). Recent data suggest that IL-23 may also contribute to psoriasis pathology because increased IL-23p19 and IL-12p40—but not IL-12p35—mRNA expression was observed in lesional skin and monocyte and dendritic cells that were isolated from psoriatic lesions (Lee et al., 2004). Overall, there was sufficient data from human psoriasis tissue to suggest IL-12 and/or IL-23 involvement in skin pathology.

There are few animal models for psoriasis, limiting the amount of preclinical information in regard to IL-12p40 antagonism. A neutralizing antibody to IL-12p40 did ameliorate formation of psoriasiform lesions in one murine model (Hong et al., 1999). Immunocompromised mice received naïve splenic CD4+CD45RB^{hi} T cells from wild-type or IFNγ knock-out mice and were challenged one day later with lipopolysaccharide plus IL-12 or staphylococcal enterotoxin B. The skin lesions that developed exhibited many of the histological features of human psoriasis, regardless of whether IFNγ could be expressed by the donor cells. These lesions were abolished when an anti-murine IL-12p40 mAb was administered on days 7 and 35 after the T cell transfer. Subsequent studies that analyzed anti-IL-12p40 mAb treatment in later stages of disease demonstrated that higher doses and prolonged treatment could almost completely resolve psoriatic lesions. These lesions, however, did return once mAb treatment ceased (Hong et al., 2001). These observations suggest that IL-12 can contribute to psoriatic lesions, and neutralization through murine IL-12p40 mAbs, provided significant protection from psoriatic disease.

The most recent *in vivo* validation of targeting IL-12p40 for psoriasis was achieved in human patients dosed with a human anti-human IL-12p40 mAb (Kauffman et al., 2004). The Phase 1 study was nonrandomized, open-label, and tested single intravenous (IV) ascending doses of IL-12p40 mAb in 18 subjects with moderate to severe psoriasis vulgaris. Twelve of 18 subjects achieved at least a 75% improvement in their disease severity after a single IV infusion of anti-IL-12p40. These benefits were sustained over the 16-week study period in many patients. Furthermore, the efficacy data suggest a correlation of clinical response with the dose of anti-IL-12p40 mAb. Further studies are necessary to determine the duration of disease suppression and the necessity and timing for multiple mAb doses. These data are very encouraging for patients with psoriasis and effectively validated IL-12p40 as an appropriate drug target for this debilitating disease.

B. Crohn's Disease

Crohn's disease (CD) is an immune-mediated disease of the gastrointestinal tract characterized by transmural infiltration of lymphocytes and macrophages, granulomas, fissuring ulceration, and submucosal fibrosis (reviewed by Bouma and Strober, 2003). Evidence from human CD tissue suggests that CD is mediated by Th1 cells, IL-12, and perhaps IL-23. IL-12p35, IL-12p40, and IL-12Rβ2 mRNA are upregulated in lamina propria mononuclear cells and ileal specimens from CD patients (Berrebi et al., 1998; Monteleone et al.,

1997; Parrello et al., 2000). In addition, *ex vivo* cell analysis demonstrated that IL-12 is produced by CD lamina propria cells, leading to increased IFNγ production (Colpaert et al., 2002; Parronchi et al., 1997). IL-23p19 expression within CD inflamed mucosa was shown to be significantly increased and correlated with lesion severity upon endoscopy (Schmidt et al., 2005; Stallmach et al., 2004). Common IL-12 and IL-23 signaling pathways were also noted to be activated in CD tissue. T cell nuclear extracts demonstrated increased levels of STAT4 and t-bet, which are indicative of IL-12 signaling (Neurath et al., 2002). Together, these human CD tissue analyses suggested that IL-12p40 is a rational target to consider for the treatment of Crohn's disease.

IL-12p40 has been studied as a therapeutic target in many mouse models of colitis. Anti-IL-12p40 antibodies administered at early or late time points in a rodent model of colitis improved clinical and histopathological changes (Neurath et al., 1995). In a separate study using IL-10-deficient mice, anti-IL-12p40 antibodies ameliorated disease and decreased CD4+ cell infiltration (Davidson et al., 1998). In a hapten-induced colitis model, IL-12p35 and IL-12Rβ1 knock-out mice demonstrated milder disease than wild-type mice (Camoglio et al., 2002). Interestingly, IL-12p40-deficient mice were more susceptible to this model of colitis, which could suggest a protective role for IL-23, or p40 monomers or homodimers. Regardless, there was sufficient evidence in animal colitis models to warrant targeting IL-12p40 in human CD.

Defining evidence for the role of IL-12p40 in CD was shown in a double-blind, placebo-controlled clinical study in 79 human CD patients. Once-weekly subcutaneous administration of an anti-IL-12p40 mAb induced long-lasting remission in anti-IL–12p40 treated patients (Mannon et al., 2004). Clinical improvement was associated with a reduction of IL-12 and IFNγ production by mononuclear cells isolated from the affected tissues. Subsequent studies are required to define further the appropriate treatment regimen; however, this is an exciting opportunity for patients with CD and serves to further validate IL-12p40 as an appropriate drug target for certain immune-mediated diseases.

C. Rheumatoid Arthritis

The pathogenesis of rheumatoid arthritis (RA) involves complex interactions between T and B lymphocytes, macrophages, fibroblast-like synoviocytes, and a network of cytokines. Cytokines are thought to mediate tissue destruction through chondrocyte and fibroblast activation, which leads to cartilage loss and bone erosion. The success of anti-TNF biological therapies and the prominence of T lymphocytes in rheumatoid synovitis suggests that targeting cytokines involved in T cell activation, such as IL-12 and IL-23, might be appropriate for RA therapy. Indeed, the prevalence of IL-17 in RA tissues implies the contribution of IL-23 because IL-23 is known to be a potent inducer of IL-17 production (Miossec, 2003). Similarly, IFNγ is predominantly expressed by re-stimulated synovial T cells, which implies contribution from IL-12 (Miossec, 2000). Indeed, IL-12 expression has been observed in RA synovial tissue (Morita et al., 1998) and RA granulomata (Hessian et al., 2003). Although not as striking as the evidence for TNFα involvement, there is evidence from RA tissue that IL-12 or IL-23 may also be involved in disease pathology.

The majority of IL-12p40 target validation resides in the animal model of RA, called collagen-induced arthritis (CIA). Chronic treatment of CIA mice with an anti-murine IL-12p40 mAb significantly attenuated the histopathological and clinical severity of disease (Malfait et al., 1998). Total collagen-specific antibody levels were decreased even though the isotype balance did not change. Interestingly, disease protection increased upon co-administration with anti-TNFα mAb (Butler et al., 1999). Perhaps the strongest evidence for the role of IL-12 and IL-23 in RA has been generated using mice genetically deficient for IL-12p40, IL-12p35, and IL-23p19. IL-12p40 and IL-23p19 knock-out mice were completely protected from CIA, though IL-12p35 knock-out mice exhibited increased disease severity (McIntyre et al., 1996; Murphy et al., 2003). This observation could even suggest a protective effect of IL-12 in this model. The role of IL-12 or IL-23 in human RA may not have been established to date in the relevant scientific literature, therefore IL-12p40 may not yet be considered a validated target for this disease.

D. Multiple Sclerosis

Multiple sclerosis (MS) is a neurological disease of the central nervous system (CNS) characterized by loss of sensorimotor and cognitive functions believed to be due to neuroinflammation, myelin antibodies, demyelination, and axonal deterioration (reviewed by Gran et al., 2004). Human-tissue analysis indicates that T cell infiltration and IL-12 or IL-23 may contribute to MS pathogenesis. In patients with MS, there is increased IL-12p40 production in MS lesions (Windhagen et al., 1995), peripheral blood mononuclear cells (van Boxel-Dezaire et al., 1999), serum (Nicoletti et al., 1996), and cerebrospinal fluid (Fassbender et al., 1998). Futhermore, levels of IL-12p40-positive monocytes in the blood were suggested to correlate with MS disease activity (Makhlouf et al., 2001). In addition, increased frequency of IL-12Rβ1- and IL-12Rβ2-positive T cells were identified in MS samples (Ozenci et al., 2001). Collectively, these observations suggest that similar immune mechanisms are involved in MS as the other immune-mediated diseases discussed thus far.

Effective targeting of IL-12p40 has been achieved using IL-12p40 neutralizing mAbs in a murine or primate demyelinating disease model that mimics human MS. This model is called experimental autoimmune encephalomyelitis (EAE). Administration of recombinant IL-12 can exacerbate EAE, whereas treatment with neutralizing anti-IL-12p40 antibodies will inhibit both murine and primate models of EAE (Brok et al., 2002; Constantinescu et al., 1998; Leonard et al., 1995). MAb studies in chronic remitting-relapsing murine EAE have allowed the investigation of temporal and specific inhibition of IL-12 and IL-23. Anti-IL-12p40 or anti-IL-23p19 mAb treatment will significantly ameliorate EAE when dosing is initiated prior to or during established disease. The level of disease suppression is greatest when mAb is dosed earlier in the course of disease. In contrast, anti-IL-12p35 mAb treatment is ineffective when dosed at any time. *Ex vivo* cellular analysis suggests that IL-12p40 targeting prevents complete activation of CD4+ T cells, thus preventing upregulation of cell-surface markers and chemokine receptors (JM Benson, unpublished results). Thus, IL-12p40 mAb antagonist studies in the EAE model have begun to elucidate downstream immunological effects of IL-12p40 inhibition.

Evidence for the role of IL-12 or IL-23 in EAE and MS has also been demonstrated using knock-out mice that are deficient in IL-12 or IL-23 subunits. IL-12p40- or IL-12Rβ1-deficient mice are completely resistant to EAE, even though mice deficient in IL-12p35 or IL-12Rβ2 are fully susceptible (Becher et al., 2002; Gran et al., 2002; Zhang et al., 2003). This suggested that alternative p40 cytokines, such as IL-23, are responsible for disease. The role of IL-23 in EAE was recently confirmed in studies using IL-23p19-deficient mice. These animals demonstrated complete resistance to disease induction, similar to IL-12p40-deficient mice (Cua et al., 2003). Subsequent studies suggest that IL-23 promotes differentiation of a unique subset of encephalitogenic CD4+ T cells that produce significant amounts of IL-17, thus suggesting a novel mechanism of immunopathology for EAE and MS (Langrish et al., 2005). Characterization of the collection of IL-12 and IL-23 knock-out mice in the EAE model has demonstrated a clear contribution of IL-12p40 and IL-23 to this autoimmune pathogenesis.

MS poses a unique challenge for therapeutic intervention given that the pathology occurs in the CNS. Because many therapeutic molecules cannot cross the blood-brain barrier (BBB), target localization is an important point to consider in MS target-validation studies. Although peripheral mAb treatment is effective in murine and primate EAE, this disease model is induced peripherally and thereby might also be successfully treated through peripheral immunomodulation. Therefore it is difficult to determine if peripheral treatment is sufficient for MS using the EAE model. Some IL-12p40 protein mimetic studies, however, have addressed the other side to this question by determining if therapeutic intervention directly within the CNS is sufficient to inhibit disease. For example, myelin basic protein–specific CD4+ T cells were transduced to express IL-12p40, then adoptively transferred into mice immunized to develop EAE. The IL-12p40 producing T cells were shown to traffic specifically into the CNS and a significantly reduce EAE clinical scores (Costa et al., 2001). Even though the question remains as to whether IL-12p40 targeted therapies must cross the BBB in MS, there are data to suggest that IL-12p40 is an appropriate target both in the CNS and periphery of MS patients.

The immune-mediated diseases discussed thus far represent those with the most evidence for contribution of IL-12 or IL-23. As the use of IL-12p40 antagonists expands to other disease models, there will certainly be additional therapeutic indications studied for IL-12p40 target validation. Already there are suggestions that IL-12p40 may be an appropriate target for autoimmune uveitis (Yokoi et al., 1997), insulin-dependent diabetes mellitus (Trembleau et al., 1997), allograft rejection (Piccotti et al., 1997), graft versus host disease (Nishimura et al., 1996), and even pulmonary fibrosis (Huaux et al., 2002). The caveat with all of these disease models is that they mostly mimic only a portion of the human disease of interest. Therefore, the only way to more conclusively verify that IL-12p40 is an appropriate therapeutic target in these indications is to introduce an IL-12p40 specific antagonist to the human condition.

IV. CONCLUSION

Results from a variety of *in vitro* and *in vivo* animal and human studies utilizing multiple IL-12p40 antagonist technologies have yielded consistent observations on the benefit of IL-12p40 inhibition. The use of neutralizing mAbs, knock-out mice, or protein mimetics in a variety of model systems all suggested that IL-12p40 inhibition would be beneficial for immune disorders. Thus, IL-12p40 was well validated as a therapeutic target. Of course, there are some limitations to the *in vitro* and *in vivo* modeling systems because they may not accurately reflect the complexity of human disease. While the contribution of IL-12p40 to immune-mediated disease seems to be true so far for psoriasis and CD, the safety and tolerability of long-term IL-12p40 targeting remains to be determined. As previously mentioned, IL-12 is thought to contribute to pathogen and tumor immunity, although the role for IL-23 in these pathways is less clear. For this reason, future learnings in the human clinical studies will ultimately dictate the success of IL-12p40 targeted strategies. With the remarkable findings made so far, however, there is a very promising future for IL-12p40 specific therapeutic entities.

RECOMMENDED RESOURCES

IL-12 and IL-23 reagents and assay protocols:

1. http://www.rndsystems.com
2. http://www.bdbiosciences.com
3. Coico, R. (2005). "Current Protocols of Immunology" (John Wiley & Sons).

Scientific literature reviews of IL-12 and IL-23:

1. Adorini, L. (1997). "IL-12", *Chemical Immunology*, Vol 68 (Basel, Karger).
2. Gately, M. K., L. M. Renzetti, et al. (1998). The interleukin-12/interleukin-12-receptor system: Role in normal and pathologic immune responses. *Annul. Rev. Immunol.* 16: 495–521.
3. Trinchieri, G. (2003). Interleukin-12 and the regulation of innate resistance and adaptive immunity. *Nat. Rev. Immunol.* 3: 133–146.

REFERENCES

Austin, L. M., M. Ozawa, et al. (1999). The majority of epidermal T cells in psoriasis vulgaris lesions can produce type 1 cytokines, interferon-gamma, interleukin-2, and tumor necrosis factor-alpha, defining TC1 (cytotoxic T lymphocyte) and TH1 effector populations: A type 1 differentiation bias is also measured in circulating blood T cells in psoriatic patients. *J. Invest. Dermatol.* 113: 752–759.

Becher, B., B. G. Durell, and R. J. Noelle (2002). Experimental autoimmune encephalitis and inflammation in the absence of interleukin-12. *J. Clin. Invest.* 110: 493–497.

Berrebi, D., M. Besnard, et al. (1998). Interleukin-12 expression is focally enhanced in the gastric mucosa of pediatric patients with Crohn's disease. *Am. J. Pathol.* 152: 667–672.

Bouma, G. and W. Strober (2003). The immunological and genetic basis of inflammatory bowel disease. *Nat. Rev. Immunol.* 3: 521–533.

Brok, H. P., M. van Meurs, et al. (2002). Prevention of experimental autoimmune encephalomyelitis in common marmosets using an anti-IL-12p40 monoclonal antibody. *J. Immunol.* 169: 6554–6563.

Butler, D. M., A. M. Malfait, et al. (1999). Anti-IL-12 and anti-TNF antibodies synergistically suppress the progression of murine collagen-induced arthritis. *Eur. J. Immunol.* 29: 2205–2212.

Camoglio, L., N. P. Juffermans, et al. (2002). Contrasting roles of IL-12p40 and IL-12p35 in the development of hapten-induced colitis. *Eur. J. Immunol.* 32: 261–269.

Colpaert, S., K. Vastraelen, et al. (2002). *In vitro* analysis of interferon gamma (IFN-gamma) and interleukin-12 (IL-12) production and their effects in ileal Crohn's disease. *Eur. Cytokine Netw.* 13: 431–437.

Constantinescu, C. S., M. Wysocka, et al. (1998). Antibodies against IL-12 prevent superantigen-induced and spontaneous relapses of experimental autoimmune encephalomyelitis. *J. Immunol.* 161: 5097–5104.

Costa, G. L., M. R. Sandora, et al. (2001). Adoptive immunotherapy of experimental autoimmune encephalomyelitis via T cell delivery of the IL-12 p40 subunit. *J. Immunol.* 167: 2379–2387.

Cua, D. J., J. Sherlock, et al. (2003). Interleukin-23 rather than interleukin-12 is the critical cytokine for autoimmune inflammation of the brain. *Nature* 421: 744–748.

Davidson, N. J., S. A. Hudak, et al. (1998). IL-12, but not IFN-gamma, plays a major role in sustaining the chronic phase of colitis in IL-10-deficient mice. *J. Immunol.* 161: 3143–3149.

Fassbender, K., A. Ragoschke, et al. (1998). Increased release of interleukin-12p40 in MS: Association with intracerebral inflammation. *Neurology* 51: 753–758.

Gately, M. K., L. M. Renzetti, et al. (1998). The interleukin-12/interleukin-12-receptor system: Role in normal and pathologic immune responses. *Annu. Rev. Immunol.* 16: 495–521.

Gillessen, S., D. Carvajal, et al. (1995). Mouse interleukin-12 (IL-12) p40 homodimer: A potent IL-12 antagonist. *Eur. J. Immunol.* 25: 200–206.

Gran, B., G. X. Zhang, and A. Rostami (2004). Role of the IL-12/IL-23 system in the regulation of T-cell responses in central nervous system inflammatory demyelination. *Crit. Rev. Immunol.* 24: 111–128.

Gran, B., G. X. Zhang, et al. (2002). IL-12p35-deficient mice are susceptible to experimental autoimmune encephalomyelitis: Evidence for redundancy in the IL-12 system in the induction of central nervous system autoimmune demyelination. *J. Immunol.* 169: 7104–7110.

Hessian, P. A., J. Highton, et al. (2003). Cytokine profile of the rheumatoid nodule suggests that it is a Th1 granuloma. *Arthritis Rheum.* 48: 334–338.

Hong, K., E. L. Berg, and R. O. Ehrhardt (2001). Persistence of pathogenic CD4+ Th1-like cells in vivo in the absence of IL-12 but in the presence of autoantigen. *J. Immunol.* 166: 4765–4772.

Hong, K., A. Chu, et al. (1999). IL-12, independently of IFN-gamma, plays a crucial role in the pathogenesis of a murine psoriasis-like skin disorder. *J. Immunol.* 162: 7480–7491.

Huaux, F., M. Arras, et al. (2002). A profibrotic function of IL-12p40 in experimental pulmonary fibrosis. *J. Immunol.* 169: 2653–2661.

Kauffman, C. L., N. Aria, et al. (2004). A phase I study evaluating the safety, pharmacokinetics, and clinical response of a human IL-12 p40 antibody in subjects with plaque psoriasis. *J. Invest. Dermatol.* 123: 1037–1044.

Kobayashi, M., L. Fitz, et al. (1989). Identification and purification of natural killer cell stimulatory factor (NKSF), a cytokine with multiple biologic effects on human lymphocytes. *J. Exp. Med.* 170: 827–845.

Langrish, C. L., Y. Chen, et al. (2005). IL-23 drives a pathogenic T cell population that induces autoimmune inflammation. *J. Exp. Med.* 201: 233–240.

Lebwohl, M. (2003). Psoriasis. *Lancet* 361: 1197–1204.

Lee, E., W. L. Trepicchio, et al. (2004). Increased expression of interleukin 23 p19 and p40 in lesional skin of patients with psoriasis vulgaris. *J. Exp. Med.* 199: 125–130.

Leonard, J. P., K. E. Waldburger, and S. J. Goldman (1995). Prevention of experimental autoimmune encephalomyelitis by antibodies against interleukin 12. *J. Exp. Med.* 181: 381–386.

Ling, P., M. K. Gately, et al. (1995). Human IL-12 p40 homodimer binds to the IL-12 receptor but does not mediate biologic activity. *J. Immunol.* 154: 116–127.

Makhlouf, K., H. L. Weiner, and S. J. Khoury (2001). Increased percentage of IL-12+ monocytes in the blood correlates with the presence of active MRI lesions in MS. *J. Neuroimmunol.* 119: 145–149.

Malfait, A. M., D. M. Butler, et al. (1998). Blockade of IL-12 during the induction of collagen-induced arthritis (CIA) markedly attenuates the severity of the arthritis. *Clin. Exp. Immunol.* 111: 377–383.

Mannon, P. J., I. J. Fuss, et al. (2004). Anti-interleukin-12 antibody for active Crohn's disease. *N. Engl. J. Med.* 351: 2069–2079.

McIntyre, K. W., D. J. Shuster, et al. (1996). Reduced incidence and severity of collagen-induced arthritis in interleukin-12-deficient mice. *Eur. J. Immunol.* 26: 2933–2938.

Miossec, P. (2000). Are T cells in rheumatoid synovium aggressors or bystanders? *Curr. Opin. Rheumatol.* 12: 181–185.

Miossec, P. (2003). Interleukin-17 in rheumatoid arthritis: If T cells were to contribute to inflammation and destruction through synergy. *Arthritis. Rheum.* 48: 594–601.

Monteleone, G., L. Biancone, et al. (1997). Interleukin 12 is expressed and actively released by Crohn's disease intestinal lamina propria mononuclear cells. *Gastroenterology* 112: 1169–1178.

Morita, Y., M. Yamamura, et al. (1998). Expression of interleukin-12 in synovial tissue from patients with rheumatoid arthritis. *Arthritis Rheum.* 41: 306–314.

Murphy, C. A., C. L. Langrish, et al. (2003). Divergent pro- and antiinflammatory roles for IL-23 and IL-12 in joint autoimmune inflammation. *J. Exp. Med.* 198: 1951–1957.

Neurath, M. F., I. Fuss, et al. (1995). Antibodies to interleukin 12 abrogate established experimental colitis in mice. *J. Exp. Med.* 182: 1281–1290.

Neurath, M. F., B. Weigmann, et al. (2002). The transcription factor T-bet regulates mucosal T cell activation in experimental colitis and Crohn's disease. *J. Exp. Med.* 195: 1129–1143.

Nicoletti, F., F. Patti, et al. (1996). Elevated serum levels of interleukin-12 in chronic progressive multiple sclerosis. *J. Neuroimmunol.* 70: 87–90.

Nishimura, T., A. Sadata, et al. (1996). The therapeutic effect of interleukin-12 or its antagonist in transplantation immunity. *Ann. N. Y. Acad. Sci.* 795: 371–374.

Oppmann, B., R. Lesley, et al. (2000). Novel p19 protein engages IL-12p40 to form a cytokine, IL-23, with biological activities similar as well as distinct from IL-12. *Immunity* 13: 715–725.

Ozenci, V., M. Pashenkov, et al. (2001). IL-12/IL-12R system in multiple sclerosis. *J. Neuroimmunol.* 114: 242–252.

Parham, C., M. Chirica, et al. (2002). A receptor for the heterodimeric cytokine IL-23 is composed of IL-12Rbeta1 and a novel cytokine receptor subunit, IL-23R. *J. Immunol.* 168: 5699–5708.

Parrello, T., G. Monteleone, et al. (2000). Up-regulation of the IL-12 receptor beta 2 chain in Crohn's disease. *J. Immunol.* 165: 7234–7239.

Parronchi, P., P. Romagnani, et al. (1997). Type 1 T-helper cell predominance and interleukin-12 expression in the gut of patients with Crohn's disease. *Am. J. Pathol.* 150: 823–832.

Piccotti, J. R., S. Y. Chan, et al. (1997). Differential effects of IL-12 receptor blockade with IL-12 p40 homodimer on the induction of CD4+ and CD8+ IFN-gamma-producing cells. *J. Immunol.* 158: 643–648.

Presky, D. H., H. Yang, et al. (1996). A functional interleukin 12 receptor complex is composed of two beta-type cytokine receptor subunits. *Proc. Natl. Acad. Sci. U. S. A.* 93: 14002–14007.

Schmidt, C., T. Giese, et al. (2005). Expression of interleukin-12-related cytokine transcripts in inflammatory bowel disease: Elevated interleukin-23p19 and interleukin-27p28 in Crohn's disease but not in ulcerative colitis. *Inflamm. Bowel Dis.* 11: 16–23.

Stallmach, A., T. Giese, et al. (2004). Cytokine/chemokine transcript profiles reflect mucosal inflammation in Crohn's disease. *Int. J. Colorectal Dis.* 19: 308–315.

Stern, A. S., F. J. Podlaski, et al. (1990). Purification to homogeneity and partial characterization of cytotoxic lymphocyte maturation factor from human B-lymphoblastoid cells. *Proc. Natl. Acad. Sci. U. S. A.* 87: 6808–6812.

Trembleau, S., G. Penna, et al. (1997). Deviation of pancreas-infiltrating cells to Th2 by interleukin-12 antagonist administration inhibits autoimmune diabetes. *Eur. J. Immunol.* 27: 2330–2339.

Trinchieri, G. (2003). Interleukin-12 and the regulation of innate resistance and adaptive immunity. *Nat. Rev. Immunol.* 3: 133–146.

van Boxel-Dezaire, A. H., S. C. Hoff, et al. (1999). Decreased interleukin-10 and increased interleukin-12p40 mRNA are associated with disease activity and characterize different disease stages in multiple sclerosis. *Ann. Neurol.* 45: 695–703.

Vandenbroeck, K., I. Alloza, et al. (2004). Inhibiting cytokines of the interleukin-12 family: Recent advances and novel challenges. *J. Pharm. Pharmacol.* 56: 145–160.

Windhagen, A., J. Newcombe, et al. (1995). Expression of costimulatory molecules B7-1 (CD80), B7-2 (CD86), and interleukin 12 cytokine in multiple sclerosis lesions. *J. Exp. Med.* 182: 1985–1996.

Yawalkar, N., S. Karlen, S., et al. (1998). Expression of interleukin-12 is increased in psoriatic skin. *J. Invest. Dermatol.* 111: 1053–1057.

Yokoi, H., K. Kato, et al. (1997). Prevention of experimental autoimmune uveoretinitis by monoclonal antibody to interleukin-12. *Eur. J. Immunol.* 27: 641–646.

Zhang, G. X., B. Gran, et al. (2003). Induction of experimental autoimmune encephalomyelitis in IL-12 receptor-beta 2-deficient mice: IL-12 responsiveness is not required in the pathogenesis of inflammatory demyelination in the central nervous system. *J. Immunol.* 170: 2153–2160.

8 THE GPIIb/IIIa ANTAGONIST ABCIXIMAB FOR ACUTE PERCUTANEOUS CORONARY INTERVENTION

ROBERT E. JORDAN

Ph.D., Centocor Research and Development, Inc., Radnor, Pennsylvania

The emergence of percutaneous catheter procedures for the restoration of blood flow in occluded atherosclerotic coronary arteries introduced acute and potentially severe local vascular injuries. These balloon-device–based procedures generated locally injured surfaces that could be highly attractive to blood platelets and carried the risk of rapid vessel closure. The available antiplatelet agents, such as aspirin, were generally weak and nonspecific and were inadequate to fully protect against this new type of provocation. An alternative pathway involving a platelet adhesion receptor, however, was identified as a promising target for the potent antiplatelet efficacy that was required for these interventional procedures. The target was an integral membrane receptor termed GPIIb/IIIa (in integrin nomenclature, $\alpha_{IIb}\beta_3$). On activated platelets, GPIIb/IIIa binds to a natural ligand, fibrinogen, to interconnect platelets into potentially thrombotic aggregates. The search for selective antagonists of GPIIb/IIIa and platelet aggregation led to the development of a potent monoclonal antibody and small-molecule competitive antagonists that mimic the natural ligand(s) of the receptor. This review will focus on the development of the monoclonal antibody abciximab. The inhibition of the GPIIb/IIIa receptor pathway was extensively studied *in vitro*, in well-established animal models of thrombosis and in several large clinical trials. More than a decade of continued research has yielded further details of GPIIb/IIIa and the structural complexity, conformational responsiveness and signaling connections that could not be envisioned during early drug development. Abciximab and other antagonists continue to provide insights into the receptor and to contribute to the effort to refine anti- GPIIb/IIIa therapy.

I. INTRODUCTION

Platelets were long suspected to be critical participants in the intravascular, thrombotic occlusions that led to heart attacks. Platelets readily bind to normally hidden elements in blood vessels to minimize blood leakage (e.g., the hemostatic plug at breaks and tears in a vessel). When platelets encounter subendothelial components, such as collagen and von Willebrand factor (vWF), in a diseased but otherwise intact atherosclerotic artery, pathological occlusive thrombi can form. Weak and relatively nonspecific inhibitors of platelet function, such as aspirin, provide a measure of protection against episodes of blockage during the gradual evolution of coronary artery disease. In advanced atherosclerosis and unstable angina, thrombotic episodes can occur repeatedly and there is need for more potent and selective therapeutic antagonists. In the early days of angioplasty it was clear that in the absence of any antiplatelet therapy, thrombosis was frequent and this could be partially reduced with aspirin (Barnathan et al., 1987). As interventional practice evolved, it became clear that thrombotic events resulting in myocardial infarctions or the need for urgent revascularization procedure or even death were still occurring, requiring more potent platelet inhibition. The platelet GPIIb/IIIa receptor appeared to be an ideal target for drug development for the prevention of acute thrombosis.

There have been several excellent and comprehensive reviews of platelet thrombosis, and the development of GPIIb/IIIa antagonists for treatment of percutaneous intervention (PCI), myocardial infarction (MI), unstable angina, and follow-on indications (Agah et al., 2002; Coller, 2001; Law and Phillips, 1999). The present overview will focus on various aspects of the GPIIb/IIIa receptor that influenced the development of abciximab and emerging information that continues to add to the interpretation of its clinical benefit.

II. RATIONALE FOR GPIIb/IIIa AS A TARGET IN CORONARY ARTERIAL DISEASE

The participation of the GPIIb/IIIa receptor in the aggregation of platelets in normal hemostasis as well as in pathological thrombosis is now well established. Arrival at this understanding and the identification of GPIIb/IIIa as a therapeutic target was the result of contributions from many scientific disciplines (Coller, 1995). These included the studies of fibrinogen biochemistry and platelet physiology, investigations on the rare naturally occurring defects of GPIIb/IIIa in Glanzmann thrombasthenia, basic studies of the cell-adhesion process and the integrin family of receptors, and the availability of predictive animal models of acute thrombosis and arterial pathology. The eventual clinical development of the monoclonal antibody abciximab, as well as other small-molecule antagonists of GPIIb/IIIa, resulted from a convergence of information from each of these diverse fields in the early 1990s.

Of particular importance, the absence of functional GPIIb/IIIa on the platelets of individuals with Glanzmann thrombasthenia was shown to underlie the inability of these platelets to bind fibrinogen. This defect was further correlated with a lack of aggregation response when these platelets were stimulated with physiological agonists that rapidly induce normal platelet aggregation (e.g., adenosine diphosphate [ADP], epinephrine, and thrombin) (Nurden, 2005). This crucial identification of fibrinogen as the physical adhesive ligand and GPIIb/IIIa as the fibrinogen receptor prompted investigators to apply the evolving technology of monoclonal antibodies to the GPIIb/IIIa target. Specifically, the goal was an antibody that would bind to the receptor in such a way as to block the association of fibrinogen to activated platelets.

III. GENERATION OF THE 7E3 MONOCLONAL ANTIBODY AGAINST GPIIb/IIIa

Although activation at the receptor level was suspected to occur, details of molecular functioning and GPIIb/IIIa structure were far from well understood in the 1980s. One approach taken for antibody generation was immunization with intact, washed human platelets (Coller, 1985). The rationale was to generate antibodies to a conformation of GPIIb/IIIa that was minimally perturbed and then to select candidate antibodies based on their functional ability to block fibrinogen binding. A screening assay based on the binding of platelets to immobilized fibrinogen pointed to several candidate antibodies, notably 10E5 and 7E3. The latter is the murine parent antibody of the chimeric therapeutic version, abciximab (Coller et al., 1983).

Because of the potential for causing Fc-mediated cell (platelet) destruction, the intact murine IgG1 kappa antibody was never intended for therapeutic use. Instead, the proteolytically cleaved $F(ab')_2$ and Fab fragments of murine 7E3 IgG1 were selected for studies in animals and for the initial human testing. As a consequence of removing the Fc domain, the $F(ab')_2$ and Fab fragments of IgG possess little or no ability to fix complement and/or cause antibody-dependent cellular cytotoxicity (ADCC). Later, the Fab fragment of the human/murine chimeric version (containing 50% human sequences in the constant domains) was chosen for final human testing—a decision motivated primarily by the expectation for minimized human immune response (Jordan et al., 1996, 1997). Steps in the progression of antibody and fragment development leading to abciximab are depicted in Figure 8.1. In recent years, the development of other therapeutic monoclonal antibodies has employed more extensive humanization of murine antibodies to foil host recognition and immune response. It is now feasible to generate fully human monoclonal antibodies using phage display and transgenic animals, and these approaches are likely to succeed in further reducing immunogenicity.

A review of the *in vitro* and *in vivo* preclinical investigations and clinical studies with abciximab will follow. The extensive validation that was conducted for the first "drug" to be approved in this class may still serve as a useful guide in the complex path of drug development to an emerging target.

IV. *IN VITRO* STUDIES

A. Quantitative Aspects of Platelet GPIIb/IIIa

The GPIIb/IIIa receptor is normally present only on the circulating platelet and its bone marrow progenitor, the megakaryocyte. One step in the validation of GPIIb/IIIa as a therapeutic target was to quantify platelet receptor numbers and the degree of blockade required for inhibition of platelet aggregation. Estimates of approximately 40,000 surface receptors per activated platelet had been recorded based on binding studies using radiolabeled fibrinogen (Bennett and Vilaire, 1979). Similar values were obtained using radiolabeled monoclonal IgG antibodies (Newman et al., 1985). Although 7E3 IgG bound to both resting and activated platelets at a similar level, we noted that 7E3 could bind to two GPIIb/IIIa receptors simultaneously because of the bivalent nature of the IgG. This led to studies that showed that the actual external membrane receptor number could be higher than 80,000 per platelet based on binding studies using the monovalent Fab fragment (Wagner et al., 1996). As data accumulated, it

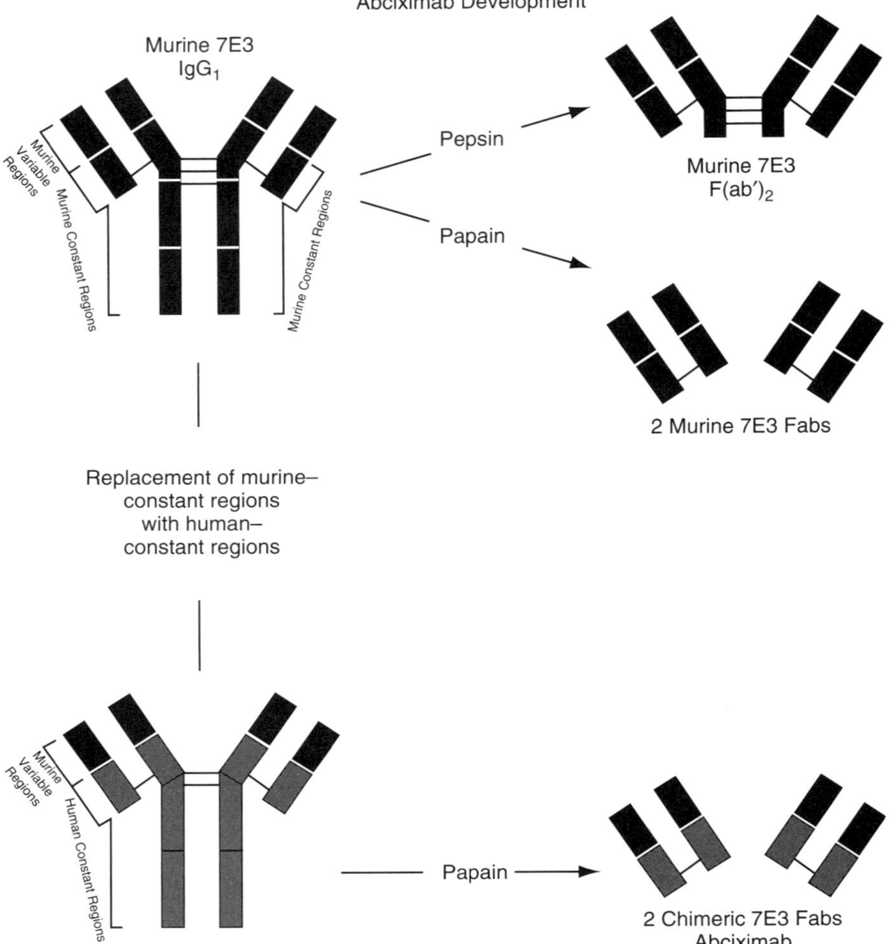

FIGURE 8.1 **Schematic representation of different clinically tested fragments of the 7E3 antibody.** The first trials employed the bivalent $F(ab')_2$ fragment of murine 7E3 resulting from pepsin digestion (upper right). Later, the monovalent Fab fragment of murine 7E3 resulting from papain digestion was tested. In the lower part of the figure, the replacement of the murine constant regions with human equivalents is depicted and this chimeric IgG was digested to Fab fragments with papain. The approximate molecular weights of IgG, $F(ab')_2$, and Fab are 160,000, 100,000, and 48,000, respectively. In the text, chimeric 7E3 Fab is generally denoted by its USAN proper name, abciximab (ReoPro®).

became clear that the GPIIb/IIIa number on platelets from different individuals could be highly variable. For *in vivo* dose finding, it was found that the percent blockade of total receptors per platelet was the most useful parameter for pharmacodynamic correlations. The quantitative GPIIb/IIIa blockade analyses in *in vitro* studies and in *ex vivo* determinations for both animal and human testing were performed using a radiolabeled version of 7E3 antibody to assess unblocked receptors before and after treatment (Coller et al., 1995). Despite the accessibility of the platelets in treated patients, there still remains a degree of uncertainty about the functioning of platelets in a test tube compared to flowing blood *in vivo*.

The ability to quantify receptor blockade showed that the optimized bolus dose of antibody was in relatively close stoichiometric equivalence to available GPIIb/IIIa receptors—on average, no more than 1.5 times the number of molecules of 7E3 relative to the total number of surface receptors on circulating and splenic pools of platelets. The access to circulating blood platelets for analysis before and after treatment greatly facilitated the pharmacological evaluations. Few instances in drug development have provided such convenient access to the actual cellular target of therapeutic intervention.

B. Functional Testing of Anti-GPIIb/IIIa Antibodies

Platelet aggregometry measures the ability of platelets in a platelet-rich plasma suspension to clump together and thereby increase the transmission of visible light in a cuvette. Platelets respond at different rates and extents depending on the activation stimulus that is used. Thus, any calibration of the degree of GPIIb/IIIa blockade required for inhibition of platelet aggregation proved to be partly a function of the strength and type of platelet agonist employed for aggregation. Adenosine diphosphate (ADP) proved to be a reproducible and consistent stimulus that could be conveniently used and compared among different laboratories. ADP, however, elicits different responses depending on the concentration used. For example, at ADP concentrations of less than 2 μM, platelets often demonstrate a weak, biphasic, and sometimes reversible response in standard aggregometry. Higher concentrations (e.g., 5 to 20 μM) caused more vigorous responses that were usually irreversible. In order to avoid overestimating the inhibitory activity of abciximab, platelet function testing with ADP was most often conducted at 5 to 20 μM. Other agonists, such as collagen and epinephrine, were found to yield variable results, especially when different preparations were compared among different laboratories. When it became available, the thrombin receptor activating peptide, TRAP, was found to be a potent stimulus for platelet aggregation and was adopted for certain specialized testing (Vu et al., 1991). Thrombin itself could not routinely be used because of the presence of heparin in some samples. (The anticoagulant heparin, widely used in cardiovascular interventions, is a cofactor for inhibition of thrombin by antithrombin III.) One unexpected and confounding variable for a different GPIIb/IIIa antagonist (eptifibatide) was the augmentation of its platelet inhibitory activity in standard citrate anticoagulant (Phillips et al., 1997). In the case of eptifibatide, a recalibration of the dose-finding was needed in noncitrate anticoagulated plasma to accurately reflect *in vivo* activity of the drug. This did not turn out to be a significant concern for abciximab.

Although turbidometric aggregometry is the most established tool for pharmacodynamic assessment of GPIIb/IIIa antagonists, a number of alternative approaches have been devised. Among these methods are shear-induced aggregation systems, including the cone and plate viscometer, the Xylum clot signature analyzer, and the PFA-100. Additionally, the Rapid Platelet Function Analyzer (RPFA) using TRAP-induced agglutination of platelets to fibrinogen-coated beads in whole blood has shown particular promise for point of care monitoring (Smith et al., 1999). Each of these methods has been applied to the pharmacodynamic monitoring of abciximab and other antagonists. Not surprisingly, there are subtle and sometimes distinct differences in the results obtained with the different techniques, further underscoring the challenge of correlating *in vitro* parameters to *in vivo* thrombotic phenomena.

C. Species Cross-Reactivity

Monoclonal antibodies often possess limited species cross-reactivity and the 7E3 antibody was no exception. This restriction made the choice of animal models problematic for the *in vivo* validation of efficacy. Cross reactivity was largely limited to nonhuman primates for the monovalent Fab fragments and to dogs and rats (as learned much later) for higher doses of the bivalent $F(ab')_2$ fragments (Sassoli et al., 2001). The finding that different antibody doses and/or variants were required in different animal species is based on the differing affinities of abciximab for the same receptor target in the respective species. The ability of 7E3 to inhibit platelet function in dogs turned out to be critically important for the preclinical validation as it enabled the use of the more available canine models of arterial thrombosis.

D. Correlations of Antibody Binding to Functional Inhibition

Figure 8.2 depicts an experiment that correlates increasing abciximab concentrations with quantitative levels of receptor blockade and inhibition of 5 μM ADP-induced platelet aggregation. This example employed human platelet-rich plasma and closely paralleled the results obtained in similar experiments using platelets from cynomolgus monkeys (Jordan et al., 1996). There are graded inhibitory effects at increasing antibody concentrations with complete inhibition of platelet aggregation at concentrations nearing 2 μg/mL. The percent receptor blockade at which complete inhibition of aggregation was achieved was approximately 80%. In the preclinical studies, nearly identical correlations were obtained when platelets were immediately prepared from animals that were treated with increasing bolus doses of abciximab. These were important comparisons since they established that the *in vitro* concentration-effect relationships applied directly to the *in vivo* setting where clearance and other factors might have influenced the binding of abciximab to platelets. Based on a number of studies of this type, the 80% receptor blockade of GPIIb/IIIa was adopted as the pharmacological target for abciximab (and later, for other GPIIb/IIIa antagonists as well).

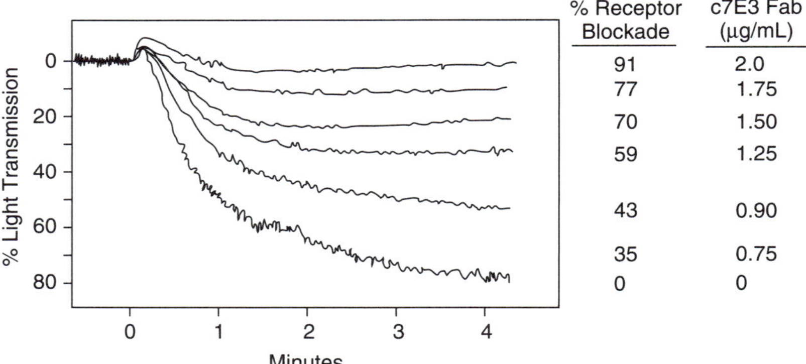

FIGURE 8.2 The dose-dependent inhibition of platelet aggregation by abciximab correlated with the blockade of platelet GPIIb/IIIa receptors. The traces represent platelet aggregation responses to 5 μM ADP. Separate aggregation responses were performed at increasing concentrations of abciximab (μg/mL) and the receptor blockade was independently assessed on parallel samples using the radiometric method described by Coller (1985). From *Adhesion Receptors as Therapeutic Targets*. Reproduced by permission of Routledge/Taylor & Francis Group, LLC.

V. ANIMAL STUDIES

A. Bolus Dose Studies

In vivo, platelet thrombosis is typically an arterial phenomenon occurring when platelets encounter vascular damage (e.g., in advanced atherosclerosis). The first human clinical indication for abciximab was anticipated to be the treatment of acute, abrupt thrombotic vessel closure that was observed to accompany some balloon angioplasty procedures in coronary arteries. The adoption of rigorous preclinical challenge models was considered a key step in the validation of the efficacy and safety of GPIIb/IIIa antagonism. In various laboratories, animal models had been developed that mirrored the acute human clinical setting and many of these models were incorporated into a systematic preclinical program of efficacy testing. Intravenous doses were assigned, depending on the animal model, to achieve the 80% receptor blockade and complete (or nearly complete) inhibition of platelet aggregation. The preclinical models have been reviewed in detail previously (Jordan et al., 1996) and a listing of the characteristics of the model and major findings in them are provided in Table 8.1. The consistent success of a single intravenous bolus dose of the antibody in a spectrum of acute models was an encouraging and necessary step in showing that a single treatment with abciximab (immediately before injury) could result in immediate, and in some cases prolonged, protection against intravascular thrombosis.

B. Extended Duration of Therapy

Anti-GPIIb/IIIa therapy for prevention of acute thrombosis was a logical first step in antithrombotic therapy with this class of drug; a therapy that could also be envisioned for preventive treatment in even more complex thrombotic settings, such as unstable angina and myocardial infarction. Even though the vessel injury resulting from balloon angioplasty was predictable with regard to its timing of onset, the requirements for duration and degree of sustained intensity of anti-GPIIb/IIIa therapy were uncertain. On this point, animal data and available models were limited. Lacking specific knowledge, it seemed reasonable to maintain the 80% receptor blockade until the vessel was sufficiently healed and the dangers of platelet recruitment and thrombus formation subsided. It should be noted that the human Glanzmann thrombasthenia condition (in which lifelong deficiency of platelet GPIIb/IIIa leads to episodic and highly variable degrees of hemorrhage risk) suggested that a limited duration of high-grade GPIIb/IIIa inhibition would be tolerated.

As a Fab fragment, abciximab could be expected to clear rapidly from the circulation by filtration in the kidneys (Yokota et al., 1993). To address the challenge of maintaining 80% receptor blockade when the free antibody Fab fragment was rapidly cleared and new platelets were continuously entering circulation from the bone marrow, a continuous infusion regimen was expected to be needed. Similar issues were to be encountered during the development of small-molecule parenteral antagonists for which clearance from the circulation was an even more rapid phenomenon. Toxicology studies with abciximab in monkeys established that bolus plus infusion regimens of as long as 96 hours could effectively prolong the intense initial platelet inhibition. An unexpected finding even in short-term treatments was that abciximab slowly re-equilibrated among circulating platelets and this redistribution continued

TABLE 8.1 The 7E3 Antibody in Preclinical Thrombosis Models

Preclinical results	Animal Model	Version of Antibody 7E3	Bolus Dose, mg/kg	Reference
Prevented cyclic flow reductions following vascular injury to the coronary artery	Dog	$F(ab')_2$	0.8	Coller et al., 1986
Prevented cyclic flow reductions following vascular injury in carotid arteries	Monkey	$F(ab')_2$	0.2	Coller et al., 1989
Blocked injury-induced coronary artery thrombosis and occlusion following electrolytic injury for up to 6 days	Dog	$F(ab')_2$	0.8	Bates et al., 1992
Prevented carotid artery thrombosis and occlusion following electrolytic injury	Monkey	Fab (abciximab)	0.20 and 0.25	Rote et al., 1994
Prevented coronary artery thrombosis and occlusion following balloon angioplasty	Dog	$F(ab')_2$	0.8	Bates et al., 1991
Recanalized thrombosed arteries in combination with tPA	Dog	$F(ab')_2$	0.8	Gold et al., 1988
Prevented coronary artery reocclusion after reperfusion with tPA	Dog	$F(ab')_2$	0.8	Yasuda et al., 1988
Lysed tPA-resistant thrombus in everted coronary artery in combination with tPA	Dog	$F(ab')_2$	0.8	Yasuda et al., 1990
Accelerated thrombolysis by rt-PA in coronary arterial models resistant to aspirin, prostacyclin, or thromboxane A_2	Dog	$F(ab')_2$	0.8	Fitzgerald et al., 1991
Prevented the reocclusion of recanalized vessels without concomitant heparin	Dog	$F(ab')_2$	0.8	Mickelson et al., 1990
Enhanced and sustained coronary thrombolysis by rt-PA	Baboon	Fab (abciximab)	0.45	Kohmura et al., 1993

long after the cessation of treatment (Mascelli et al., 1998). The continuous re-equilibration of abciximab (visualized by flow cytometry) yielded a profile in which circulating platelets all possessed a comparable, but slowly declining, occupancy of external membrane GPIIb/IIIa by abciximab. This resulted in a pharmacodynamic profile of tapered recovery of platelet function, a phenomenon that was found to also apply to human pharmacology (Mascelli et al., 1998). Although partial recovery of platelet aggregation was seen within 4 to 6 hours after cessation of treatment, full recovery was generally not restored for up to 24 to 48 hours. This additional period of slow progressive recovery avoids an abrupt restoration of platelet function and may have contributed to the effectiveness of the relatively short 12-hour infusion ultimately adopted for acute PCI indications.

VI. PLATELET PHARMACODYNAMIC PHENOMENA RECOGNIZED LATER

A. Internalization Within Platelets

An additional and unforeseen targeting consideration came with the realization that abciximab gained access to an internal alpha granule pool of GPIIb/IIIa receptors. This latter pool of receptors may comprise an additional 50% of those on the external membrane (i.e., ~40, 000 receptors) (Law and Phillips, 1999). Upon platelet activation, the alpha granule populations of GPIIb/IIIa receptors become expressed on the external membrane, where they may participate in ligand binding and aggregation (Kleiman et al., 1995; Woods et al., 1986). Based on electron microscopic analysis of platelet sections from treated patients, abciximab was localized by immunogold labeling within the cannilicular system and in some, but not all, of the internal alpha granules (Nurden et al., 1999). Interestingly, the translocation from the surface to those internal granules was rapid and could be visualized at the earliest observations (3 hours). The results suggested that a secretion-dependent residual aggregation could be supported by unbound internal receptors and the ability of drugs to block all of the pools of GPIIb/IIIa was a potential consideration for drugs that target this receptor. To attempt to integrate this finding with the earlier relationships among dose duration, recovery, distribution, and *in vivo* benefit underscored the complexity of the pharmacokinetic and pharmacodynamic parameters for GPIIb/IIIa antagonists.

Those points were especially evident in the attempts to develop orally active inhibitors of platelet GPIIb/IIIa (a topic that is covered elsewhere in this volume). This class of drug appeared to promise long-acting therapies for prevention of cardiovascular thrombotic diseases. In a number of late-stage clinical trials, however, there was disappointing efficacy and an unexpected rise in the incidence of adverse events and mortality (Quinn et al., 2002). The reason for these troubling results remains uncertain, and there was some speculation that the cycles of high and low inhibition associated with twice-daily oral administration might be involved. Even 48-hour administration of an intravenous GPIIb/IIIa antagonist was found to be associated with increased adverse thrombotic endpoints in certain patient subgroups (The GUSTO IV-ACS Investigators, 2001). These perplexing clinical results prompted numerous attempts to identify potential mechanisms and conditions in which GPIIb/IIIa antagonists could cause paradoxical thrombosis and activation of platelets.

B. Potential Activating Actions of GPIIb/IIIa Antagonists

Among the proposed mechanisms for platelet pro-aggregatory activity by GPIIb/IIIa antagonists is a report that fibrinogen binding to platelets may be induced under highly specific conditions of low-level receptor blockade (Peter et al., 1998). These conclusions have been disputed and the results alternately attributed to artifactual thrombin generation (Frelinger et al., 2001). In other studies, GPIIb/IIIa antagonists were found to potentiate platelet alpha granule release even at doses/concentrations at which fibrinogen binding to the receptor was blocked (Schneider et al., 2000). It remains uncertain if these phenomena are directly relevant to the clinical actions of abciximab. They serve as examples of the continued growth in the understanding of the complexity of platelet biology. Many of these parameters were not known or appreciated during the initial stages of GPIIb/IIIa target validation. Future generations of GPIIb/IIIa receptor antagonists will certainly need to address them in depth. An additional phenomenon has gained particular recent attention: the potentiation of expression of the platelet granule membrane component, CD40 ligand (CD40L).

CD40L is a molecule normally associated with immune signaling when T cell CD40L interacts with its receptor CD40 on B cells to promote immune responses. CD40L and CD40 are both present in and on other cells. Ironically, CD40L is also present within platelets in the alpha granule, where it is normally hidden from the circulation. Upon platelet activation, alpha granules and their contents, including CD40L, become exposed on the platelet surface. The CD40L on platelets was shown to have the potential to promote inflammatory reactions of endothelial cells by engaging the CD40 receptor on those cells (Henn et al., 1998). Normally, the blockade of GPIIb/IIIa at optimum doses/concentrations of antagonists such as abciximab blocks platelet aggregation, alpha granule release, and expression of granule contents. Recent findings, however, suggest that at suboptimal inhibitory levels (<20%), at which a fraction of receptors are blocked, granule release of CD40L may actually be promoted (Nannizzi-Alaimo, et al., 2003). These findings suggest that anti-GPIIb/IIIa agents could to be pro-inflammatory under certain and perhaps briefly occurring conditions (e.g., during recovery from higher levels of receptor blockade). This hypothesis, while of unproven clinical association, may be pertinent to the oral GPIIb/IIIa compounds where chronic dosing was typically aimed at a lower level of platelet inhibition and for which recurrent inhibitory cycles were normal. In any case, these results further highlight the complexity and diversity of interconnected functions of the GPIIb/IIIa receptor—a degree of subtlety that was clearly not appreciated at the outset of the validation of the GPIIb/IIIa target.

GPIIb/IIIa has been termed the responsive integrin in recognition of its ability to transmit outside-in as well as inside-out signals during platelet activation (Hato et al., 2002; Hynes, 2002). These functional attributes were recognized well in advance of detailed knowledge of the receptor structure (Du et al., 1991). With the recent delineation of the crystal structures of the extracellular domains of GPIIb/IIIa and avB3, it has now become more clear how the conformation of the heterodimeric receptor responds to signals and to the binding of ligands (Xiao et al., 2004; Xiong et al., 2002). Indeed, even the binding of certain small-molecule mimetics of fibrinogen, for example, RGD-containing peptides, can initiate a substantial conformational alteration of the receptor. It was known for some time that antibodies could be elicited to altered or ligand-occupied forms of the receptor and to what were termed ligand-induced binding sites (LIBS) (Du et al., 1991, Kouns et al., 1990). In fact, the occupancy of the GPIIb/IIIa receptor by certain small-molecule antagonists could only be monitored through the use of monoclonal LIBS-directed reagents (Jennings et al., 2000). The ability to generate such reagents caused concerns that human patients might develop similar immune reactivity after exposure to GPIIb/IIIa antagonists. Indeed, LIBS-like human immune responses have been reported and have been linked to acute thrombocytopenia in rare patients (Bougie et al., 2003).

C. Immunogenicity

Although immunogenicity may not usually be considered a pharmacodynamic topic, it is included here because of the potential effects such responses could have on *in vivo* platelet function. Unfortunately, animal studies were of limited utility for predicting host immune responses in humans and were not a major factor in the validation of the target (although in the case of abciximab, the specific decision regarding the human/murine chimeric structure was done with the expectation of minimizing human immunogenicity).

Human immune responses to abciximab were shown to develop primarily to the antibody itself in approximately 6% of patients on first exposure (Knight et al, 1995). Adverse consequences have generally been rare—in part because of the acute nature of the therapy and the fact that host antibodies tend to appear only 7 days or later after exposure. Delayed immune responses have been associated with thrombocytopenia at 7 or more days after first exposure for abciximab (Curtis et al., 2004; Nurden et al., 2004), and tirofiban (Bosco et al., 2005). Repeated treatments with abciximab have been observed to cause higher rates of immune responses. The incidence of thrombocytopenia is increased when a positive immune titer is present at the time of re-administration (Dery et al., 2004). Thus, GPIIb/IIIa antagonists must be added to the list of examples in the platelet immunology literature in which drug-induced immune responses are capable of causing platelet clearance (Aster, 2002). Clearly, the avoidance of immune responses of any type is a goal for future drug development against targets on vascular cells unless the clearance or removal of that cell type is advantageous. Avoidance of host LIBS-like immune responses will be a more difficult, but not impossible, task (Kouns et al., 1992).

VII. CROSS REACTIVITY WITH OTHER INTEGRINS

A. The $\alpha_v\beta_3$ Receptor

Abciximab turned out not to be specific for the platelet GPIIb/IIIa receptor. Work in several laboratories established that the antibody bound to two other members of the extended integrin family of cell-surface receptors. One of these is the $\alpha_v\beta_3$ vitronectin receptor present on many cell types, including endothelium, smooth muscle cells, and leukocytes (Byzova et al., 1998). The $\alpha_v\beta_3$ receptor shares the same β subunit with GPIIb/IIIa (also called $\alpha_{IIb}\beta_3$) and the recent crystal structural information available on both receptors, makes it evident that abciximab binds to epitopes largely defined by the β_3 subunit (Artoni et al., 2004). There was not a clear prospective advantage for the blockade of $\alpha_v\beta_3$ in the treatment of arterial thrombosis and most of the alternative small-molecule antagonists were developed with a strict specificity for GPIIb/IIIa. Nevertheless, evidence was gained from the earliest abciximab trial, the EPIC study, that patients required fewer follow-up revascularizations—an observation that could be consistent with an inhibition of $\alpha_v\beta_3$-mediated cell migration/proliferation and decreased restenosis in the treated vessels. Although not universally found in later trials in all subjects, in certain patient subsets (e.g., diabetics), an inhibition of restenosis has been observed (Lincoff, 1999) and was prospectively confirmed in a trial entirely conducted in diabetic subjects (Mehilli et al., 2004).

B. The Mac-1 Receptor

Abciximab was also shown to bind to a second integrin, the Mac-1 receptor (otherwise termed $\alpha_M\beta_2$). The Mac-1 receptor is present on the surfaces of many leukocytes, including activated immune effector cells (e.g., monocytes and macrophages). The binding of abciximab to this transiently expressed receptor is of lower affinity than to the β_3 integrins (Altieri and Edgington, 1988). Potential clinical benefit was theorized as an inhibition of inflammatory cell function in the area of atherosclerotic disease and localized vessel damage. Indeed, measurement of a set of inflammatory markers that typically rise following angioplasty (including tissue necrosis factor [TNF], interleukin-6 [IL-6] and C-reactive protein [CRP]) showed a suppressed increased with abciximab treatment (Lincoff, 2001). It is nevertheless unclear if the anti-inflammatory effect of abciximab derives from a direct interaction with Mac-1 on inflammatory cells or is due to secondary mechanisms due to platelet GPIIb/IIIa blockade leading to an inhibition of adhesion of cells such as monocytes to platelets and/or inhibition of the release of platelet granule components such as CD40L (Nannizzi-Alaimo, 2003).

The clinical advantages or possible disadvantages of $\alpha_v\beta_3$ and Mac-1 inhibition by abciximab currently remain obscured by the complex biology and pathology of severe coronary artery disease. Although a detailed understanding of the target specificity of any new agent is critical to its validation path, in the case of abciximab there were new findings at multiple points along this path. New data continue to provide intriguing insights into the pathology of acute coronary artery disease and the diverse roles of different integrin receptors.

VIII. INTEGRATION OF CLINICAL PHARMACOLOGY AND PRECLINICAL STUDIES

The human pharmacodynamic and dose response studies of abciximab were conducted in normal healthy subjects and in patients with stable coronary disease in trials in Europe, the United States, and Japan. Some studies included the concomitant medications heparin and aspirin to gain information on any potential interactions with these two common agents. By extension of the preclinical programs, the Phase I studies were designed to identify a bolus dose that would achieve the 80% receptor blockade target and profoundly inhibit *ex vivo* platelet aggregation. These goals were met at a weight-adjusted bolus dose of 0.25 mg per kilogram of body weight. Because the effects of a single bolus dose began to wane within 4 to 6 hours, the next goal was to define the strategy to sustain the inhibition for at least 12 hours. This goal was achieved in early clinical trials with a non–weight adjusted infusion of 10 μg/ min and was later refined to a weight-adjusted regimen of 0.125 μg/kg/ min for 12 hours (to a maximum of 10 μg/ min). An additional pharmacodynamic measurement was applied in some early Phase I studies. This additional test was the template bleeding time method, a method purportedly unique in its ability to assess *in vivo* platelet function and thus, perhaps, the effects of abciximab treatment. The bleeding time measurement (at a defined cut in the skin) proved highly variable and did not yield consistent results among different clinical sites. As a result, the inhibition of platelet aggregation was established as the primary methodology for quantitative assessment of abciximab activity.

The Phase I studies provided the needed information to define the dose regimen for sustained anti-platelet activity for at least 12 hours. The early clinical programs in percutaneous coronary arterial intervention (primarily balloon angioplasty) built upon this foundation.

IX. CLINICAL STUDIES

A. Early Studies in Man

Many of the individual clinical trials of abciximab in PCI have been systematically reviewed (Lincoff, 1999; van den Brand and Simoons, 1999). In general, these trials were conducted in a setting of acute interventional cardiology in balloon angioplasty and, as time progressed, in an evolving setting of more aggressive balloon expansion and stent placement. The widespread use of high-dose heparin together with abciximab was gradually supplanted by low-dose heparin regimens as data accumulated for its effectiveness and decreased hemorrhagic risk. In addition, the thienopyridine platelet ADP receptor antagonists, ticlopidine and clopidogrel, were not yet available. In the first years, few patients received the newly available metallic stents—sheaths placed within areas of arterial narrowing following balloon expansion. Drug-coated stents were not then available and did not become so until well after the introduction of abciximab. Thus, the 12 years since abciximab was first approved have seen substantial advances in the interventional treatment of coronary atherosclerotic lesions. Despite these improvements in practice and pharmacology, abciximab has continued to provide substantial benefit for the prevention of ischemic complications in high-risk PCI procedures. A brief overview of the validation of abciximab and the GPIIb/IIIa target in selected clinical trials follows. The open-label Phase I and Phase II trials supported both the antithrombotic potential and safety of abciximab. For example, the cyclic flow variations resulting from transient formation of platelet thrombi at sites of vascular injury were shown to be effectively blocked by abciximab in patients undergoing balloon angioplasty (Anderson et al., 1994). Similar positive results were gained in patients with severe (refractory to other therapy) unstable angina in which abciximab reduced the occurrence of recurrent ischemic events during and after angioplasty (Simoons et al., 1994). These early indicators of improved clinical outcome supported the feasibility of GPIIb/IIIa blockade for controlling acute coronary thrombotic events and led to the larger Phase III, randomized and placebo-controlled efficacy trials.

B. Phase III Efficay Trials

A sequence of pivotal Phase III trials in acute coronary interventions was then conducted over a span of several years and eventually encompassed more than 8000 patients. These trials included EPIC, CAPTURE, EPILOG and EPISTENT (The EPIC Investigators, 1994, Lincoff, 1999; van den Brand and Simoons, 1999). The EPIC trial in high-risk patients, (Evaluation of c7E3 Fab for the Prevention of Ischemic Complications), compared bolus abciximab only (0.25 mg/kg) to the same bolus plus 12-hour infusion regimen (10 μg/min) and a placebo group. The treatment was initiated immediately prior to the coronary intervention and benefit was assessed as a combined 30-day clinical endpoint of 1) the need for repeat vascular intervention, 2) myocardial infarction, and 3) death. The results showed the bolus plus infusion regimen to be the most effective with a 35% reduction of the combined clinical endpoints from

12.8% of patients in the placebo group to 8.3% in the abciximab bolus plus infusion arm. Follow-up of these patients indicated that the clinical benefits were maintained at 6 months, at 1 year, and longer, and established that the prevention of acute events could have long-lasting clinical benefits. The clinical benefits in EPIC, however, were gained at the risk of excess bleeding and this issue was addressed by lowering the heparin anticoagulant dose in the subsequent EPILOG trial.

The EPILOG trail (Evaluation in PTCA to Improve Long-term Outcome with abciximab GPIIb/IIIa blockade) investigated abciximab in a broad patient population with different risk levels and undergoing various interventional procedures. EPILOG addressed the hemorrhagic problem by limiting heparin use to 70 U/Kg with a targeted, activated clotting time (ACT) of >200 s. This was a significant reduction compared to the standard high-dose regimens of the period. Another change was the adoption of a weight-adjusted 12-hour abciximab infusion of 0.125 μg/kg/ min. While planned for 4800 patients, this trial was terminated at 2792 enrolled patients after an unexpectedly strong clinical benefit was seen at the first interim analysis. The combined incidence of death, MI or urgent revascularization at 30 days was 11.7% in the placebo group but was reduced in the abciximab groups to 5.2% (with low-dose heparin) or 5.4% (with high-dose heparin). The early suppression of ischemic events was maintained at 6 months and 1 year, again confirming the long-term benefits of short-term anti-GPIIb/IIIa treatment.

The EPISTENT trial addressed the adoption of coronary stenting as the increasingly dominant mode of coronary intervention. EPISTENT (Evaluation of Platelet Inhibition in STENTing) enrolled 2399 patients and compared the following three groups; 1) stent plus placebo, 2) stent plus abciximab (0.25 mg/kg bolus plus 0.125 μg/kg/ min for 12 hours) and, 3) balloon angioplasty plus the same abciximab regimen. The standard 30-day composite endpoint was reached in 10.8% of placebo patients, 6.9% of abciximab patients undergoing only angioplasty and 5.3% of abciximab patients receiving a stent. As with the previous trials, the statistical significance was highly positive ($p < 0.001$ for the abciximab plus stent group compared to placebo). Thus, these three trials provided compelling evidence that blockade of platelet GPIIb/IIIa by abciximab at the time of coronary intervention yielded significant and sustained clinical benefits. The Kaplan-Meier analyses of the primary endpoints of EPIC, EPILOG and EPISTENT for the key abciximab arms compared to placebo are depicted in Figure 8.3.

Another Phase III trial was CAPTURE; a European based study that was designed for 1400 patients. CAPTURE (Chimeric 7E3 AntiPlatelet Therapy in Unstable Angina REfractory to Standard Treatment) studied patients with refractory unstable angina who were scheduled to undergo balloon angioplasty. These patients received placebo or abciximab 18 to 24 hours before the procedure and it was continued for 1-hour post procedure. The clinical endpoints were analogous to those in the EPIC/EPILOG/EPISTENT series described above. At an interim analysis (1050 patients), the efficacy was found to have exceeded a predetermined positive statistical stopping threshold and the trial was terminated with 1265 patients treated. The composite endpoint was reduced from 15.5% in the placebo group to 10.8% in the abciximab group. Thus, the prepeated Phase III trail results in acute interventional settings validated GPIIIb/IIIa as a therapeutic target and abciximab as a valuable therapeutic antibody.

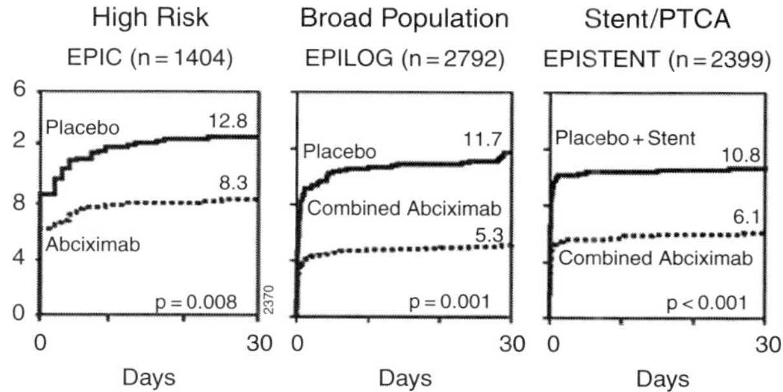

FIGURE 8.3 **Kaplan-Meier event rates for death, myocardial infarction, or urgent intervention through 30 days.** The values associated with each curve denote the total percent of patients in the respective treatment groups with a clinical event. The reduction of acute events in these three separate trials was also found to be extended to at least 3 years in follow-up analyses.

X. CONCLUSION

The introduction of platelet GPIIb/IIIa receptor blockade has been termed one of the most significant advances in the practice of interventional cardiology (Lincoff, 1999). During the period that abciximab and other GPIIb/IIIa antagonists have been available, many additional improvements in interventional devices and pharmacological agents have come about. Despite these many advances, the inhibition of platelet GPIIb/IIIa remains an important component in the treatment of acute coronary arterial disease. There is also reason to believe that similar benefits could apply to other and more chronic thrombotic settings. The medical treatment of unstable angina, myocardial infarction, and stroke are active areas of continued investigation. In addition, inhibition of GPIIb/IIIa has been considered for the treatment of peripheral arterial disease, Kawasaki disease, heparin-induced thrombocytopenia, sickle cell disease, tumor angiogenesis, and other conditions associated with GPIIb/IIIa- or $\alpha_v\beta_3$–mediated conditions (Coller, 1995, 2001). The validation of GPIIb/IIIa for new and different diseases may prompt new avenues of basic research and development similar to the extensive path that led to the treatment of coronary arterial disease. The growing utility of this treatment has its roots in the basic science and pharmacology studies that led to the identification and validation of the GPIIb/IIIa receptor as a drug target.

RECOMMENDED RESOURCES

http://www.rxlist.com/cgi/generic2/abcixim.htm
http://www.drugs.com/reopro.html

REFERENCES

Agah, R., E. F. Plow, and E. J. Topol (2002). GPIIb-IIIa antagonists. *In* "Platelets" (A. D. Michelson, ed). pp. 769–785. Academic Press, New York.

Altieri, D. C. and T. W. Edgington (1988). A monoclonal antibody reacting with distinct adhesion molecules defines a transition in the functional state of the receptor CD11b/CD18 (Mac-1)[1]. *J. Immunol.* 8: 2656–660.

Anderson, H. V., R. L. Krikeeide, et al. (1994). Cyclic flow variations after coronary angioplasty in humans: Clinical and angiographic characteristics and elimination with 7E3 monoclonal antiplatelet antibody. *JACC.* 23: 1031–1037.

Artoni, A., H. Li, et al. (2004). Integrin β3 regions controlling binding of murine mAb 7E3: Implications for the mechanism of integrin αIIbβ3 activation. *PNAS.* 101: 13,114–13, 120.

Aster, R. H. (2002). Drug-induced thrombocytopenia. *In* "Platelets" (A.D. Michelson, ed.), pp. 593–606. Academic Press. New York.

Barnathan, E. S., J. S. Schwartz, et al. (1987). Aspirin and dipyridamole in the prevention of acute coronary thrombosis complicating coronary angioplasty. *Circulation.* 76: 125–134.

Bates, E. R., M. J. McGillem, et al. (1991). A monoclonal antibody against the platelet glycoprotein IIb/IIIa receptor complex prevents platelet aggregation and thrombosis in a canine model of coronary antioplasty. *Circulation.* 84: 2463–2469.

Bates, E. R., D. G. Walsh, et al. (1992). Sustained inhibition of the vessel wall-platelet interaction after deep coronary artery injury by temporary inhibition of the platelet glycoprotein IIb/IIIa receptor. *Coron. Artery. Dis.* 3: 67–76.

Bennett, J. S. and G. Vilaire (1979). Exposure of platelet fibrinogen receptors by ADP and epinephrine. *J. Clin. Invest.* 64: 1393–1401.

Bosco, A., G. Kidson-Gerber, and S. Dunkley (2005). Delayed tirofiban-induced thromobcytopenia: Two case reports. *J. Thromb. Haemost.* 2: 1109–1110.

Bougie, D. W., P. R. Wilker (2003). Acute thrombocytopenia after treatment with dirofiban or eptifibatide is associated with antibodies specific for ligand-occupied GPIIb/IIIa. *Blood.* 15: 2071–2076.

Byzova, T. V., R. Rabbani (1998). Role of integrin $\alpha_v\beta_3$ in vascular biology. *Thromb. Haemost.* 80: 726–34.

Coller, B. S., E. I. Peerschke, et al. (1983). A murine monoclonal antibody that completely blocks the binding of fibrinogen to platelets produces a thrombasthenic-like state in normal platelets and binds to glycoproteins IIb and/or IIIa. *J. Clin. Invest.* 72: 325–338.

Coller, B. S. (1985). A new murine monoclonal antibody reports an activation-dependent change in the conformation and/or microenvironment of the plaetlet glycoprotein IIb/IIIa complex. *J. Clin. Invest.* 76: 101–108.

Coller, B. S., J. D. Folts, et al. (1986). Antithrombotic effet of a monoclonal antibody to the platelet glycoprotein IIb/IIIa receptor in an experimental animal model. *Blood* 68: 783–786.

Coller, B. S., J. D. Folts, et al. (1989). Abolition of in vivo platelet thrombus formation in primates with monoclonal antibodies to the platelet GPIIb/IIIa receptor: Correlation with bleeding time, platelet aggregation, and blockade of GPIIb/IIIa receptors. *Circulation* 80: 1766–1774.

Coller, B. S. (1995). Blockade of platelet GPIIb/IIIa receptors as an antithrombotic strategy. *Circulation* 92: 2373–2380.

Coller, B. S. (1999). Glycoprotein IIb/IIIa antagonists: development of abciximab and pharmacology of select agents. *In* "Contempoary Cardiology: Platelet Glycoprotein IIb/IIIa Inhibitors in Cardiovascular Disease" (A. M. Lincoff and E. J. Topal, eds.), pp 67–89. Humana Press Inc., Totowa, NJ.

Coller, B. S. (2001). Anti-GPIIb/IIIa Drugs: Current strategies and future directions. *Throm. Haemost.* 86: 427–443.

Curtis, B. R., J. Swyers, et al. (2002). Thrombocytopenia after second exposure to abciximab is caused by antibodies that recognize abciximab-coated platelets. *Blood* 99: 2054–2059.

Dery, J. P., G. A. Braden, et al. (2004). Final results of the ReoPro readministration registry. *Am. J. Cardiol.* 93: 979–984.

Du, X., E. F. Plow, et al. (1991). Ligands "activate" integrin $\alpha_{IIb}\beta_3$ (platelet GPIIb-IIIa). *Cell* 65: 409–416.

Fitzgerald, D. J., M. Hanson, and G. A. FitzGerald (1991). Systemic lysis protects against the effects of platelet activation during coronary thrombolysis. *J. Clin. Invest.* 88: 1589–1595.

Frelinger, A. L., M. I. Furman, et al. (2001). Dissociation of glycoprotein IIb/IIIa antagonists from platelets does not result in fibrinogen binding or platelet aggregation. *Circulation* 104: 1374–1379.

Gold, H. K., B. S. Coller, et al. (1988). Rapid and sustained coronary artery recanalization with combined bolus injection of recombinant tissue-type plasminogen activator and monoclonal antiplatelet GPIIb/IIIa antibody in a canine preparation. *Circulation* 77: 670–677.

Hato, T., M. H. Ginsberg, and S. J. Shattil (2002). Integrin $\alpha_{IIb}\beta_3$. *In* "Platelets" (A.D. Michelson, ed.), 105–116. Academic Press, Boston.

Henn, V., J. R. Slupsky, et al. (1998). CD40 ligand on activated platelets triggers an inflammatory reaction of endothelial cells. *Nature* 391: 591.

Hynes, R. O. (2002). Integrins: Bidirectional, allosteric signaling machines. *Cell* 110: 673–667.

Jennings, L. K., J. H. Haga, and S. M. Slack (2000). Differential expression of a ligand induced binding site (LIBS) by GPIIb-IIIa ligand recognition peptides and parenteral antagonists. *Thromb. Haemost.* 84: 1095–1102.

Jordan, R. E., M. A. Mascelli, et al. (1997). Pharmacology and clinical development of abciximab (c7E3 Fab, ReoPro): A monoclonal antibody inhibitor of GPIIb/IIIa and $\alpha_v\beta_3$. *In* "New Therapeutic Agents in Thrombosis and Thrombolysis" (A.A. Sasahara, J. Loscalzo, eds.), 291–333. Marcel Dekker, Inc., New York.

Jordan, R. E., C. L. Wagner, et al. (1996). Preclinical development of c7E3 Fab: A mouse/human chimeric monclonal antibody fragment that inhibits platelet function by blockade of GPIIb/IIIa receptors with observations on the immunogenicity of c7E3 Fab in humans. *In* "Adhesion Receptors as Therapeutic Targets" (M.A. Horton, ed.), 281–305. CRC Press, New York.

Kleiman, N. S., A. E. Raizner, et al. (1995). Differential inhibition of platelet aggregation induced by adenosine diphosphate or a thrombin receptor-activating peptide in patients treated with bolus chimeric 7E3 Fab: Implications for inhibition of the internal pool of GPIIb/IIIa receptors. *JACC.* 26: 1665–1671.

Knight, D. M., C. Wagner, et al. (1995). The immunogenicity of the 7E3 murine monoclonal Fab antibody fragment variable region is dramatically reduced in humans by substitution of human for murine constant regions. *Mol. Immunol.* 32: 1271–1281.

Kohmura, C., H. K. Gold, et al. (1993). A chimeric murine/human antibody fab fragment directed against the platelet GPIIb/IIIa receptor enhances and sustains arterial thrombolysis with recombinant tissue-type plasminogen activator in baboons. *Arterioscler. Thromb.* 13: 1837–1842.

Kouns, W. C., C. D. Wall, et al. (1990). A conformation-dependent epitope of human platelet glycoprotein IIIa. *J. Biol. Chem.* 33: 20,594–20, 601.

Kouns, W., T. Wller, et al. (1992). Identification of a peptidomimetic inhibitor with minimal effects on the conformation of GPIIb-IIIa. *Blood* Suppl. 80: 165a.

Law, D. A. and D. R. Phillips (1999). Glycoprotein IIb-IIIa in platelet aggregation and acute arterial thrombosis. *In* "Contempoary Cardiology: Platelet Glycoprotein IIb/IIIa Inhibitors in Cardiovascular Disease" (A. M. Lincoff and E. J. Topal, eds.), 35–66. Humana Press Inc., Totowa, NJ.

Lincoff, A. M. (1999). Abciximab during percutaneous coronary intervention: The EPIC, EPILOG, and EPISTENT trials. *In* "Contemporary Cardiology: Platelet Glycoprotein IIb/IIIa Inhibitors in Cardiovascular Disease" (A. M. Lincoff, E. J. Topol, eds.), 93–113. Humana Press Inc., Totowa.

Lincoff, A. M., R. M. Califf, et al., for the Evaluation of Platelet IIb/IIIa Inhibition in Stenting Investigators. (1999). Complementary clinical benefits of coronary-artery stenting and blockade of platelet glycoprotein IIb/IIIa receptors. *N. Engl. J. Med.* 341: 319–327.

Lincoff, A. M., D. J. Kereiakes, et al. (2001). Abciximab suppresses the rise in levels of circulating inflammatory markers after percutaneous coronary revascularization. *Circulation* 104: 163–167.

Mascelli, M. A., E. T. Lance, et al. (1998). Pharmacodynamic profile of short-term abciximab treatment demonstrates prolonged platelet inhibition with gradual recovery from GP IIb/IIIa receptor blockade. *Circulation* 97: 1680–1688.

Mehilli, J., A. Kastrati, et al. (2004). Randomised clinical trial of abciximab in diabetic patients undergoing elective percutaneous coronary interventions after treatment with a high loading dose of clopidogrel. *Circulation* 110: 24, 3627–3635.

Mickelson, J. K., P. J. Simpson, et al. (1990). Antiplatelet antibody [7E3 $F(ab')_2$] prevents rethrombosis after recombinant tissue-type plasminogen activator-induced coronary artery thrombolysis in a canine model. *Circulation* 81: 617–627.

Nannizzi-Alaimo, L., et al. (2003). Inhibitory effects of glycoprotein IIb/IIIa antagonists and aspirin on the release of soluble CD40 ligand during platelet stimulation. *Circulation* 107: 1123–1128.

Newman, P. J., R. W. Allen, et al. (1985). Quantitation of membrane glycoprotein IIIa and intact human platelets using the monoclonal antibody, AP-3. *Blood* 65: 227–232.

Nurden, A. T. (2005). Qualitative disorders of platelets and megakaryocytes. *J. Thromb. Haemost.* 3: 1773–1782.

Nurden, P., C. Poujol, et al. (1999). Labeling of the internal pool of GP IIb-IIIa in platelets by c7E3 Fab fragments (abciximab): Flow and endocytic mechanisms contribute to the transport. *Blood* 93: 1622–1633.

Nurden, P., G. Clofent-Sanchez, et al. (2004). Delayed immunologic thrombocytopenia induced by abciximab. *Thromb. Haemost.* 92: 820–828.

Peter, K., M. Schwarz, et al. (1998). Induction of fibrinogen binding and platelet aggregation as a potential intrinsic property of various glycoprotein IIb/IIIa ($\alpha_{IIb}\beta_3$) inhibitors. *Blood* 92: 3240–3249.

Phillips, D. R., W. Teng, et al. (1997). Effect of Ca^{2+} on GP IIb-IIIa interactions with Integrilin. *Circulation* 96: 1488–1494.

Quinn, M. J., E. F. Plow, and E. J. Topol (2002). Platelet glycoprotein IIb/IIIa inhibitors: Recognition of a two-edged sword? *Circulation* 106: 379–385.

Rote, W. E., M. A. Nedelman, et al. (1994). Chimeric 7E3 prevents carotid artery thrombosis in cynomolgus monkeys. *Stroke* 25: 1223–1233.

Sassoli, P. M., E. L. Emmell, et al. (2001). 7E3 $F(ab')_2$, an effective antagonist of rat αIIbβ3 and $\alpha_v\beta_3$, blocks *in vivo* thrombus formation and *in vitro* angiogenesis. *Thromb. Haemost.* 85: 896–902.

Schneider, D. J., D. J. Taatjes, and B. E. Sobel (2000). Paradoxical inhibition of fibrinogen binding and potentiation of α-granule release by specific types of inhibitors of glycoprotein IIb-IIIa. *Cardiovasc. Res.* 42: 437–446.

Simoons, M. L., M. J. de Boer, et al. (1994). Randomized trial of GPIIb/IIIa platelet receptor blocker in refractory unstable angina. *Circulation* 89: 596–603.

Smith, J. W., S. R. Steinhubl, et al. (1999). Rapid platelet-function assay: An automated and quantitative cartridge-based method. *Circulation* 99: 620–625.

The GUSTO Investigators (2001). Reperfusion therapy for acute myocardial infarction with fibrinolytic therapy or combination reduced fibrinolytic therapy and platelet glycoprotein Iib/IIIa inhibition: The GUSTO V randomised trial. *Lancet* 357: 1905–1914.

The EPIC Investigators (1994). Use of a monoclonal antibody directed against the platelet glycoprotein IIb/IIIa receptor in high-risk coronary angioplasty. *NEJM*. 330: 956–961.

van den Brand, M. J. B. M. and M. L. Simoons (1999). *In* "Contemporary Caridiology: Platelet Glycoprotein IIb/IIIa Inhibitors in Cardiovascular Disease" (A. M. Lincoff, E. J. Topol, eds.), 143–168. Humana Press Inc., Totowa.

Vu, T-K. H., D. T. Hung, et al. (1991). Molecular cloning of a functional thrombin receptor reveals a novel proteolytic mechanism of receptor activation. *Cell* 64: 1057–1068.

Wagner, C. L., M. A. Mascelli, et al. (1996). Analysis of GPIIb/IIIa receptor to number by quantification of 7E3 binding to human platelets. *Blood* 88: 907–914.

Woods, V. L., L. E. Wolff, and D. M. Keller (1986). Resting platelets contain a substantial centrally located pool of glycoprotein IIb-IIIa complex which may be accessible to some but not other extracellular proteins. *J. Biol. Chem.* 261: 15, 242–15, 251.

Xiao, T., J. Takagi, et al. (2004). Structural basis for allostery in integrins and binding to fibrinogen-mimetic therapeutics. *Nature* 432: 59–67.

Xiong, J-P., T. Stehle, et al. (2002). Crystal structural of the extracellular segment of integrin $\alpha_V\beta_3$ in complex with an Arg-Gly-Asp ligand. *Science* 296: 151–155.

Yasuda, T., H. K. Gold, et al. (1988). Monoclonal antibody against the platelet glycoprotein (GP) IIb/IIIa receptor prevents coronary artery reocclusion after reperfusion with recombinant tissue-type plasminogen activator in dogs. *J. Clin. Invest.* 81, 1284–1291.

Yasuda, T., H. K. Gold, et al. (1990). Lysis of plasminogen activator-resistant platelet-rich coronary artery thrombus with combined bolus injection of recombinant tissue-type plasminogen activator and antiplatelet GPIIb/IIIa antibody. *J. Am. Coll. Cardiol.* 16: 1728–1735.

Yokota, T., D. E. Milenic, et al. (1993). Microautoradiographic analysis of the normal organ distribution of radioiodinated single-chain Fv and other immunoglobulin forms. *Cancer Res.* 53: 3776–3783.

III

VALIDATING TARGETS OF SMALL MOLECULE APPROACHES

9

EPIDERMAL GROWTH FACTOR RECEPTOR (EGFR) INHIBITOR FOR ONCOLOGY: DISCOVERY AND DEVELOPMENT OF ERLOTINIB

KENNETH K. IWATA,* SHANNON E. BEARD,† AND JOHN D. HALEY*

* *Ph.D., OSI Pharmaceuticals, Farmingdale, New York*
† *Ph.D., OSI Pharmaceuticals, Melville, New York*

Molecular-targeted drug therapies are based upon our understanding of tumor-cell biology to identify oncogenic targets and develop therapeutic approaches to these targets using small-molecular-weight inhibitors, antibodies, and antisense. Unlike traditional chemotherapeutics, which rely upon differential toxicities to tumor cells versus normal cells, molecular-targeted therapy approaches identify key molecular drivers of tumorigenicity and develop drug therapies against the target. The goal of molecular-targeted therapies is to develop agents with better efficacy and tolerability than current chemotherapeutic drugs. During the past two decades, the increased understanding of the involvement of growth factors, receptors, and tyrosine kinases in tumor biology has made these oncogenes the focus of new, targeted cancer-drug-discovery efforts. The extensive literature on epidermal growth factor receptor (EGFR) and its ligands in tumor cell lines and clinical specimens has made EGFR an important molecular target for drug discovery. The focus of this chapter is to provide a background on EGFR as a cancer target, the development of erlotinib (a potent selective inhibitor of EGFR tyrosine kinase), and preclinical data supporting erlotinib's advancement into the clinic. The concluding section will present some of the issues faced by molecular-targeted therapies such as erlotinib. These will include the role of biomarkers in the identification of patients who might best respond to treatment and efforts to optimize the use of molecular-targeted therapies.

I. INTRODUCTION

Tyrosine kinases are among the earliest molecular targets identified for cancer-drug discovery (Gschwind et al., 2004). Determination that the DNA sequence

Target Validation in Drug Discovery

TABLE 9.1 EGFR Over-Expression and Clinical Prognosis

Tumor Type	Cases/Year[a]	Percent[b,c] Over-expression	Related to Poor Prognosis	References
Prostate	232,090	40–70	Yes	Kim et al. 2001 Scher et al. 1995
Breast	212,930	35–70	Yes	Nicholson et al. 2001
NSCLC	172,570	50–90	Yes	Hirsch et al. 2003 Selvaggi et al. 2004
Colon	104,950	45–80	Yes	Resnik et al. 2004 Nicholson et al. 2001 O'Dwyer et al. 2002
Bladder	63,210	30–90	Yes	Neal et al. 1990
Pancreatic	32,180	30–50	Yes	Xiong et al. 2002 Papouchado et al. 2005
Head & Neck	29,370	70–100	Yes	Pomerantz and Grandis 2004
Ovarian	22,220	35–60	Yes	Scambia et al. 1992

[a]American Cancer Society 2005.
[b]Ciardiello, F. et al. *Expert Opin. Investig. Drugs* (2002) 11: 755–768.
[c]Salomon, D. S. et al. *Crit. Rev. Oncol. Hematol.* (1995) 19: 183–232.

for a virally associated oncogene v-sis has a proto-oncogene homologue c-sis, which encodes for the growth factor PDGFRβ (reviewed Heldin and Westermark, 1990), provided evidence that growth factors and growth factor receptors can drive tumorigenesis. The observation that growth factors, growth factor receptors, and kinases play a major role in oncogenesis, along with advances in molecular biology and automated high-throughput drug-screening technologies, has revolutionized approaches to small-molecule cancer-drug discovery resulting in the emergence of molecular targeted therapeutics. Unlike traditional chemotherapeutic drug discovery, which relies upon differential toxicities of normal cells versus tumor cells, targeted drug discovery identifies key molecular mediators of tumorigenicity against which small-molecule inhibitors are developed to generate cytostatic and/or cytotoxic effects on tumors with minimal negative effects on normal tissue and functions. Targeted approaches to anti-cancer therapy have the potential to further reduce toxicity and provide a better quality of life for the patient. EGFR and its ligands have long been observed to be elevated in many tumor cell lines (Arteaga, 2001; Salomon et al., 1995) and in tumor biopsy specimens and associated with poor clinical prognosis (Table 9.1). These observations made EGFR an ideal early target for molecularly targeted drug discovery.

II. EPIDERMAL GROWTH FACTOR RECEPTOR AND LIGANDS

Epidermal growth factor (EGF) is a highly conserved small molecular weight 6000 Dalton polypeptide that was first identified and purified by Stanley Cohen (reviewed Carpenter and Cohen, 1990).

EGFR is a 170-kDa transmembrane receptor tyrosine kinase. The EGFR gene encodes a 1186 amino acid polypeptide that is highly glycosylated in the extracellular ligand binding portion of the receptor. The extracellular domain consists of 621 amino acids including two cysteine-rich domains, to which N-linked carbohydrate is attached. The transmembrane domain contains 23 amino acids and the cytoplasmic domain, containing intrinsic tyrosine kinase activity, comprises of 542 amino acids.

Seven ligands for EGFR—EGF, TGFα, amphiregulin, betacellulin, epigen, epiregulin and heparin binding EGF-like growth factor (HB-EGF)—have been identified (Burgess et al., 2003). Crystallographic data suggest that the extracellular regions of EGFR function as a negative regulators of intrinsic kinase activity, which is relieved by interaction with ligand (Burgess et al., 2003, Ferguson et al., 2003, Schlessinger 2002). Binding of EGFR with ligand results in a conformational change, followed by receptor dimerization and activation of the tyrosine kinase domain. As shown in Figure 9.1, upon receptor activation, the EGFR transphosphorylates the tyrosine residues on the adjacent EGFR dimer.

More than 12 tyrosines in the cytoplasmic domain of EGFR have been shown to be phosphorylated and provide binding sites for SH2-containing proteins linked to various signal transduction pathways (Sordella et al., 2004; Thelemann et al., 2005). The outcome from some of these activated EGFR pathways is shown in Figure 9.2. These activated pathways are associated with oncogenic activities such as increased proliferation, invasion, metastasis, angiogenesis, decreased sensitivity to chemotherapeutic agents, and decreased apoptosis (Ciardiello and Tortora, 2002). In addition to signaling through EGFR receptor homodimers, EGFR can form heterodimers with other members of the human EGF receptor (HER) family.

Shown in Figure 9.3 are the other members of the HER family with their respective molecular sizes, ligands, and nomenclatures. The amino-acid sequence of the kinase domains of HER2, HER3, and HER4 are related by 82%, 59%, and 79%, respectively (Arteaga, 2001). The extracellular domains of HER2, HER3, and HER4 are much less related by 44%, 36%, and 48%, respectively, and differ significantly in their ligand selectivity. EGFR and HER4 both bind ligands and have active tyrosine kinase catalytic activity. HER3 can bind ligand, but the kinase domain is enzymatically inactive. HER2 has an enzymatically active kinase domain, but has no known ligand.

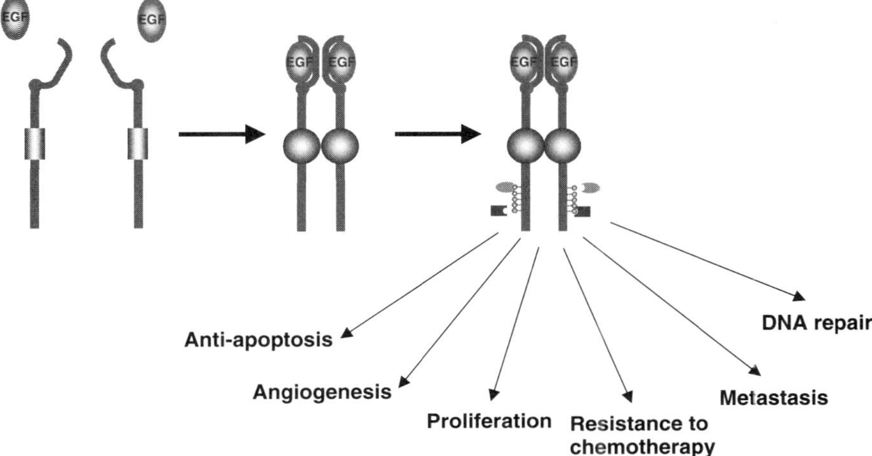

FIGURE 9.1 **A model of EGFR activation.** The left figure shows EGFR with the extracellular region of EGFR functioning as negative regulator of activity. The middle figure shows EGF bound to EGFR relieving the negative interaction resulting in EGFR dimerization and receptor tyrosine kinase activation. The right figure shows some of the activities resulting from activated EGFR.

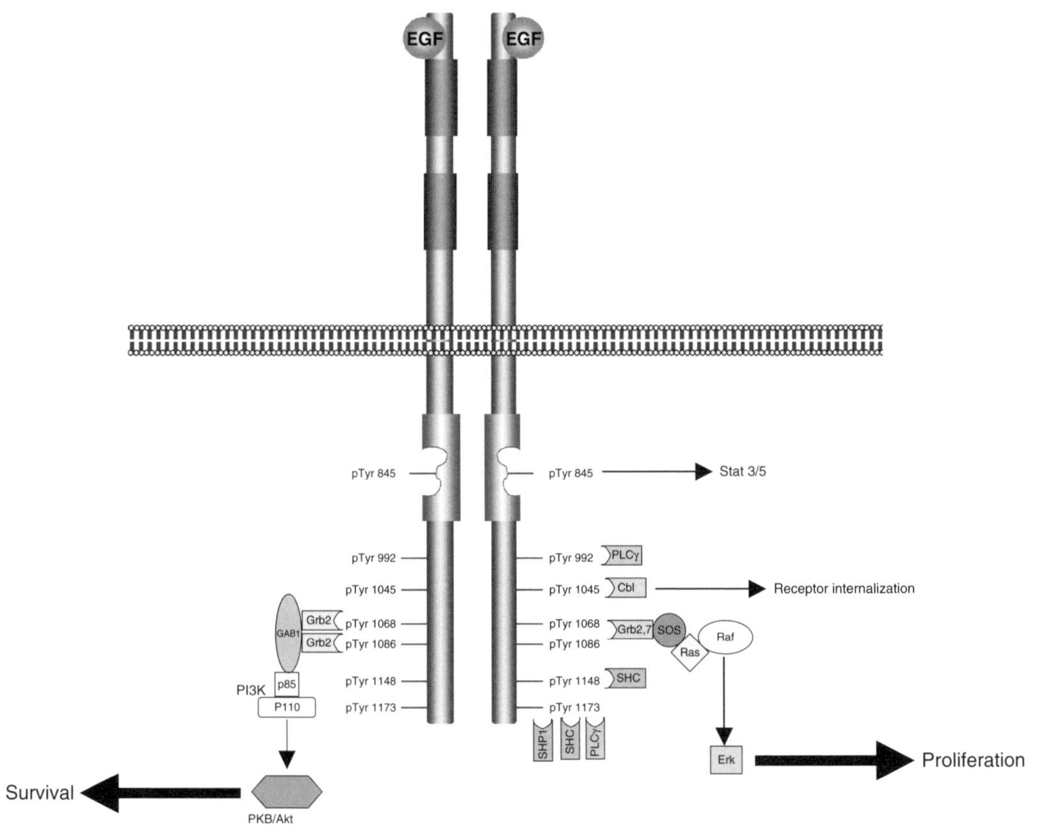

FIGURE 9.2 **A model of EGFR signal transduction and activities.** Activated EGFR transphosphorylates tyrosines on the adjacent receptor providing docking sites for phosphotyrosine binding proteins that form signal transduction complexes.

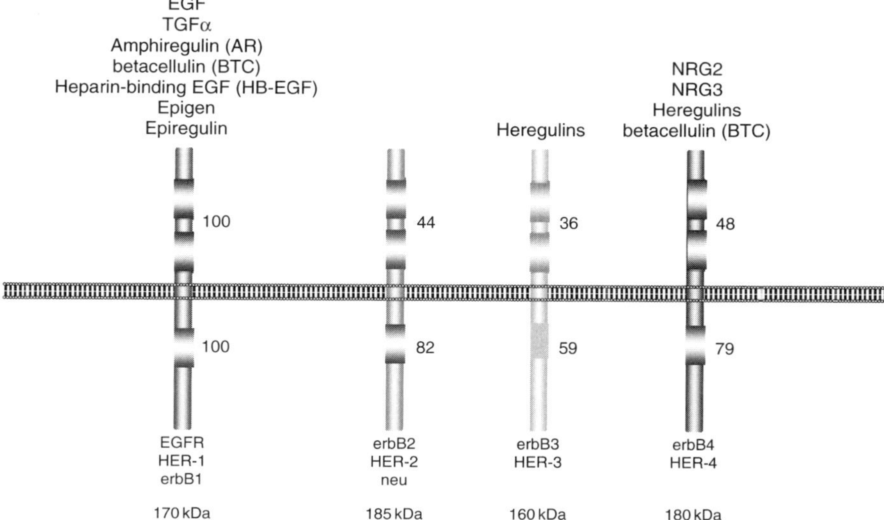

FIGURE 9.3 HER family members and ligands.

As illustrated in Figure 9.4, the dimer status (e.g., homodimer or heterodimer) can determine the intensity of signal generated by ligand binding as indicated in the relative proliferative index with higher numbers reflecting a higher proliferative capability (Pinkas-Kramarski et al., 1996). The formation of heterodimers with EGFR allows a receptor with an active kinase but no ligand (i.e., HER2), or a receptor that can bind ligand but lacks an enzymatically active kinase domain (i.e., HER3), to activate signal transduction pathways or receptor internalization (Yarden, 2001; Yarden and Sliwkowski, 2001). EGFR homodimers can bind an intracellular protein Cbl, which targets EGFR for internalization and degradation (Thien, 2001). The EGFR heterodimer with HER2, however, fails to sufficiently recruit Cbl allowing EGFR to be internalized and recycled to the cell surface (Yarden, 2001). Consequently, the dimer status of EGFR can determine the signaling pathways activated, the intensity of the signal, and the fate of EGFR on the membrane surface. Quantitating the EGFR homodimer and heterodimer status might, therefore, be useful as a diagnostic biomarker of tumor responsiveness to EGFR inhibitors as a single agent or in combination with other biological and chemotherapeutic agents.

A. EGFR in Normal Physiological Function

In considering an oncogene as a cancer drug discovery target, it is important to understand the normal physiological roles and functions to anticipate potential toxicities. If a molecular target serves a critical role in normal biological or physiological functions, the potential toxicities could obviate the target from consideration as a drug-discovery effort.

Tyrosine kinase activity has been shown to be essential for EGFR-mediated biochemical effects following ligand binding on target cells (Chen et al., 1987). In normal epithelial cells there are approximately 100,000 receptors expressed on the cell surface. EGFR is expressed by a wide range of cell types (Carpenter

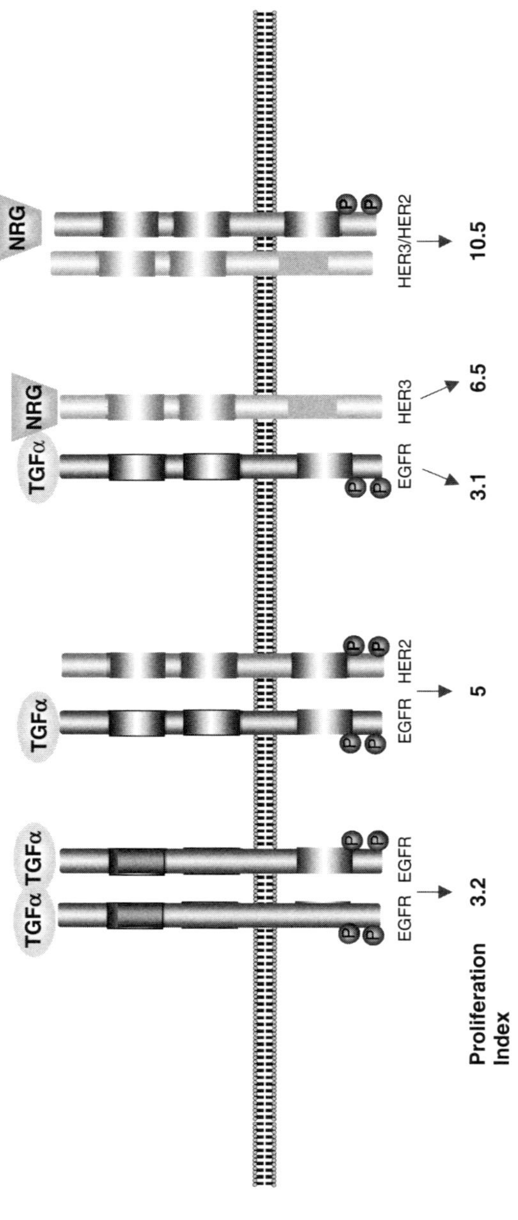

FIGURE 9.4 HER homodimerization and heterodimerization determines proliferative response to ligand binding.

et al., 1979; De Luca et al., 2000; Nagao et al., 2002). EGFR and its ligands can activate a multitude of signaling pathways (see Figure 9.2), which results in the stimulation of a number of biological activities, including proliferation, migration, and angiogenesis, through induction of VEGF mRNA transcription; increased anti-apoptotic signals; and increased DNA repair pathways (Ciardiello and Tortora 2002; Magné et al., 2003).

EGFR is expressed on the plasma membrane of cells derived from all three embryonic germ layers and transduces a signal produced by binding any of the previously described EGF-like family of molecules. In cells that express the EGFR, binding of the ligands can result in mitogenesis and cell survival (reviewed Yarden and Sliwkowski, 2001). In mammalian cells the receptor also controls the process of epithelial cell differentiation (Haigler et al., 1986; Boonstra et al., 1985).

Although EGFR is involved in signal transduction pathways for important cellular activities, it does not appear crucial for cell survival. Mice expressing defective EGFR with <10% wild-type EGFR activity or mice deficient in TGFα expression do not appear to have severe consequences (Luetteke et al., 1994). The major phenotype of these genetically EGFR-defective mice is curly hair and slight myopia. These observations suggested that EGFR inhibitors would be well tolerated with minimal or manageable toxicity.

The EGFR gene is localized on chromosome 7 (Davies et al., 1980; Shimizu et al., 1980). The EGFR gene contains a polymorphic dinucleotide (CA) repeat in intron 1, which showed *in vitro* and *in vivo* that EGFR gene transcription activity was inversely correlated to the number of CA dinucleotide repeats. An ethnic association has been observed whereby a larger number of repeats is prevalent in the Asian population and a shorter number of repeats in Caucasians (Amador et al., 2004; Liu et al., 2003). How CA dinucleotide repeats affect sensitivity to EGFR inhibitors by normal tissue or tumors and how it relates to tumor response or toxicities are currently unclear.

B. EGFR Signal Transduction

EGFR signal transduction is a broad area that is well covered in several review articles (Schlessinger 2000; Hynes and Lane, 2005; Yarden and Sliwkowski, 2001). Briefly, binding of ligand results in the transphosphorylation of the receptor at a number of tyrosine residues (Downward et al., 1984a; Margolis et al., 1989) located in the EGFR C-terminal phosphorylation domain, where mutation of these tyrosine residues has only subtle effects on the kinase activity (Honegger et al., 1988). EGFR phosphorylation occurs in trans by an intermolecular phosphorylation of oligomerized receptors (Haley et al., 1989; Honegger et al., 1989). Ligand binding was shown to result in EGFR dimerization and tyrosine kinase activation (Schlessinger, 2002) leading to phosphorylation of cellular substrates by the ligand-activated EGFR tyrosine kinase. A large number of functionally defined, intracellular molecules have been identified as substrates for the EGFR tyrosine kinase, such as Erk, SHC, PLC-γ, p85 PI-3 kinase, Grb2 (Blagoev et al., 2004). As shown in Figure 9.2, activated EGFR forms complexes with other proteins involved in downstream signal transduction (Thelemann et al., 2005). The existence of a signal transduction complex was postulated (Ullrich and Schlessinger, 1990) and complexes of this type were shown to coordinate the diverse responses elicited by EGF binding (Pawson and Nash, 2003).

C. EGFR and Tumorigenesis

The over-expression of proto-oncoproteins is a mechanism by which the neoplastic transformation of normal cells into their tumor counterparts can occur (Bishop, 1987). The earliest association of EGFR signaling pathway and tumorigenicity was observed in 1976 when virally transformed cells and many tumor cell lines were observed to secrete peptide factors, which competed for binding to EGFR. These peptides were subsequently identified as TGFα (Todaro et al., 1980), which is one of the ligands for EGFR. Additionally, links of EGFR as an oncogene came when it was discovered that the avian erythroblastosis virus (AEV) contains the v-erbB oncogene that was found to encode the avian homologue of the human EGFR (Downward et al., 1984b).

EGFR over-expression is frequently observed in human carcinomas, particularly those that express ligands, such as TGFα in an autocrine manner (Ozanne et al., 1986; Sporn and Todaro, 1980). The constitutive activation of the receptor tyrosine kinase has been shown to result in phenotypic transformation of immortalized mammalian fibroblasts in culture (DiFiore et al., 1987a; Stern et al., 1987; Velu et al., 1987).

Over-expression of EGFR or its ligands (e.g., TGFα) has been observed in a high percentage of tumor biopsy samples and has been associated with poor prognosis in prostate, breast, colon, bladder, pancreatic, head and neck, and ovarian cancers (see Table 9.1). As shown in Table 9.1, the reported percentage of tumors over-expressing EGFR within a given tumor type covers a broad range. This wide range reflects the various methodologies, such as ligand binding, immunohistochemistry (IHC) and fluorescent in situ hybridization (FISH) used to quantitate EGFR over-expression. Even within the same methodology (e.g., IHC), standardized analysis such as percent cells scoring positive, intensity of staining or some combination of the two (Hirsch and Witta, 2005) have not been used in the literature, resulting in a diversity of reports on EGFR over-expression.

In human carcinoma, EGFR over-expression frequently correlates with amplification of its gene. However, over-expression of EGFR protein does not always accompany or require amplification of the receptor gene and it was postulated that several mechanisms exist by which receptor over-expression may occur: altered truncated mRNA stabilization, gene amplification, and decreased EGFR degradation (Haley and Waterfield, 1991).

Over-expression of the full-length EGFR alone is insufficient for phenotypic transformation, suggesting additional molecular events are required. Receptor activation, for instance by binding of ligand, is an absolute requirement for EGFR-mediated transformation of immortalized rodent fibroblasts. Tyrosine kinase activity has been shown to be essential for the transforming properties of the truncated chicken EGFR v-erbB (Ng and Privalsky, 1986). In contrast to the ligand dependence for transformation shown by EGFR, over-expression of c-erbB2/neu alone is able to transform NIH3T3 cells in culture (DiFiore et al., 1987b). Human EGFR with truncation of the ligand-binding domains alone, however, is sufficient for transformation of immortalized rodent fibroblasts. N-terminally truncated EGFR stimulates growth of transfected cells under conditions of anchorage-independent growth (Haley et al., 1989; Khazaie et al., 1988; Wells and Bishop, 1988). Increasing the expression of N-terminally truncated receptor elevates phenotypic transformation, correlating with increased tyrosine kinase activity in the absence of ligand (Haley et al., 1989). EGFR activation status as determined by ligand binding, dimerization status, truncations, or mutations appears to be the important determinant of tumorigenicity rather than expression level. Tyrosine kinase activation following truncation of extracellular ligand binding domains has been shown for the EGFRvIII, v-ros, v-kit and c-erbB2/neu (Blume-Jensen and Hunter, 2001; Peles et al., 1991).

Normal cells and tissue appear to tolerate significantly reduced levels of EGFR signaling. Cells of the hematopoietic lineage do not express EGFR (DeLuca et al., 2000) and viable cells have been retrieved from EGFR knock-out animals (Woodworth et al., 2000). Conversely, tumor cells may have a greater dependence/oncoaddiction on activated oncogenic signal transduction pathways than normal cells (Jain et al., 2002; Weinstein, 2002).

Several approaches that reduce EGFR activity in tumor cell lines have demonstrated the potential of EGFR inhibitors as antitumor drugs. A dominant negative mutant form of EGFR (Redemann et al., 1992) resulted in reduced proliferation and invasiveness of human colon tumor. The downregulation of EGFR expression using antisense reduced proliferation and invasiveness of human rhabdomyosarcoma cells. Antibodies that bind the extracellular EGF binding domain have been shown to effectively inhibit tumor cell proliferation *in vitro* in cell culture studies and *in vivo* using human tumor xenografts. Inhibition of EGFR in the colorectal cell line DiFi (Baselga, et al., 1993; Fan et al., 1993) and the head and neck tumor cell line HN5 (Modjtahedi et al., 1998) resulted in the significant induction of apoptosis. The dramatic induction of apoptosis following inhibition of EGFR shows the dependence of some tumor cells on EGFR activity.

D. Strategy for EGFR as a Cancer Drug Target

The goal of the erlotinib discovery team was to identify a small molecular weight, potent, highly selective, orally available inhibitor of EGFR that would be effective against a wide range of human neoplasms. The therapeutic objective was to develop a safe, orally active, potent selective inhibitor of the human EGFR tyrosine kinase capable of increasing the survival and quality of life of cancer patients. Before embarking on a drug-discovery effort, a drug profile was developed. One of the considerations was whether to develop a highly selective inhibitor of EGFR tyrosine kinase activity or to develop a multikinase inhibitor that inhibited several members of the HER family or other kinases. A highly selective inhibitor has the possible advantage of limiting side effects, which might allow for a greater spectrum of combinations with other targeted agents as well as chemotherapeutic agents by minimizing the chances of overlapping toxicities. A multikinase inhibitor has the potential advantage of being more efficacious by targeting several signaling pathways in tumor cells. A drug that inhibits multiple pathways could, however, have a broader range of toxicities and perhaps a more limited capability for combining with other agents due to overlapping toxicities. The screening goal was to find a potent screening "hit" compound with a potency <1 μM in a cell-based assay with selectivity against HER2 kinase activity and a >100-fold selectivity against other purified tyrosine kinases.

EGFR as a drug discovery target can be approached several ways. EGFR drug discovery high-throughput screening can use purified EGFR kinase in an *in vitro* kinase assay, an intact cell-based assay measuring inhibition of EGFR, or a ligand-binding competition assay. The approach used for erlotinib drug discovery was a cell-based assay.

III. DISCOVERY OF AN EGFR INHIBITOR FOR DRUG DEVELOPMENT

A cell-based assay was developed using a human cell line that expressed high levels of EGFR. The assay used cells seeded in 96-well plates. The cells were incubated with screening compounds and then EGF was added to activate the receptor. Inhibition of phosphorylation was quantitated using a horseradish peroxidase–conjugated antibody against phosphotyrosine. OSI Pharmaceutical's automated high-throughput screening systems were used to screen several hundred thousand compounds. From this screen, several lead compounds were identified that inhibited EGFR tyrosine phosphorylation. As a cell-based screen, what is not known is whether the compound is inhibiting ligand binding, receptor dimerization, or receptor tyrosine kinase activity. The nature of inhibition was determined by testing lead compounds in an *in vitro* kinase assay using purified EGFR. EGFR was purified from A431 cells using an anti-EGFR affinity column eluted with EGF. Using purified, full-length EGFR in an *in vitro* kinase assay, the lead compounds from screening were confirmed as inhibitors of EGFR tyrosine kinase activity. One of the advantages of a cell-based assay is that it identifies only those active compounds that are able to cross the cell membrane. The lead compounds were, therefore, already selected for being potent inhibitors of cellular EGFR tyrosine kinase activity. Of the lead compounds, a quinazolinamine was selected for further development (Morin, 2000; Moyer et al., 1997).

Hundreds of analogues of the lead compound were synthesized and tested in cell-based EGFR kinase assays and *in vitro* assays using purified EGFR to determine structure activity relationships (SAR). A cell-based assay was also used to determine effect on HER2 kinase activity using a human breast cell construct that expressed high levels of activated HER2. Analogues were also tested in a MTT (3-(4,5)-dimethylthiazol-2-yl)-2,5-diphenyl tetrazolium bromide) proliferation/toxicity assay to identify toxic compounds.

The lead compounds that showed promising *in vitro* kinase and cell-based kinase activities were tested in *in vivo* models using human cell lines that express high levels of EGFR (>1,000,000 receptors/cell) such as the cervical carcinoma cell line A431 or the human head and neck carcinoma cell line HN5. A phosphoEGFR capture enzyme-linked immunosorbent assay (ELISA) was developed and used to evaluate lead compounds for *in vivo* activity in tumor xenografts (Pollack et al., 1999). Once the cells formed a palpable tumor, a lead compound was administered by oral gavage; the tumor xenograft harvested after 1 hour and the tumor homogenate were analyzed for phosphoEGFR levels relative to total EGFR (Pollack et al., 1999). Those compounds that demonstrated potent inhibition of EGFR in tumor xenografts were evaluated for tumor growth inhibitory (TGI) activity using HN5 and/or A431 cell lines. Compounds were administered by oral gavage or intraperitoneally once daily to identify a potent, orally available inhibitor of EGFR kinase that was a potent inhibitor of tumor growth. These efforts ultimately lead to the synthesis and selection of erlotinib as the drug candidate (Figure 9.5).

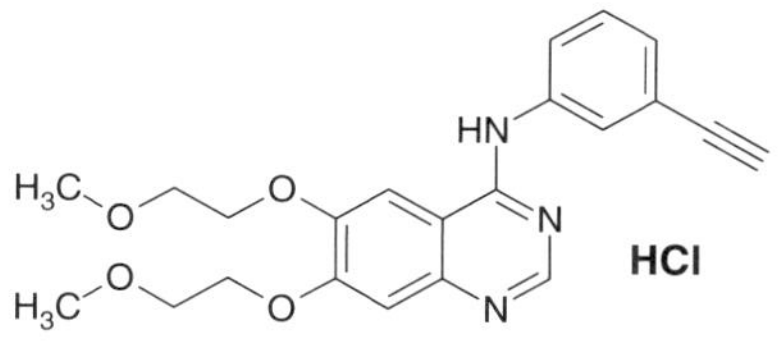

N-(3-ethynylphenyl)-6,7-bis(2-methoxyethoxy)-4-quinazolinamine, monohydrochloride

FIGURE 9.5 The structure and chemical nomenclature of the hydrochloride salt form of erlotinib. Erlotinib is also known as Tarceva®, OSI-774 and CP-358,774.

A. Preclinical Characterization of Erlotinib

As shown in Figure 9.5, erlotinib is a quinazolinamine class of compound with the chemical name N-(3-ethynylphenyl)-6,7-bis(2-methoxyethoxy)-4-quinazolinamine, with a formula molecular weight of 393.4 and is also known as OSI-774 or CP-358,774. Tarceva®, the pharmaceutical preparation of erlotinib, is a formulation containing the hydrochloride salt. Erlotinib is a potent inhibitor of purified EGFR tyrosine kinase activity with an IC_{50} of 2nM. As shown in Table 9.2, erlotinib is also a highly selective inhibitor of EGFR with some activity on HER2 and VEGFR, though at much higher drug concentrations. The *in vitro* kinase assays shown in Table 9.2 used the Km concentration of ATP for each kinase. In the literature, different potency values have been reported for erlotinib and other EGFR inhibitors, which may be the result of differences in assay conditions, including the ATP concentration used in the kinase assay. In a cell-based EGFR activity assay, erlotinib potently inhibited EGFR phosphorylation in an intact HN5 human head and neck tumor cell line with an IC_{50} of 20 nM. In MTT-proliferation assays using HN5 or DiFi (colorectal tumor cell line), erlotinib showed IC_{50} values ranging between 70 and 100 nM. Erlotinib was further tested for EGFR pathway selectivity using a mitogenesis assay. In the mitogenesis assay, cells were serum-starved, treated with compound, and mitogenically stimulated with growth factors that use receptor tyrosine kinases including EGFR.

TABLE 9.2 Erlotinib Kinase Selectivity

Kinase	IC_{50} (nM)
HER 1	2
HER 2	800
HER 4	>10,000
VEGFR	600
PDGFRβ	>10,000
FAK	>10,000
TrkA	4,500
IGF-1R	>10,000
Met	>10,000
CSF-1R	>10,000
Src	1,300
Abl	1,500
Ret	800
Insulin R	>10,000
Lck	>10,000

As shown in Figure 9.6, erlotinib was shown to be highly selective in inhibiting the EGFR mitogenic activity with an IC_{50} ranging between 70 and 200 nM, requiring >1 μM to show activity in the other receptor tyrosine kinase pathways involving IGF-1, basic FGF and PDGFβ.

In vivo, erlotinib potently inhibited tumor growth of HN5 xenografts in athymic mice with ED_{50}s of 10 mg/kg (Pollack et al., 1999). A plot of the percentage inhibition of EGFR phosphorylation with tumor growth inhibition showed a linear relationship, establishing a mechanistic link between EGFR kinase inhibition and antitumor efficacy. *In vivo* combination studies with xenografts of erlotinib with cisplatin, doxorubicin, 5-FU, paclitaxel, vinorelbine, or gemcitabine showed no increased toxicity (Table 9.3). Though the goal of the study was to investigate whether combinations of erlotinib with chemotherapeutic agents could be tolerated, some additive efficacious effects were observed in combinations of erlotinib with cisplatin, doxorubicin, paclitaxel, and gemcitabine (Table 9.3) supporting the potential clinical combination of erlotinib with chemotherapeutic drugs.

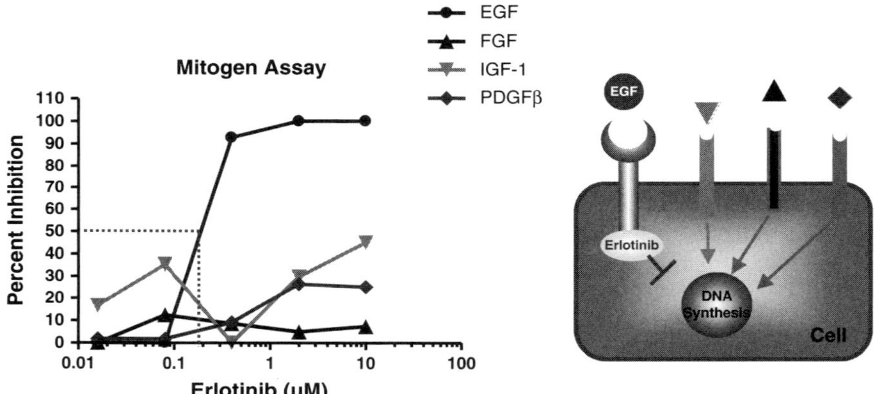

FIGURE 9.6 The selective inhibition of the EGFR signaling pathway by erlotinib is demonstrated in this mitogenesis experiment. The figure on the left shows the dose response effect of erlotinib on the mitogenic stimulation of serum-starved FRE cells stimulated by EGF, basic FGF, IGF-1, and PDGFβ. The figure on the right diagramatically shows the selective inhibition of the EGFR mitogenic pathway by erlotinib.

TABLE 9.3 Summary of Erlotinib and Chemotherapy Agent Combinations *In Vivo*

Agent	Dose/Route/Regimen	Lethality	Efficacy
Cisplatin	10 mg/kg IV qd × 1	No increase	Additive Effects
Doxorubicin HCl	15 mg/kg IV qd × 1	No increase	Additive Effects
5-Fluorouracil	200 mg/kg IP qd × 1	No increase	No interaction
Paclitaxel	10 mg/kg IP qd × 5	No increase	Additive Effects
Vinorelbine tartrate	25 mg/kg IV qid × 3	No increase	No interaction
Gemcitabine HCl	100 mg/kg IP tid × 4	No increase	Additive Effects

In responsive tumor cells (e.g., HN5), erlotinib was shown to block cell-cycle progression at the G1 phase (Moyer et al., 1997). Erlotinib-induced G1 arrest of some tumors suggests that the sequence of addition of erlotinib and a chemotherapeutic agent may determine the efficacy of a combination. *In vitro* studies combining erlotinib with docetaxel using A549, a non small–cell lung cancer (NSCLC) cell line (Kimura et al., 2004) and *in vivo* studies of erlotinib with cisplatin using HN5 xenografts (Pollack et al., 1999) both showed that the sequence of a chemotherapeutic agent added first followed by erlotinib showed the most effective activity versus the reverse order or the simultaneous combination of both agents. The applicability of sequence of addition of erlotinib with chemotherapeutic agents beyond NSCLC, head and neck tumors, as well as tumors in the clinic, has yet to be determined.

B. Erlotinib in the Clinic

A Phase I dose escalation study of orally administered erlotinib given once daily showed that it was well tolerated with a maximum tolerated dose (MTD) of 150 mg. The dose-limiting toxicity (DLT) and main side effects were rash and diarrhea (Hidalgo et al., 2001). Additional Phase I studies explored the safety of combinations of erlotinib with chemotherapy (e.g., paclitaxel/carboplatin, gemcitabine/cisplatin, and docetaxel) and showed that increased toxicity was not observed above what was expected for the agents when administered as single agents (Grunwald and Hidalgo 2003; Herbst, 2003).

Three Phase II single-agent studies with erlotinib on NSCLC, head and neck, and ovarian cancer patients showed that erlotinib was well tolerated and demonstrated patient responses in a number of different types of cancers (Grunwald and Hidalgo, 2003; Perez-Soler, 2004). A Phase II study on breast cancer showed erlotinib was well tolerated but had minimal activity in this heavily pretreated population with locally advanced or metastatic breast cancer. The positive outcomes from the NSCLC Phase II studies supported the initiation of the pivotal Phase III studies.

One large, placebo-controlled Phase III study (BR.21) was conducted on NSCLC patients after failure of at least one prior chemotherapeutic treatment (Shepherd et al., 2005). In BR.21, a robust survival advantage was observed in the group of patients receiving erlotinib. Based upon this trial, erlotinib was approved in 2004 for advanced NSCLC patients after failure of at least one prior round of chemotherapy. Two other Phase III studies, Tribute and Talent, were conducted on NSCLC patients who had not previously been treated with chemotherapy. These studies evaluated combinations of erlotinib with carboplatin/paclitaxel and cisplatin/gemcitabine, respectively. The outcomes of both of these trials were disappointing, with no statistically significant benefit observed in the group that received both chemotherapy with erlotinib over the treatment group treated with the chemotherapy alone (Perez-Soler, 2004). A Phase III study, PA.3, was conducted on patients with locally advanced, unresectable, or metastatic pancreatic cancer using erlotinib in combination with gemcitabine. In PA.3, the patients receiving erlotinib in combination with gemcitabine showed a statistically significant survival benefit over the control group that received gemcitabine alone (Moore et al., 2005). BR.21 and PA.3 demonstrated that erlotinib was effective on more than one tumor type and that erlotinib could be combined with a chemotherapeutic drug (i.e., gemcitabine) and be effective.

C. EGFR and Biomarkers

Trastuzumab is an antibody targeted to the extracellular domain of HER2. HER2 over-expression is an example of a biomarker used to select breast-cancer patients who would benefit the most from treatment with trastuzumab. Patients with high levels of HER2 expression showed a larger benefit to trastuzumab treatment as compared with patients with low expression of HER2 (Fornier et al., 2002).

Thus far, a strong correlation has not been observed between high expression levels of EGFR and response to erlotinib, gefitinib, and cetuximab or other EGFR inhibitors (Arteaga, 2002; Chung et al., 2005; El-Rayes and LoRosso, 2004; Hirsch and Witta, 2005; Perez-Soler, 2004), which may be because of the selective mode of action of these particular EGFR inhibitors. As shown in Figure 9.1, EGFR can activate multiple signal transduction pathways. EGFR signal transduction pathways are also linked to other signaling pathways activated by oncogenes, other receptors, and signaling molecules. These other pathways may be able to circumvent EGFR inhibition by signaling downstream or in parallel with EGFR.

As suggested in Figure 9.2, activated EGFR results in the increased phosphorylation of PKB/Akt. In an *in vitro* assay, it was shown that a high concentration of IGF-1 was able to block erlotinib-induced apoptosis in DiFi cells (Moyer et al., 1997). This observation suggests that the PKB/Akt pathway activated by IGF-1 was able to circumvent apoptosis induced by erlotinib inhibition of EGFR signaling in this colorectal cell line. EGFR inhibitors have been shown to potently inhibit PKB/Akt phosphorylation in responsive cell lines (Amann, et al., 2005; Moasser et al., 2001). MDA MBA468, a human-breast-tumor cell line with mutated inactive PTEN was resistant to growth inhibition and reduction of PKB/Akt phosphorylation when treated with the EGFR inhibitor gefitinib (Moasser et al., 2001). Restoration of PTEN activity in MDA MB468 resulted in cells sensitive to gefitinib (Bianco et al., 2003; She et al., 2003). Patients with glioblastoma multiforme (GBM) tumors with high levels of EGFR expression and low levels of phosphorylated PKB/PKB/Akt had a better response to erlotinib treatment than those with low levels of EGFR expression and high levels of phosphorylated PKB/Akt (Haas-Kogan et al., 2005). Tumors may utilize EGFR-independent activation of PKB/Akt to resist EGFR inhibitors. These observations suggest that tumor cells can use pathways such as PKB/Akt to overcome the effects of EGFR inhibitors and the activation state of PKB/Akt may determine the responsiveness of tumor cells to EGFR inhibitors. Understanding the underlying mechanism by which a tumor is resistant to EGFR inhibitors may identify biomarkers of response/resistance and also allow for the rational selection of combinations with other modalities (e.g., chemotherapeutic drugs, biologicals, radiation) to overcome tumor insensitivity to EGFR inhibitors.

I. EGFR Mutations

EGFR with mutations in the kinase domain were reported to predict responsiveness of tumors in NSCLC patients to EGFR kinase inhibitors erlotinib and gefitinib (Lynch et al., 2004; Paez et al., 2004; Pao et al., 2005). In the BR.21 Phase III study of NSCLC patients treated with erlotinib, there was a trend of EGFR mutation showing a greater response rate, but patients with wild-type or mutated EGFR both showed similar survival benefit when treated with erlotinib (Tsao et al., 2005). A cell line expressing the L858R mutation in the EGFR kinase domain was shown to be more sensitive to gefitinib inhibition than three cell lines expressing wild-type EGFR, but the wild-type EGFR-expressing cell lines did show some response at higher gefitinib concentrations (Paez et al., 2004). The above observations suggest that while tumors expressing mutated EGFR are more likely to respond to EGFR inhibitors than tumors expressing wild-type EGFR, a potent EGFR inhibitor administered at effective doses could be efficacious on tumors expressing either wild-type or mutated EGFR.

Although most of the EGFR mutations in the kinase domain have been associated with greater sensitivity by the tumors to EGFR inhibitors, a new mutation, T790M, has been identified which appears to be insensitive to gefitinib (Haber and Settleman, 2005; Pao et al., 2005). The prevalence of this T790M mutation in tumor cell lines and in patient tumors as well as its response to other EGFR inhibitors is actively being investigated.

Another mutant form of EGFR, EGFRvIII, which is missing a major portion of the extracellular ligand binding domain, has been observed as prevalent in malignant gliomas (Learn et al., 2004). Though cells expressing EGFRvIII have been reported to be resistant to some EGFR inhibitors (Learn et al., 2004), erlotinib has been shown to inhibit tumorigenic cells expressing EGFRvIII *in vitro* (Iwata et al., 2002).

II. EGFR Expression as a Biomarker

Colorectal cancer patients with tumors that were determined to be EGFR negative were reported to have responded to treatment with cetuximab (Erbitux®), which is an antibody targeted against the extracellular domain of EGFR (Chung et al., 2005). This study demonstrates a limitation of the currently available EGFR tests and the analysis of IHC slides (Hirsch and Witta, 2005). There may be tumors that express a low level of EGFR, which are sensitive to EGFR inhibitors, but score EGFR negative due to the limitation of the current EGFR tests and analytical methods. Although EGFR over-expression may not be predictive of response to EGFR inhibitors, such as erlotinib, gefitinib, and cetuximab (Arteaga, 2002; Cortes-Funes and Parra, 2003; El-Rayes and LoRosso, 2004; Soulieres et al., 2004; Parra et al., 2004), more sensitive and/or combination of analytical methods, such as IHC and FISH, may identify patients with tumors lacking functional levels of EGFR expression who may not respond to EGFR inhibitors.

III. Epithelial Mesenchymal Transition

Recent proteomic analysis of erlotinib sensitive and insensitive NSCLC cell lines identified several biomarkers of responsiveness (Thompson et al., 2005). E-cadherin was strongly associated with the erlotinib-sensitive cell lines, and vimentin expression was strongly associated with erlotinib insensitivity. Epithelial tumor cells are known to express E-cadherin. When the epithelial tumor cells become metastatic, they undergo a process known as epithelial mesenchymal transition (EMT). During this process, the epithelial tumor cells transition to a mesenchymal phenotype expressing biomarkers associated with the mesenchymal phenotype (Thiery, 2003). These observations also suggest that tumor cells may utilize and be dependent upon the EGFR pathway differently, depending upon which phase of EMT they occupy (Yauch et al., 2005). These observations suggest potential biomarkers to identify patients who may benefit the most from erlotinib treatment. The EMT link of sensitivity to EGFR inhibitors also suggests future areas of research to understand EGFR signaling pathways used by tumor cells.

D. Erlotinib in Combination Therapeutic Modalities

The potential for combining an EGFR inhibitor with chemotherapeutic drugs was early demonstrated using an anti-EGFR antibody with doxorubicin and cisplatin (Baselga et al., 1993; Fan et al., 1993). Small-molecule EGFR inhibitors have also been shown to be synergistic with chemotherapeutic drugs in *in vivo* studies (Ciardiello et al., 2000; Sirotnak et al., 2000). As previously described, preclinical *in vivo* studies showed that erlotinib could be combined with a broad spectrum of cytotoxic agents without increased toxicity and additive antitumor activity was observed with some combinations (Table 9.3). *In vitro* studies have shown additive activity when erlotinib was combined with biological agents such as the mTOR inhibitor rapamycin (Birle et al., 2003), histone deacetylase (HDAC) inhibitors (Witta et al., 2005), and the proteasome inhibitor bortezomib (Piperdi et al., 2004). Erlotinib has also been shown to have additive activity when combined with the anti-EGFR antibody cetuximab, demonstrating how two targeted approaches to EGFR may yield greater potency (Huang, 2004).

There may be circumstances that drive tumor cells to acquire a greater dependence or addiction to the EGFR pathway. The breast tumor cell line MCF-7 made resistant to fulvestrant or tamoxifen showed a significantly greater sensitivity to EGFR inhibitors and expressed elevated levels of EGFR relative to the parental MCF-7 (Hutcheson et al., 2002; Nicholson et al., 2002). It was also shown that tamoxifen or fulvestrant in combination with the EGFR inhibitor gefitinib delayed the onset of resistance to the anti-estrogens (Nicholson et al., 2002). Breast (MCF-7), ovarian (A2780), and cervical (ME180) tumor cell lines made resistant to doxorubicin, paclitaxel, and cisplatin, respectively, showed significantly greater sensitivity to the EGFR inhibitor erlotinib and expressed higher levels of EGFR relative to the parental cell lines (Dai et al., 2005). These studies show how some tumor cells may become more dependent upon the EGFR pathway in response to external stress such as exposure to chemotherapeutic agents, thereby making them more sensitive to EGFR inhibitors. These observations suggest that in some circumstances administration of EGFR inhibitors in sequence following treatment with chemotherapeutics might be effective.

EGFR inhibitors, such as erlotinib, block cells in the G1 phase of the cell-cycle, which could reduce the effectiveness of chemotherapeutics that depend upon cells actively transiting through the cell-cycle to maximally exert their antitumor effect. *In vitro* (Kimura et al., 2004) and *in vivo* experiments (Pollack et al., 1999) have demonstrated that intermittent dosing of tumor cells with chemotherapeutic drugs followed by erlotinib was more effective than the reverse order or when both drugs were dosed concurrently. Intermittent dosing of erlotinib can also allow higher doses of erlotinib to be administered than can be given on a daily schedule. Intermittent scheduling of erlotinib and docetaxel in a Phase I study showed that such a schedule is well tolerated and showed some clinically beneficial possibilities (Davies et al., 2005).

Understanding threshold and duration effects may be important in designing scheduling combinations of EGFR inhibitors with chemotherapeutics.

In combination with chemotherapeutics, transient high concentrations of EGFR inhibitors may affect tumors differently than chronic daily exposure at lower concentrations. High transient concentrations of small-molecule EGFR inhibitors may be effective in inhibiting anti-apoptotic pathways, whereas chronic lower doses may exert greater effects on cell-cycle activities (Solit et al., 2005).

EGFR functions through a multitude of signaling pathways and is influenced by linkages with other network signal transduction pathways (Yarden and Sliwkowski, 2001). Understanding which signaling pathways and network connections are being used by tumor cells would help in the selection of appropriate molecular targeted therapies and chemotherapies to use in combination with EGFR inhibitors. A better understanding of the functions of the EGFR pathways and interactions with other signaling pathways will also allow the optimization of scheduling/timing or sequencing of EGFR inhibitors with chemotherapy, biological agents, or radiotherapy.

E. EGFR Inhibitors

As mentioned earlier in this chapter, EGFR provides an example of how multiple approaches can be directed against a single oncogene target. Drug-discovery efforts have targeted the EGFR tyrosine catalytic domain in the intracellular part of the receptor using small-molecule inhibitors and the extracellular ligand-binding domain using monoclonal antibodies. Both of these approaches have yielded clinically approved drugs (e.g., erlotinib and cetuximab). Erlotinib was approved for patients with advanced NSCLC after failure of at least one prior round of chemotherapy. The EGFR antibody cetuximab has been approved for patients with irinotecan-refractory metastatic colorectal cancer.

Antibodies are large (e.g., 150,000 Dalton) molecular weight proteins that bind the extracellular domain of EGFR, thereby preventing ligand binding, sterically affecting EGFR dimer formation, inducing receptor internalization and possible stimulation of an antibody-dependent cell-mediated cytotoxicity (ADCC) response. Small-molecule inhibitors may not stimulate an ADCC response, but may be able to penetrate tissue more effectively than antibodies. Antibody inhibitors of EGFR have long half-lives and are administered intravenously on a weekly basis (Herbst and Langer, 2002), whereas small-molecule reversible inhibitors of EGFR have shorter half-lives allowing daily oral administration. The shorter half-life of small-molecule inhibitors may allow more flexibility in pulsatile dosing and scheduling in combination with other agents. Differences in targeting EGFR using antibodies (e.g., cetuximab) versus small-molecule inhibitors (e.g., erlotinib) were also apparent in a NSCLC cell line selected for resistance to cetuximab. This cell-line variant was resistant to growth inhibition by cetuximab, but was still sensitive to erlotinib (Huang et al., 2004). The different properties of erlotinib and cetuximab offers the possibility of combining the two different EGFR drugs, which may allow a more effective "shut-down" of the EGFR pathways. This was demonstrated when erlotinib combined with cetuximab showed additive inhibition of proliferation both *in vitro* and *in vivo* (Huang et al., 2004).

There are a number of small-molecule inhibitors of EGFR in the clinic (Figure 9.7). These include reversible selective inhibitors (e.g., erlotinib, gefitinib), irreversible inhibitors (e.g., canertinib), and dual HER family member

Erlotinib (HER 1 specific, reversible)

Gefitinib (HER 1 specific, reversible)

Canertinib (HER specific, irreversible)

Lapatinib (HER specific, reversible)

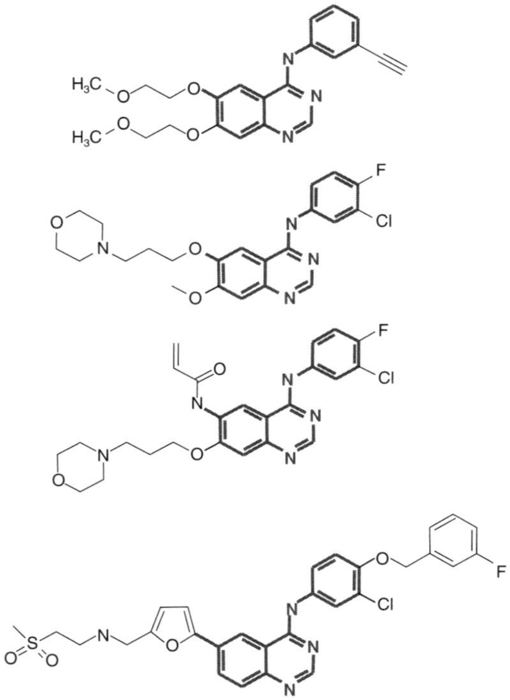

FIGURE 9.7 The structure and activity of several EGFR kinase inhibitors. The chemical features that are identical between the structures are highlighted in red. In parenthesis are the reported selectivity and activities of the EGFR inhibitors. It is apparent that minor differences in chemical structure can result in significant differences in target kinase selectivity.

inhibitors (e.g., lapatinib). Each of these drugs inhibits EGFR and has similarities in chemical structure, highlighted in red. As is apparent in Figure 9.7, small differences in chemical structure result in dramatic differences in selectivity and reversibility. The observation that small differences in chemical structure can result in major differences in biological and clinical activity has been reported for the chemotherapeutic drugs paclitaxel and docetaxel. Both have the identical primary mechanism of action and very similar chemical structures (Figure 9.8), but in a clinical study on metastatic breast tumor patients,

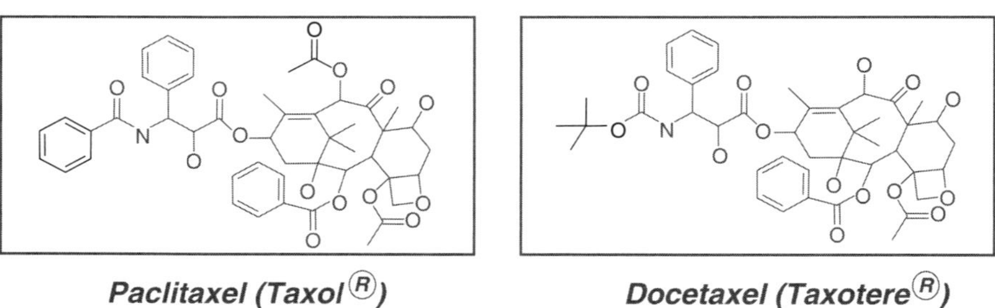

FIGURE 9.8 A comparison of the structures of paclitaxel and docetaxel provides an example of how small differences in chemical structure of two drugs with similar mechanisms of action can result in different clinical outcomes. The chemical features that are identical between the two drugs are highlighted in red.

docetaxel showed greater efficacy (Jones et al., 2005). What is learned and attributed to one member of a drug class may not always be applicable to all members.

IV. CONCLUSIONS

EGFR has been shown to be an important target for cancer drug discovery owing to its oncogenic activities in a broad spectrum of tumor types. Erlotinib provides an example of how a molecular-targeted cancer drug can be developed from concept to clinic. Erlotinib demonstrated that a small-molecule EGFR tyrosine kinase inhibitor can provide survival benefit to NSCLC patients after failure of at least one prior round of chemotherapy, and to pancreatic cancer patients, in large Phase III clinical trials. The challenge is to understand why some patients did not benefit from treatment. Response to erlotinib may be affected by patient variables such as drug metabolism or tumor variables such as the signal transduction or mutational status of the tumor. Patient variables may be managed through dose and scheduling. Managing tumor variables requires a better understanding of tumor utilization of EGFR and the interactions between EGFR signaling pathways and other signal transduction networks. Understanding the tumor variables of erlotinib responsiveness would help identify patients who would best respond to treatment, provide for the rational selection of combination agents and improve administration of combinations whether through sequencing or scheduling of treatments.

Discovery was a first step in the development of a molecular targeted therapy, such as an EGFR tyrosine kinase inhibitor for oncology. Identifying the patients most likely to benefit from treatment and optimizing its use remain future challenges.

RECOMMENDED RESOURCES

pubmed
http://www.ncbi.nlm.nih.gov/
Science Magazine
http://www.sciencemag.org/
Science Magazine Signal Transduction
http://stke.sciencemag.org/
Science Magazine Functional Genomics
http://www.sciencemag.org/feature/plus/sfg/
Plos Open Action Journal
http://medicine.plosjournals.org/perlserv/?request=index-html&issn=1549-1676
Clinical Trials
http://www.oncolink.upenn.edu/
Protein Database
http://www.ihop-net.org/UniPub/iHOP/gs/91035.html
General Biotechnology Search Database
http://www.psb.ugent.be/links/biotech.htm
Chemistry Database
http://www.chemdex.org/
News Resource
http://www.biospace.com/
News Resource

http://www.genengnews.com/
Disease Websites
American Cancer Society
http://www.cancer.org/docroot/home/index.asp
Lung Cancer
http://www.lungcanceronline.org/
http://www.lungcancer.org/
http://www.lungusa.org/site/pp.asp?c=dvLUK9O0E&b=22542
Pancreatic Cancer
http://www.pancan.org/
http://www.lustgartenfoundation.org/

REFERENCES

Amador, L. M., D. Oppenheimer, et al. (2004). An epidermal growth factor receptor intron 1 polymorphism mediates response to epidermal growth factor receptor inhibitors. *Cancer Res.* 64: 9139–9143.

Amann, J., S. Kalyankrishna, et al. (2005). Aberrant epidermal growth factor receptor signaling and enhanced sensitivity to EGFR inhibitors in lung cancer. *Cancer Res.* 65: 226–235.

Arteaga, C. L. (2001). The epidermal growth factor receptor: From mutant oncogene in nonhuman cancers to therapeutic target in human neoplasia. *J. Clin. Oncol.*19: 32s–40s.

Arteaga, C. L. (2002). Epidermal growth factor receptor dependence in human tumors: More than just expression? *Oncologist* 7: 31–39.

Baselga, J., L. Norton, et al. (1993). Antitumor effects of doxorubicin in combination with anti-epidermal growth factor receptor monoclonal antibodies. *J. Natl. Cancer Inst.* 85: 1327–1333.

Bianco, R., I. Shin, et al. (2003). Loss of PTEN/MMAC1/TEP in EGF receptor-expressing tumor cells counteracts the antitumor action of EGFR tyrosine kinase inhibitors. *Oncogene* 22: 2812–2822.

Birle D. C., C. Y. F. Yau, et al. (2003). Interactions between rapamycin and the HER1/EGFR inhibitor erlotinib (Tarceva™, OSI-774) on downstream signaling pathways and tumor growth in cervical carcinoma xenografts. *Proc. Am. Assoc. Cancer Res.* 44: 1076.

Bishop, J. M. (1987). The molecular genetics of cancer. *Science* 235: 305–311.

Blagoev, B., S. E. Ong, et al. (2004). Temporal analysis of phosphotyrosine-dependent signaling networks by quantitative proteomics. *Nat. Biotechnol.* 22: 1139–1145.

Blume-Jensen, P. and T. Hunter. (2001). Oncogenic kinase signalling. *Nature* 411: 355–365.

Boonstra, J., S. W. De Laat, and M. Ponec (1985). Epidermal growth factor receptor expression related to differentiation capacity in normal and transformed keratinocytes. *Exp. Cell Res.* 161: 421–433.

Burgess, A. W., H. Cho, et al. (2003). An open-and-shut case? Recent insights into the activation of EGF/ErbB receptors. *Mol. Cell* 12: 541–552.

Carpenter, G. and S. Cohen (1990). Epidermal growth factor. *J. Biol. Chem.* 265: 7709–7712.

Carpenter, G., L. King, and S. Cohen (1979). Rapid enhancement of protein phosphorylation in A-431 cell membrane preparations by epidermal growth factor. *J. Biol. Clin.* 254: 4884–4891.

Chen, W. S., C. S. Lazar, et al. (1987). Requirement for intrinsic protein tyrosine kinase in the immediate and late actions of the EGF receptor. *Nature* 328: 820-823.

Chung, K. Y., J. Shia, et al. (2005). Cetuximab shows activity in colorectal cancer patients with tumors that do not express the epidermal growth factor receptor by immunohistochemistry. *J. Clin. Oncol.* 23: 1803–1810.

Ciardiello, F., R. Caputo, et al. (2000). Antitumor effect and potentiation of cytotoxic drugs activity in human cancer cells by ZD-1839 (Iressa), an epidermal growth factor receptor-selective tyrosine kinase inhibitor. *Clin. Cancer Res.* 6: 2053–2063.

Ciardiello, F. and G. Tortora (2002). Anti-epidermal growth factor receptor drugs in cancer therapy. *Expert Opin. Investig. Drugs* 11: 755–768.

Cortes-Funes, H. and H. S. Parra (2003). Extensive experience of disease control with gefitinib and the role of prognostic markers. *Br. J. Cancer* 89: S3–S8.

Dai, Q., Y. H. Ling, et al. (2005). Enhanced sensitivity to the HER1/epidermal growth factor receptor tyrosine kinase inhibitor erlotinib hydrochloride in chemotherapy-resistant tumor cell lines. *Clin. Cancer Res.* 11: 1572–1578.

Davies, A. M., V. A. Grosse, et al. (1980). Genetic analysis of epidermal growth factor action: Assignment of human epidermal growth factor receptor gene to chromosome 7. *Proc. Natl. Acad. Sci. U. S. A.* 77: 4188–4192.

Davies, A. M., P. N. Lara, et al. (2005). Intermittent erlotinib in combination with docetaxel (DOC): Phase I schedules designed to achieve pharmacodynamic separation. *J. Clin. Oncol.* 23: 630s.

De Luca, A., S. Pignata, et al. (2000). Detection of circulating tumor cells in carcinoma patients by a novel epidermal growth factor receptor reverse transcription-PCR assay. *Clin. Cancer Res.* 6: 1439–1444.

Di Fiore, P. P., J. H. Pierce, et al. (1987a). Overexpression of the human EGF receptor confers an EGF-dependent transformed phenotype to NIH 3T3 cells. *Cell* 51: 1063–1070.

Di Fiore, P. P., J. H. Pierce, et al. (1987b). ErbB-2 is a potent oncogene when overexpressed in NIH/3T3 cells. *Science* 237: 178–182.

Downward, J., P. Parker, and M. D. Waterfield (1984a). Autophosphorylation sites on the epidermal growth factor receptor. *Nature* 311: 483–485.

Downward, J., Y. Yarden, et al. (1984b). Close similarity of epidermal growth factor receptor and v-erb-B oncogene protein sequences. *Nature* 307: 521-527.

El-Rayes, B. F. and P. M. LoRosso (2004). Targeting the epidermal growth factor receptor. *Br. J. Cancer* 91: 418–424.

Fan, Z., J. Baselga, et al. (1993). Antitumor effect of anti-epidermal growth factor receptor monoclonal antibodies plus cis-diamminedichloroplatinum on well established A431 cell xenografts. *Cancer Res.* 53: 4637–4642.

Ferguson, K. M., M. B. Berger, et al. (2003). EGF activates its receptor by removing interactions that autoinhibit ectodomain dimerization. *Mol. Cell* 11: 507–517.

Fornier, M., M. Risio, et al. (2002). HER2 testing and correlation with efficacy of trastuzumab therapy. *Oncology* 16: 1340–1342.

Grunwald, V. and M. Hidalgo (2003). Developing inhibitors of the epidermal growth factor receptor for cancer treatment. *J. Natl. Cancer Inst.* 95: 851–867.

Gschwind, A., O. M. Fischer, and A. Ullrich (2004). The discovery of receptor tyrosine kinases: Targets for cancer therapy. *Nat. Rev. Cancer* 4: 361–370.

Haas-Kogan, D. A., M. D. Prados, et al. (2005). Epidermal growth factor receptor, protein kinase B/Akt, and glioma response to erlotinib. *J. Natl. Cancer Inst.* 97: 880–887.

Haber, D. A. and J. Settleman (2005). Overcoming acquired resistance to Iressa/Tarceva with inhibitors of a different class. *Cell Cycle* 4: e40–e42.

Haigler, H., R. Gordon, and P. Chiang (1986). Down-regulation of EGF receptors. *Ann. N. Y. Acad. Sci.* 463: 102–104.

Haley, J. D., J. J. Hsuan, and M. D. Waterfield (1989). Analysis of mammalian fibroblast transformation by normal and mutated human EGF receptors. *Oncogene* 4: 273–283.

Haley, J. D. and M. D. Waterfield (1991). Contributory effects of de novo transcription and premature transcript termination in the regulation of human epidermal growth factor receptor proto-oncogene RNA synthesis. *J. Biol. Chem.* 266: 1746–1753.

Heldin, C. and B. Westermark (1990). Platelet-derived growth factor: Mechanism of action and possible *in vivo* function. *Cell Regul.* 1: 555–566.

Herbst R. S. and C. J. Langer (2002). Epidermal growth factor receptors as a target for cancer treatment: The emerging role of IMC-C225 in the treatment of lung and head and neck cancers. *Semin. Oncol.* 29: 27–36.

Herbst R. S. (2003). Erlotinib (Tarceva): An update on the clinical trial program. *Semin. Oncol.* 30: 34–46.

Hidalgo, M., L. L. Siu, et al. (2001). Phase I and pharmacologic study of OSI-774, an epidermal growth factor receptor tyrosine kinase inhibitor, in patients with advanced solid malignancies. *J. Clin. Oncol.* 19: 3267–3279.

Hirsch, F. R., M. Varella-Garcia, et al. (2003). Epidermal growth factor receptor in non-small-cell lung carcinomas: Correlation between gene copy number and protein expression and impact on prognosis. *J. Clin. Oncol.* 21: 3798–3807.

Hirsch, F. R. and S. Witta (2005). Biomarkers for prediction of sensitivity to EGFR inhibitors in non-small cell lung cancer. *Curr. Opin. Oncol.* 17: 118–122.

Honegger, A., T. J. Dull, et al. (1988). Kinetic parameters of the protein tyrosine kinase activity of EGF-receptor mutants with individually altered autophosphorylation sites. *EMBO J.* 7: 3053–3060.

Honegger, A. M., R. M. Kris, et al. (1989). Evidence that autophosphorylation of solubilized receptors for epidermal growth factor is mediated by intermolecular cross-phosphorylation. *Proc. Natl. Acad. Sci. U. S. A.* 86: 925–929.

Huang, S., E. A. Armstrong, et al. (2004). Dual-agent molecular targeting of the epidermal growth factor receptor (EGFR): Combining anti-EGFR antibody with tyrosine kinase inhibitor. *Cancer Res.* 64: 5355–5362.

Hutcheson, I. R., R. A. McClelland, et al. (2002). Increased sensitivity of tamoxifen- and fulvestrant-resistant MCF-7 breast cancer cells to the growth inhibitory action of ZD1839 ('Iressa'). 1st *Internation Symposium on Signal Transduction Modulators in Cancer Therapy.*

Hynes, N. E. and H. A. Lane (2005). ERBB receptors and cancer: The complexity of targeted inhibitors. *Nat. Rev. Cancer* 5: 341–354.

Iwata, K. K., K. Provoncha, and N. Gibson (2002). Inhibition of mutant EGFRvIII transformed cells by tyrosine kinase inhibitor OSI-774 (Tarceva). *J. Clin. Oncol.* 21: 79.

Jain, M., C. Arvanitis, et al. (2002). Sustained loss of a neoplastic phenotype by brief inactivation of MYC. *Science* 297: 102–104.

Jones, S. E., J. Erban, et al. (2005). Randomized phase III study of docetaxel compared with paclitaxel in metastatic breast cancer. *J. Clin. Oncol.* 23: 5542–5551.

Khazaie, K., T. J. Dull, et al. (1988). Truncation of the human EGF receptor leads to differential transforming potentials in primary avian fibroblasts and erythroblasts. *EMBO J.* 7: 3061–3071.

Kim E. S., F. R. Khuri, and R. S. Herbst (2001). Epidermal growth factor receptor biology (IMC-C225). *Curr. Opin. Oncol.* 13: 506–513.

Kimura, T., C. M. Mahaffey, et al. (2004). Apoptotic effects of the docetaxel: OSI-774 combination in non-small cell lung carcinoma (NSCLC) cells. *J. Clin. Oncol.* 23: 649.

Learn, C. A., T. L. Hartzell, et al. (2004). Resistance to tyrosine kinase inhibition by mutant epidermal growth factor receptor variant III contributes to the neoplastic phenotype of glioblastoma multiforme. *Clin. Cancer Res.* 10: 3216–3224.

Liu, W., F. Innocenti, et al. (2003). Interethnic difference in the allelic distribution of human epidermal growth factor receptor intron1 polymorphism. *Clin. Cancer Res.* 9: 1009–1012.

Luetteke, N. C., H. K. Phillips, et al. (1994). The mouse waved-2 phenotype results from a point mutation in the EGF receptor tyrosine kinase. *Genes Dev.* 8: 399–413.

Lynch, T. J., D. W. Bell, et al. (2004). Activating mutations in the epidermal growth factor receptor underlying responsiveness of non-small-cell lung cancer to gefitinib. *N. Engl. J. Med.* 350: 2129–2139.

Magne', M., J. Fischel, et al. (2003). Molecular mechanisms underlying the interaction between ZD1839 ('Iressa') and cisplatin/5-fluorouracil. *Br. J. Cancer* 89: 585–592.

Margolis, B. L., I. Lax, et al. (1989). All autophosphorylation sites of epidermal growth factor (EGF) receptor and HER2/neu are located in their carboxyl-terminal tails: Identification of a novel site in EGF receptor. *J. Biol. Chem.* 264:, 10,667–10,671.

Moasser, M. M., A. Basso, et al. (2001). The tyrosine kinase inhibitor ZD1839 ("Iressa") inhibits HER2-driven signaling and suppresses the growth of HER2-overexpressing tumor cells. *Cancer Res.* 61: 7184–7188.

Modjtahedi, H., K. Affleck, et al. (1998). EGFR blockade by tyrosine kinase inhibitor or monoclonal antibody inhibits growth, directs terminal differentiation and induces apoptosis in the human squamous cell carcinoma HN5. *Int. J. Oncol.* 13: 335–342.

Moore, M. J., D. Goldstein, et al. (2005). Erlotinib plus gemcitabine compared to gemcitabine alone in patients with advanced pancreatic cancer: A phase III trial of the National Cancer Institute of Canada Clinical Trials Group [NCIC-CTG]. *J. Clin. Oncol.* 23: 1s.

Morin, M. J. (2000). From oncogene to drug: development of small molecule tyrosine kinase inhibitors as anti-tumor and anti-angiogenic agents. *Oncogene* 19: 6574–6583.

Moyer, J. D., E. G. Barbacci, et al. (1997). Induction of apoptosis and cell cycle arrest by CP-358,774, an inhibitor of epidermal growth factor receptor tyrosine kinase. *Cancer Res.* 57: 4838–4848.

Nagao, K., H. Hisatomi, et al. (2002). Expression of molecular marker genes in various types of normal tissue: Implication for detection of micrometastases. *Int. J. Mol. Med.* 10: 307–310.

Neal, D. E., L. Sharples, et al. (1990). The epidermal growth factor receptor and the prognosis of bladder cancer. *Cancer* 65: 1619–1625.

Ng, M. and M. L. Privalsky (1986). Structural domains of the avian erythroblastosis virus erbB protein required for fibroblast transformation: Dissection by in-frame insertional mutagenesis. *J. Virol.* 58: 542–553.

Nicholson, R. I., J. M.W. Gee, and M. E. Harper (2001). EGFR and cancer prognosis. *Eur. J. Cancer* 37: S9–S15.

Nicholson, R. I., I. R. Hutcheson, et al. (2002). Modulation of epidermal growth factor receptor in endocrine-resistant, estrogen-receptor-positive breast cancer. *Ann. N. Y. Acad. Sci.* 963: 104–115.

O'Dwyer, P. J. and A. B. Benson (2002). Epidermal growth factor receptor-targeted therapy in colorectal cancer. *Semin. Oncol.* 29: 10–17.

Ozanne, B., C. S. Richards, et al. (1986). Over-expression of the EGF receptor is a hallmark of squamous cell carcinomas. *J. Pathol.* 149: 9–14.

Paez, J. G., P. A. Janne, et al. (2004). EGFR mutations in lung cancer: Correlation with clinical response to gefitinib therapy. *Science* 304: 1497–1500.

Pao, W., V. A. Miller, et al. (2005). Acquired resistance of lung adenocarcinomas to gefitinib or erlotinib is associated with a second mutation in the EGFR kinase domain. *PLoS Med.* 2: 1–11.

Papouchado, B., L. A. Erickson, et al. (2005). Epidermal growth factor receptor and activated epidermal growth factor receptor expression in gastrointestinal carcinoids and pancreatic endocrine carcinomas. *Mod. Pathol.* advance online publication 6 May 2005.

Parra, H. S., R. Cavina, et al. (2004). Analysis of epidermal growth factor receptor expression as a predictive factor for response to gefitinib ('Iressa', ZD1839) in non-small-cell lung cancer. *Br. J. Cancer* 91: 208–212.

Pawson, T. and P. Nash (2003). Assembly of cell regulatory systems through protein interaction domains. *Science* 300: 445–452.

Peles, E., R. B. Levy, et al. (1991). Oncogenic forms of the neu/HER2 tyrosine kinase are permanently coupled to phospholipase C gamma. *EMBO J.* 10: 2077–2086.

Perez-Soler, R. (2004). The role of erlotinib (Tarceva, OSI 774) in the treatment of non-small cell lung cancer. *Clin. Cancer Res.* 10: 4238s–4240s.

Pinkas-Kramarski, R., L. Soussan, et al. (1996). Diversification of Neu differentiation factor and epidermal growth factor signaling by combinatorial receptor interactions. *EMBO J.* 15: 2452–2467.

Piperdi B., Y. Ling, and R. Perez-Soler (2004). Schedule-dependent interaction between the proteosome inhibitor, bortezomib, and the EGFR-PTK inhibitor, erlotinib, in human non-small cell lung cancer cell lines. *Proc. Am. Assoc. Cancer Res.* 45: 4010.

Pollack, V. A., D. M. Savage, et al. (1999). Inhibition of epidermal growth factor receptor-associated tyrosine phosphorylation in human carcinomas with CP-358,774: Dynamics of receptor inhibition in situ and antitumor effects in athymic mice. *J. Pharmacol. Exp. Ther.* 291: 739–748.

Pomerantz, R. G. and J. R. Grandis (2004). The epidermal growth factor receptor signaling network in head and neck carcinogenesis and implications for targeted therapy. *Semin. Oncol.* 31: 734–743.

Redemann, N., B. Holzmann, et al. (1992). Anti-oncogenic activity of signalling-defective epidermal growth factor receptor mutants. *Mol. Cell. Biol.* 12: 491–498.

Resnick, M. B., J. Routhier, et al. (2004). Epidermal growth factor receptor, c-MET, β-catenin, and p53 expression as prognostic indicators in stage II colon cancer: A tissue microarray study. *Clin. Cancer Res.* 10: 3069–3075.

Salomon, D. S., R. Brandt, et al. (1995). Epidermal growth factor-related peptides and their receptors in human malignancies. *Crit. Rev. Oncol. Hematol.* 19: 183–232.

Scambia, G., P. Benedetti Panici, et al. (1992). Significance of epidermal growth factor receptor in advanced ovarian cancer. *J. Clin. Oncol.* 10: 529–535.

Scher, H. I., A. Sarkis, et al. (1995). Changing pattern of expression of the epidermal growth factor receptor and transforming growth factor alpha in the progression of prostatic neoplasms. *Clin. Cancer Res.* 1: 545–550.

Schlessinger, J. (2000). Cell signaling by receptor tyrosine kinases. *Cell* 103: 211–225.

Schlessinger, J. (2002). Ligand-induced, receptor-mediated dimerization and activation of EGF receptor. *Cell* 110: 669–672.

Selvaggi, G., S. Novello, et al. (2004). Epidermal growth factor receptor overexpression correlates with a poor prognosis in completely resected non-small-cell lung cancer. *Annals Oncol.* 15: 28–32.

She, Q., D. Solit, et al. (2003). Resistance to gefitinib in PTEN-Null HER-overexpressing tumor cells can be overcome through restoration of PTEN function or pharmacologic modulation of constitutive phosphatidylinositol 3-Kinase/Akt pathway signaling. *Clin. Cancer Res.* 9: 4340–4346.

Shepherd, F. A., P. J. Rodrigues, et al. (2005). Erlotinib in previously treated non-small-cell lung cancer. *N. Engl. J. Med.* 353: 123–132.

Shimizu, N., M. A. Behzadian, and Y. Shimizu (1980). Genetics of cell surface receptors for bioactive polypeptides: binding of epidermal growth factor is associated with the presence of human chromosome 7 in human-mouse cell hybrids. *Proc. Natl. Acad. Sci. U. S. A.* 77: 3600–3604.

Sirotnak, F. M., M. F. Zakowski, et al. (2000). Efficacy of cytotoxic agents against human tumor xenografts is markedly enhanced by coadministration of ZD1839 (Iressa), an inhibitor of EGFR tyrosine kinase. *Clin. Cancer Res.* 6: 4885–4892.

Solit, D. B., Y. She, et al. (2005). Pulsatile administration of the epidermal growth factor receptor inhibitor gefitinib is significantly more effective than continuous dosing for sensitizing tumors to Paclitaxel. *Clin. Cancer Res.* 11: 1983–1989.

Sordella, R., D. W. Bell, et al. (2004). Gefitinib-sensitizing EGFR mutations in lung cancer activate anti-apoptotic pathways. *Science* 305: 1163–1167.

Soulieres, D., N. N. Senzer, et al. (2004) Multicenter phase II study of erlotinib, an oral epidermal growth factor receptor tyrosine kinase inhibitor, in patients with recurrent or metastatic squamous cell cancer of the head and neck. *J. Clin. Oncol.* 22: 77–85.

Sporn, M. B. and G. J. Todaro (1980). Autocrine secretion and malignant transformation of cells. *N. Engl. J. Med.* 303: 878–880.

Stern, D. F., D. L. Hare, et al. (1987). Construction of a novel oncogene based on synthetic sequences encoding epidermal growth factor. *Science* 235: 321–324.

Thelemann, A., F. Petti, et al. (2005). Phosphotyrosine signaling networks in epidermal growth factor receptor overexpressing squamous carcinoma cells. *Mol. Cell. Proteomics* 4: 356–376.

Thien, C. B. and W. Y. Langdon (2001). CBL: Many adaptations to regulate protein tyrosine kinases. *Nat. Rev.* 2: 294-305.

Thomson, S., E. Buck, et al. (2005). Epithelial-mesenchymal transition is a determinant of sensitivity of non-small cell lung carcinoma cell lines and xenografts to EGF receptor inhibition. *Cancer Res.* 65: 9455–9462.

Thiery, J. P. (2003). Epithelial-mesenchymal transitions in development and pathologies. *Curr. Opin. Cell Biol.* 15: 740–746.

Todaro, G. J., C. Fryling, and J. E. De Larco (1980). Transforming growth factors produced by certain human tumor cells: Polypeptides that interact with epidermal growth factor receptors. *Proc. Natl. Acad. Sci. U. S. A.* 77: 5258–5262.

Tsao, M. S., A. Sakurada, et al. (2005). Erlotinib in lung cancer - molecular and clinical predictors of outcome. *N. Engl. J. Med.* 353, 133–144.

Ullrich, A. and J. Schlessinger (1990). Signal transduction by receptors with tyrosine kinase activity. *Cell* 61: 203–212.

Velu, T. J., L. Beguinot, et al. (1987). Epidermal-growth-factor-dependent transformation by a human EGF receptor proto-oncogene. *Science* 238: 1408–1410.

Weinstein, I. B. (2002). Cancer. Addiction to oncogenes the Achilles heal of cancer. *Science* 297: 63–64.

Wells, A. and J. M. Bishop (1988). Genetic determinants of neoplastic transformation by the retroviral oncogene v-erbB. *Proc. Natl. Acad. Sci. U. S. A.* 85: 7597-7601.

Witta, S. E., B. Helfrich, et al. (2005). Overcoming resistance to EGFR inhibitors in NSCLC cell lines by sequential treatment with histone deacetylase inhibitors. *J. Clin. Oncol.* 23: 641S.

Woodworth, C., D. Gaiotti, et al. (2000). Targeted disruption of the epidermal growth factor receptor inhibits development of papillomas and carcinomas from human papillomavirus-immortalized keratinocytes. *Cancer Res.* 60: 4397–4402.

Xiong, H. Q. and J. L. Abbruzzese (2002). Epidermal growth factor receptor-targeted therapy for pancreatic cancer. *Semin. Oncol.* 29, 31–37.

Yarden, Y. and M. X. Sliwkowski (2001). Untangling the ErbB signalling network. *Nat. Rev. Mol. Cell. Biol.* 2: 127–137.

Yarden, Y. (2001). The EGFR family and its ligands in human cancer: Signalling mechanisms and therapeutic opportunities. *Eur. J. Cancer* 37: Suppl 4, S3–S8.

Yauch, R. L., T. Januario, et al. (2005). Epithelial versus mesenchymal phenotype determines in vitro sensitivity and predicts clinical activity of erlotinib in lung cancer patients. *Clin. Cancer Res.* 11: 8686–8698.

10 PROGRESS IN ACHIEVING PROOF OF CONCEPT FOR p38 KINASE INHIBITORS

JERRY L ADAMS,* AND JOHN C. LEE†

*Ph.D., GlaxoSmithKline Pharmaceuticals, Collegeville, Pennsylvania
†Ph.D., GlaxoSmithKline Pharmaceuticals, King of Prussia, Pennsylvania

The year 1994 marked the discovery by three independent research groups of a novel mammalian stress-induced kinase (variously named as p38, RK and CSBP), which was subsequently elucidated to be central to a signaling pathway parallel to—but distinct from—the classical growth factor–induced MAP kinase cascade. This discovery, along with the identification of the JNK family of kinases, paved the way for intense research in the molecular and cellular aspects of these important signaling events in health and disease. In the ensuing years, an impressive list of discoveries was made, ranging from the identification of p38 homologues to a detailed understanding of the complex cell-signaling pathways and the biological significance of p38. These findings were facilitated by a pivotal study linking inflammatory cytokine biosynthesis to the p38 pathways using p38 inhibitors as chemical tools. Advances in our understanding of the kinome and other technology platforms have further facilitated the rapid progress in our understanding of p38 MAP kinases as drug targets. This chapter will review important achievements towards the development of p38 kinase inhibitors to treat disease.

I. INTRODUCTION

In 1988, Lee and co-workers at SmithKline and French Labs reported that SK&F 86002 (1), one member of a larger class of anti-inflammatory dihydroimidazo[2,1-b]thiazoles, inhibited LPS-stimulated IL-1 production in human monocytes (Lee, et al., 1988). The design rationale leading to the synthesis of SK&F 86002 was to prepare a compound that possessed the structural elements of the NSAID flumizole with that of the immunomodulator levamisole. The desired outcome of this strategy was the discovery of a compound that possessed both properties and therefore would be of potential utility for the treatment of chronic inflammatory disease, specifically rheumatoid arthritis. The initial hybrid molecules, while pharmacologically of interest, were poorly soluble. In an effort to improve aqueous solubility, a series of 4-pyridyl-dihydroimidazo[2,1-b]thiazoles were prepared. One of these compounds, SK&F 86002 (1), possessed a unique profile of anti-inflammatory and immunomodulatory properties. Preclinical studies revealed several shortcomings of SK&F 86002, and clinical development was not pursued. Subsequent studies identified the inhibition of both the cyclooxygenase (COX-1) and 5-lipoxygenase (5-LO) pathways as the potential mechanism for the potent anti-inflammatory activity seen with 1. With this mechanism in mind, further optimization of this series resulted in the identification of a new development candidate, SK&F 105809 (2). The clinical development of 2 progressed to Phase I clinical trials in 1990, but was later discontinued because of safety concerns arising from longer preclinical safety studies. Additional studies with SK&F 86002, however, subsequently demonstrated the inhibition of LPS-induced IL-1 production in human monocytes. This finding led us to switch our focus from dual 5-LO/COX inhibition to cytokine inhibition as the pharmacologically more important mechanism (Adams et al., 2001).

In 1990 we began a concerted effort to discover the mechanistic basis for the inhibition of IL-1 and TNF production by this compound class. It was readily determined that cytokine inhibition was independent of dual 5-LO/COX inhibition and that the compounds did not have nonspecific effects on DNA, RNA or protein synthesis. These and other studies established that the CSAID (Cytokine-Suppressive Anti-Inflammatory Drug) properties of (1) were not the result of nonselective toxicity, and that the effects were mechanistically distinct from those of corticosteroids. A more complete understanding of the cytokine suppressive mechanism required identification of the molecular target. To this end, radiolabeled ligands were prepared and used as probes to search for the molecular target in THP.1 cells. The THP.1 cells are a human monocytic cell line in which LPS-stimulated TNF production is inhibited by (1) in a manner identical to that of human monocytes. One such tool compound was radiolabelled SB 202190 (3), a potent triaryl imidazole inhibitor of TNF production ($IC_{50} = 50\,nM$). The uptake of (3) into THP.1 cells was found to be time and temperature dependent, saturable, and competitive with unlabeled (3) but not with an inactive analogue. These results suggested that a compound-specific binding site was present in THP-1 cell lysates. A binding assay was configured to quantitate the binding of ^{3}H-SB 202190 to the cytosolic fraction of THP.1 cells. The binding was specific, time dependent, reversible, and of high affinity (KDa $\simeq 50\,nM$). When a series of structural analogues was examined, a high degree of correlation was established between cytokine biosynthesis inhibition

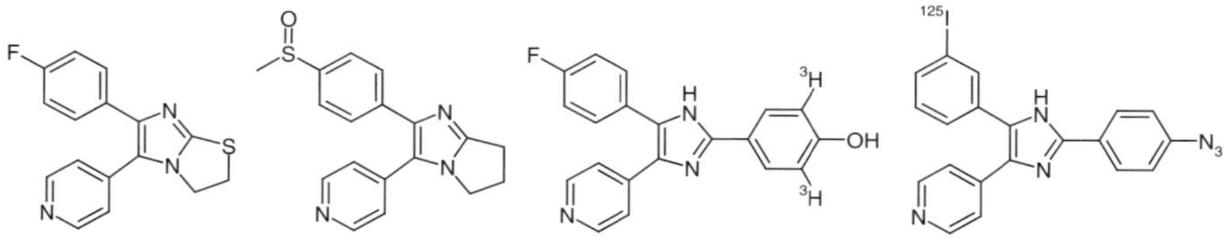

1 (SK&F 86002) 2 (SK&F 105809) 3 (SB 202190) 4 (SB 206718)

FIGURE 10.1 Early pyridinylimidazoles inhibitors of p38.

and competition in the binding assay, thus confirming that the cytosolic binding activity was related to cytokine inhibition. In order to identify the binding partner, the radio-iodinated aryl azide ^{125}I-SB 206718 (4) was synthesized as an irreversible photoaffinity label. Irradiation of this radiolabel in the presence of the partially purified THP.1 cytosol resulted in predominant labeling of a 43 kDa protein, which was identified as a pair of closely related novel serine/threonine protein kinases of the MAP kinase family. These kinases were termed CSAID binding proteins 1 and 2 (CSBP1, CSBP2). CSBP2 is the human orthologue of the murine p38α kinase (Lee, et al., 1994).

Over the ensuing years, advances in bioinformatic and genetic technologies have provided a new paradigm, linking targets to signaling pathways and to disease association. Our understanding of the MAP kinases, along with the kinome, has dramatically increased through target validation studies using knock-out and knock-down approaches, as well as more recently, *in vitro* and *in vivo* siRNA. The human kinome is now known to be composed of almost 500 protein kinases. With few exceptions, kinase inhibitors target the ATP binding site common to all kinases, and therefore selectivity rather than potency has proven to be the more difficult goal to achieve. This review of p38 kinase touches on recent developments in each of the above aspects, highlighting the value of advanced technologies and systems biology.

A. Biology of p38 Kinase

The p38 MAPK signaling pathway has been one of the most intensely studied topics in biology since its initial identification more than 10 years ago. The level of interest in this pathway has been primarily driven by two factors. First, this signaling pathway is activated by a wide array of stimuli, and is implicated in numerous diseases, most notably inflammation. Secondly, the early availability of selective p38 inhibitors provided the critical tools required to further delineate the role protein kinases play in signaling pathways, and the means to pursue the therapeutic potential of p38 inhibition. Indeed, during the last 5 years, a large number of p38 inhibitors have entered clinical trials.

At the time of the discovery of p38 MAP kinase, the first member of the MAP kinase family, extracellular signal-regulated kinase (ERK), had already been identified. It was not appreciated, however, that there were two additional sub-families of dual specificity threonine/tyrosine kinases (p38 and JNK). In 1994, several research groups independently identified a novel kinase activity (Freshney et al., 1994; Han et al., 1994; Rouse et al., 1994;) and subsequently, cloning of the human cDNA led to the identification of p38 α (Lee et al., 1994). Shortly thereafter, three other splice variants of the p38 family, p38β, P38γ, and p38δ were identified (Jiang et al., 1996, Jiang et al., 1997, Kumar et al., 1997). Two members of the family, p38α and β, are ubiquitously expressed but differentially regulated in different cell types, while the other two are more restricted in tissue distribution. To date, p38α, which has been implicated in a host of diseases, is the best understood member of the family (reviewed in Kumar et al., 2003 and Saklatvala, 2004).

Activation of p38 has been observed in various organisms as a response to many stimuli. The p38α orthologues in yeast, worm, and frog have been implicated in osmoregulation, stress responses, and cell-cycle regulation. Regulation of p38α in mammalian cells has also been well studied (reviewed in Zarubin and Han, 2005). It is now clear that the p38α signaling pathway is complex, influenced not only by stimuli and cell type, but also by various regulators and combinations of upstream activating kinases. It is well known that there are two main upstream activating kinases for p38α, MKK3 and MKK6. In addition, there is a MKK-independent mechanism of p38α activation involving transforming growth factor activated protein kinase 1 (TAK1)-binding protein (TAB) (Ge et al., 2002). The activation of p38α can be achieved by autophosphorylation after interaction with TAB1. It is also suggested that p38 negatively regulates TAK1 signaling by phosphorylating TAB1 (Cheung et al., 2003). Downstream of TAK1 signaling is IKK, which serves as an essential activation step of the Tpl2 kinase and its downstream targets, MEK1 and ERK (Waterfield et al., 2004). Thus, inhibition of p38 results in TAK1 upregulation, which leads to ERK activation. This explains why activation of ERK is frequently observed in cells treated with p38α inhibitors. This finding highlights the need to understand potential nonlinear crosstalk between kinase signaling pathways. Apart from TAK1, other upstream kinases (MAP3K) are also implicated in the activation of p38α and its close neighbor, JNK. Also contributing to the activation process are small GTP-binding proteins such as Rac1 and Cdc42 and their interaction with PAK (p21-activated kinases) and MLK1.

Downstream substrates of p38 MAP kinases are MAPKAPK2 (Kotlyarov et al., 2002) and MAPKAPK3 (McLaughlin et al., 1996), which phosphorylate various substrates—including small heat shock protein 27 (HSP27), lymphocyte-specific protein 1(LSP1), cAMP response element–binding protein (CREB), transcription factor (ATF1), SRF, and tyrosine hydroxylase. Of special interest is the MAPKAPK2 substrate tritetraprolin, a protein that destabilizes mRNA (Tchen et al., 2004). MNK is another p38 substrate that appears to be involved in translational initiation, as it phosphorylates eIF-4E (Waskewicz et al., 1997). In addition, p38 activated kinase (PRAK) and mitogen and stress-activated protein kinase (MSK1) are also known to be activated by p38α, although MSK1 is also activated by ERK (Deak et al., 1998). Not surprisingly, there are a large number of transcription factors that are regulated by p38 (reviewed in Zarubin and Han, 2005). Some examples include activating transcription factors 1, 2, and 6, SRF accessory protein (Sap1), GADD153, p53, c/EBPb, myocyte enhancing factor 2C (MEF2C), MEF2A, DIT3, ELK1, NFAT, and high mobility group-box protein (HBP1). Other unrelated proteins, such as cPLA1, Na+/H+ exchanger isoform-1 (NHE-1), tau, keratin 8, and stathmin have also been shown to be substrates for p38α.

The mechanism by which p38 MAP kinase inhibitors suppress expression of inflammatory cytokines has been elusive. Inflammatory gene expression is highly regulated at both the transcriptional and post-transcriptional levels. Early studies examining the effects of p38 inhibitors in human monocytes suggested that the regulation of inflammatory cytokine biosynthesis (primarily IL-1 and TNF) occurred at the post-transcriptional level. Subsequently, it became evident that mRNA stability can be negatively influenced by p38 pathway inhibition (Frevel et al., 2003). This observation prompted further studies to show that other inflammatory response proteins, such as COX-2, were affected in a similar fashion (Lasa et al., 2000). Other mRNAs that are stabilized by p38 include MIP-1a, gm-CSF, VEGF, and MMP-1 and -3 (Dean et al., 2004). Interestingly, MAPKAPK-2, a substrate for p38 is also involved in mRNA stabilization by p38. Catalytically active forms of MAPKAPK-2 stabilize reporter mRNA, while dominant negative MAPKAPK-2 blocks its expression (Winzen et al., 1999).

A common characteristic of structural features that influence mRNA stability is the AU-rich motif in the extended 3′UTR. This motif was first described by Shaw and Kamen (1986). There are three distinct classes of such AU-rich elements (AREs): one contains a small number of AREs (such as c-Fos), the second contains a larger number of AREs possessing multiple pentamers (such as TNFα, COX-2, etc.), and the third class contains AREs lacking the pentamers, but containing U-rich regions. The AREs target mRNA for rapid deadenylation in cells. In general, p38-regulated AREs have similar structural motifs with multiple, overlapping pentamers in the 3′UTRs. There are exceptions, however, such as MMP-1 and -3 (Reunanen et al., 2002) and tristetraprolin (Mahtani et al., 2001, Tchen et al., 2004), mRNAs that contain at least one pentameric, and U-rich sequences. The precise mechanism by which p38 regulates mRNA stability remains unclear. It is thought that the downstream kinase MAPKAPK-2 is involved, as well as an elusive ARE-binding protein. There are a number of candidates (Dean et al., 2004), but none fulfill all of the criteria for being the protein that links p38 pathways and ARE-containing mRNA. Of these, tristetraprolin is an interesting possibility, although it serves mainly as an "off-switch" in mRNA stabilization.

A large body of data from preclinical studies indicates a central role of p38 in immunological and inflammatory responses (Dong et al., 2002; Kracht and Saklatvala, 2002; Kumar et al., 2003). It is now known that the p38 pathway is selectively activated in Th1 effector T cells in response to IL-12 and IL-18. The production of Th1 cytokines, such as interferon gamma is inhibited by p38 inhibitors, while production of IL-4, a Th2 cytokine, is not. In macrophages, a number of inflammatory cytokines—such as TNF, IL-1, IL-6, and IL-8—are regulated through p38 pathways. The MKK3 knock-out mouse embryo fibroblast responds to TNF, but not to IL-1, UV, or sorbitol, thereby indicating a role for MKK3 in TNF, but not IL-1 action. In p38α knock-out mouse embryo fibroblasts, however, IL-1 induced IL-6 production is severely compromised. These data suggest that different ligands will elicit the function of different kinases in the p38 pathway. Together with a large body of genetic and *in vivo* pharmacological evidence, these results support p38 as a valid target, whose inhibition could provide therapeutic benefit in a variety of inflammatory diseases, particularly rheumatoid arthritis (Foster et al., 2000). Other possibilities for pharmacological intervention include cardiac hypertrophy, Alzheimer's disease, vascular injury, psoriasis, and inflammatory bowel disease.

II. INHIBITORS

As documented by the number of publications in this field, the design and synthesis of p38 kinase inhibitors has been an intense area of research in the pharmaceutical industry. In part, this is because of the discovery of p38 kinase as the target of the pyridinylimidazoles and the attendant keen interest of the pharmaceutical industry in exploiting the clinical potential of this target. Interest in p38α and this compound class was further heightened by the finding that SB 203580 (5), despite being an ATP competitive kinase inhibitor, possessed high kinase selectivity (Cuenda et al., 1995). This discovery confounded the prevailing view that high selectivity would not be achievable for compounds that bound in the ATP site. Industrial interest has remained high in this area, attesting to the potential therapeutic benefits of an orally active anti-cytokine agent. The obstacles to realizing this goal have proven significant and have required researchers to become increasingly more sophisticated in their search for safe and effective p38 inhibitors.

A. First Generation Inhibitors

Early examples of the pyridinylimidazoles class of p38 inhibitors were SK&F 86002 (1) and SB 203580 (5), both of which contain a 4-pyridyl group and an adjacent *p*-F-phenyl moiety appended to an imidazole core. While the initial medicinal chemistry done at SK&F established these groups as key features required for inhibition of p38 (Gallagher et al., 1997), an understanding at a molecular level was unknown until the publication of a p38/inhibitor co-crystal structure by Tong et al. (1997). Important features observed in the Tong structure and additional examples from subsequent publications by others were: 1) a hydrogen bond between the 4-pyridyl nitrogen and the amide N-H of Met 109, 2) aryl binding pocket which is behind the site normally occupied by the adenine ring of ATP formed by Thr-106 and Lys-53, and 3) a hydrogen bond between Lys-53 and the unalkylated imidazole nitrogen atom. These features are illustrated in Figure 10.2B for SB 203580. The hydrogen bond between the nitrogen of the pyridine [or pyrimidine in the case of SB 220025(6)] and the N-H of Met 109 is analogous to that seen for the N-1 adenine of ATP (Figure 10.2A). This hydrogen bond acceptor-donor interaction between the kinase backbone in the linker region and inhibitors is an almost universal feature of inhibitor-protein kinase crystal structures. On the other hand, the interaction of the imidazole nitrogen atom with the conserved lysine (Lys-53) is seen in some, but not all, structures.

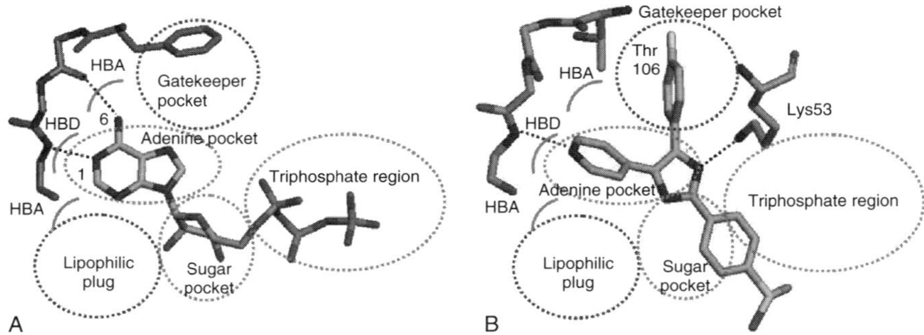

FIGURE 10.2 Traxler kinase pharmacophore model illustrated with ATP in CDK2 and SB 203580 in p38. A) Complex of human CDK2 bound to cyclin A and ATP: The perspective given is a top down view from the N-terminal to C-terminal domain in which the plane of the adenine ring matches that of the page. A short segment of the kinase linker region is shown which includes the hydrogen bonding residues (minus side chains) plus the gatekeeper residue. In this orientation, the top of the graphic corresponds to the back of the pocket and the bottom of the figure to the front or solvent. In the Traxler model, five sites were proposed of which three (adenine, sugar, and phosphate binding sites) can be directly related to ATP, and two lipophilic sites which lay outside of the region occupied by ATP (Traxler and Furet, 1999). B) The graphics were generated using DSVeiwerPro from Accelrys and the indicated PDB files. A) Complex of ATP, cyclin A and human CDK2 (PDB.1FIN), B) Complex of SB-203580 in p38αgPDB.1a9u).

Importantly, this structure did suggest the molecular basis for the selective inhibition of p38α by this structural class. Specifically, Tong proposed a key role for residue Thr-106, which formed one side of the fluorophenyl binding pocket that lay behind and orthogonal to the adenine of ATP. An alignment of the available kinase sequences revealed that p38 (both α and β) was one of a handful of serine/threonine kinases having such a sterically small residue at this position, and as such could accommodate a bulky fluorophenyl group. Whereas both p38α and β were potently inhibited by the pyridinylimidazoles, the other p38 family members (γ and δ), which possess a larger methionine gatekeeper, were not inhibited. Furthermore, the closely related ERK2, which has the larger glutamine side chain at position 106, is 1000-fold less sensitive to SB 220025 (6) ($IC_{50} = 20\,\mu M$) relative to p38α The publication of Tong et al. was quickly followed by similar findings for several tyrosine kinases that possess a sterically small residue at this position. These publications highlighted the significance of this aryl-binding pocket. Further compelling evidence that a single amino-acid residue, which is often termed the "gatekeeper" because it controls access to this binding pocket, was the primary factor governing kinase selectivity, was provided by mutagenesis studies with p38α and p38γ (Gum et al., 1998) and the tyrosine kinase, c-Src (Liu et al., 1998).

Although the primary role of the central imidazole is to serve as a scaffold to orient the vicinal aryl and 4-pyridyl rings, this heterocycle does interact with the p38 binding site and can profoundly affect inhibitor potency. Investigators at Merck were first to publish results demonstrating the nonessential nature of the central imidazole (de Laszlo et al., 1998). They investigated three sets of regioisomeric heterocyclic core replacements related to the triaryl imidazole 5. Little difference in potency was observed for the two regioisomeric furans (7 and 8) (p38 IC_{50} of 0.63 and 0.53 μM, respectively). A clear regioisomeric preference was seen with the pyrroles favoring 9 over 10 (p38 IC_{50} of 0.20 and 1.4 μM, respectively), and this preference was even more pronounced with the pyrazolones (11 and 12) (p38 IC_{50} of >0.50 and 0.035 μM, respectively). These data guided Merck chemists to synthesize the pyrrole 13 (p38 IC_{50} of 0.005 μM) analogue of SB 203580. The SAR for a series of ~40 additional pyrrole analogues paralleled that seen for the pyridinylimidazoles, suggesting a similar mode of binding. In the 10 years following the discovery of p38α as the target for the pyridinylimidazoles, the search for five-membered ring heterocyclic replacements for the core imidazole has been exhaustive (Jackson and Bullington, 2002). While a comprehensive rationalization of the SAR of these replacements is not readily discernable, most potent inhibitors retain a

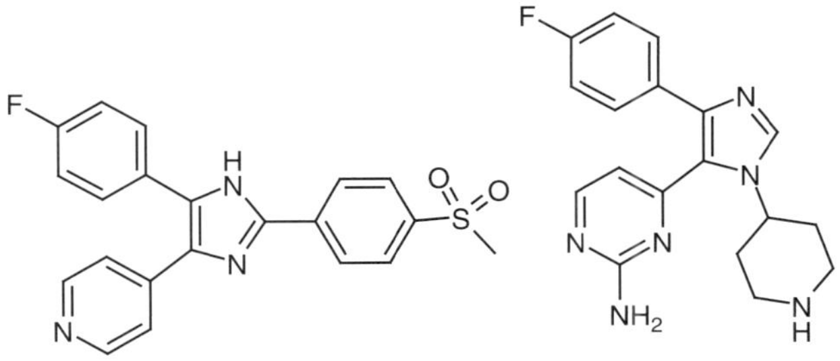

5 (SB 203580)

6 (SB 220225)

FIGURE 10.3 Pyridinylimidazoles used for p38 X-ray crystallography.

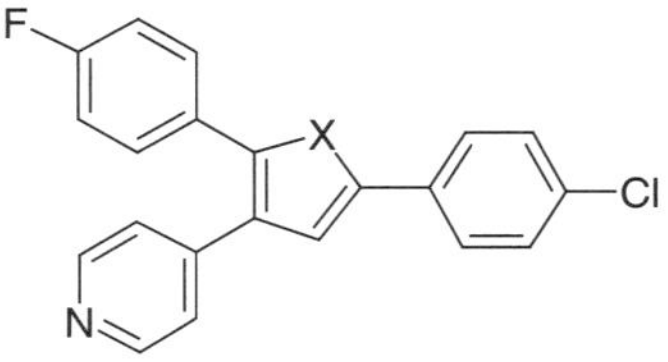

7, X = O
9, X = NH

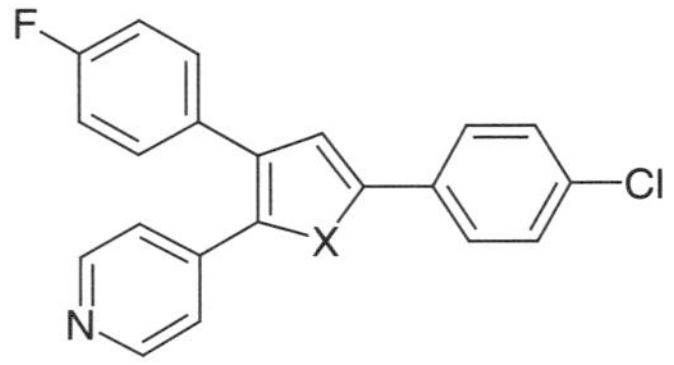

8, X = O
10, X = NH

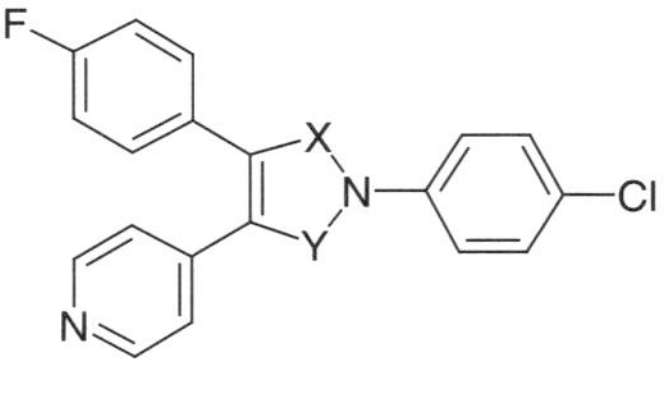

11, X = NH, Y = CO
12, X = CO, Y = NH

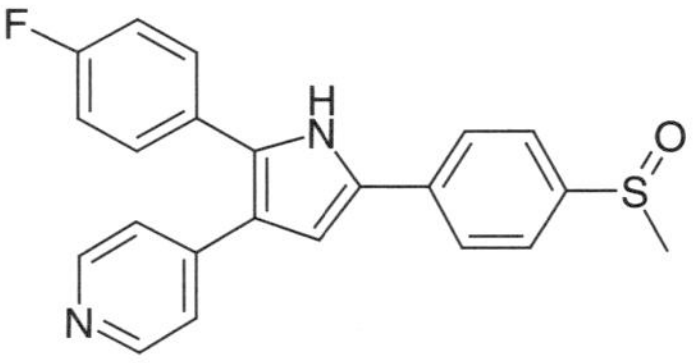

13, (L-167307)

FIGURE 10.4 Early examples of imidazole replacements from Merck.

14, X = O
15, X = NH

16 (SB 239063)

17 (SB 242235), R = OMe
18 (SB 267030), R = NH(2-MePh)

19

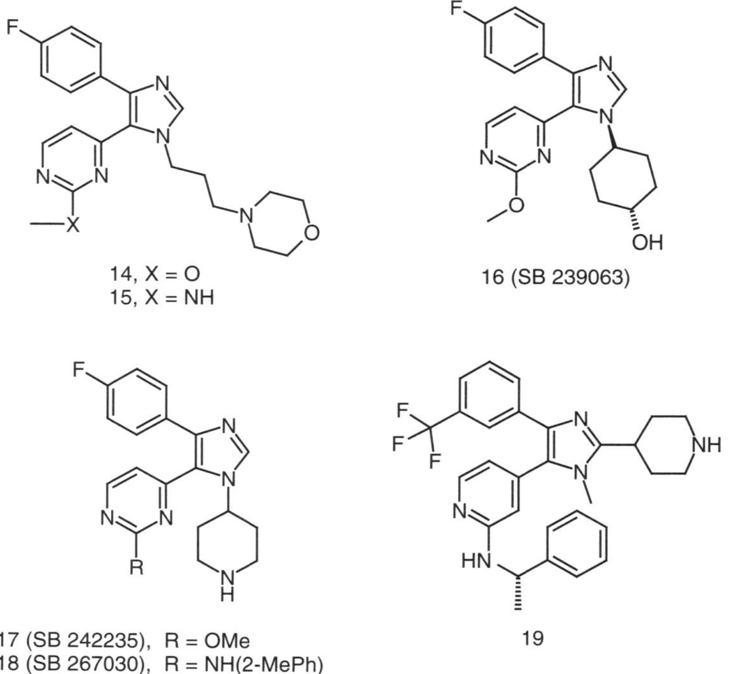

FIGURE 10.5 Pyridinylimidazoles optimized for reduced p450 inhibition and improved oral potency.

heterocycle having a hydrogen bond acceptor or donor adjacent to the aryl ring that binds in the gatekeeper pocket. This SAR suggests that maintaining an interaction with Lys-53 is generally required for potent inhibition.

As noted above, the 4-pyridyl group is essential for p38 inhibition. Unfortunately, the nitrogen lone pair of pyridine is known to ligate the ferric heme iron of cytochrome P450, and compounds containing sterically accessible pyridyl nitrogens are often potent inhibitors of P450 enzymes. This proved to be the case for the pyridinylimidazoles, such as 1 and 5. Thus the 4-pyridyl group, in addition to being required for binding to p38, was an important factor contributing to P450 inhibition. This limitation could be overcome by substitution of the pyridine with a pyrimidine ring, a weaker P450 ligand. Pyrimidin-4-yl analogues, which retained the hydrogen bonding ability of the pyridyl nitrogen, yielded compounds with the desired profile (Adams et al., 1998). Compounds containing either a 2-methoxy or a 2-aminopyrimidine (14 and 15) demonstrated improved p38α inhibition and reduced P450 inhibition. The p38α potency and oral activity of the pyrimidinylimidazoles was further improved by attachment of a cyclic group to the N-1 position of the imidazole ring (6, 16, and 17). Further optimization of the imidazole scaffold afforded the potent 2-anilinopyrimidine 18 (SB 267030, p38 Ki = 0.070 nM). Similarly, researchers at Merck (Liverton et al., 1999) have reported the potent, orally active tetra-substituted imidazole 19 containing an α-methylbenzylamine adjacent the 4-pyridyl nitrogen (p38 $IC_{50} = 0.19$ nM).

B. Second Generation Inhibitors: BIRB 796

In 2000 Dumas et al. at Bayer reported the discovery of new class of p38 inhibitors based upon a diaryl urea scaffold. Independently, in 2002 Regan et al. at Boehringer Ingelheim reported their efforts to develop a diaryl urea screening hit (20) obtained from a high-throughput screen. The activity of the initial BI compound was modest, and this finding might have not been pursued if not for the novel binding mode revealed in the co-crystal structure of 20 with p38 (Pargellis et al., 2002). Unlike the pyridinylimidazoles and inhibitors of other kinases for which X-ray structures had been determined, this N-pyrazole-N′-aryl urea did not occupy the ATP binding site. Instead the structure revealed a binding mode in which the pyrazole t-butyl group resided in the site normally occupied by the aryl group of the phenylalanine sidechain of the conserved DFG motif that marks the beginning of the activation loop. This binding mode required the outward (toward solvent) displacement of the phenylalanine residue by ~10 angstroms, thereby creating much of the binding pocket for 20. The X-ray structure also revealed that both nitrogen atoms and the oxygen atom of the urea were engaged in hydrogen bonds to the protein, a result consistent with SAR studies that showed these structural features to be essential for potent binding. In common with the pyridinylimidazoles, the 4-Cl-phenyl group of 20 occupied the gatekeeper binding pocket adjacent to Thr-106.

Efforts to optimize 20 began with the pyrazole ring. Attempts to replace the t-butyl group with both larger and smaller lipophilic groups reduced potency, as did the introduction of more polar groups. Therefore, the t-butyl group was retained in all future analogues. A striking ~100 fold enhancement in binding affinity was realized by the replacement of the pyrazole N-methyl with a phenyl or a p-tolyl moiety (21). As demonstrated by a co-crystal structure, the success of this change depends upon a lipophilic interaction with the side-chain methylenes of Glu-71, which are favorably disposed for this interaction due to the formation of hydrogen bonds between Glu-71 and the urea nitrogen atoms. Turning their attention to the 4-Cl-phenyl group, the BI group noted that the gatekeeper-binding pocket created by the small Thr-106 gatekeeper of p38α was only partially occupied. Several analogues possessing larger lipophilic groups were found to bind equivalent to, or better than phenyl. Among the most potent of these was the 1-naphthyl analogue 22. While the potency of 22 was only marginally better than 21 in a fluorescence-based binding assay a more sensitive assay suggested a 20-fold improvement in binding (Table 10.1). To their disappointment, this improved binding did not translate into more potent inhibition of TNF in THP-1 cells for 22 versus 21.

The X-ray co-crystal structure of a naphthyl analogue revealed that the distal aryl ring of the naphthyl group did fully occupy the gatekeeper pocket. Unexpectedly, binding of the added aryl ring of the naphthyl group into the gatekeeper pocket repositioned the aryl ring attached to the urea in such a manner that the *para* position was now directed toward the adenine binding pocket. To exploit this potential binding site, an inhibitor possessing an ethoxymorpholine group was prepared (24) and found to be 10 to 20 times more potent (compare 22 to 24). Addition of this same substituent to the phenyl urea resulted in a 15- to 30-fold loss in potency (compare 21 to 23), a result which strongly supports the importance of the naphthyl moiety for gaining access to the adenine pocket. In this instance (24, BIRB-796) the increase in binding ($K_d = 0.097\,nM$) did translate into potent inhibition of LPS-stimulated TNF inhibition ($IC_{50} = 18\,nM$). The X-ray structure of BIRB-796 verified the placement of the morpholine oxygen in the region of N-1 of adenine, well positioned as a hydrogen bond acceptor. Not surprisingly, the hydrogen-bond-acceptor role of the morpholinyl group of 24, which is the same interaction engaged by the pyridyl nitrogen of the pyridinylimidazoles, could be substituted by alternative hydrogen-bond acceptors, including the 4-pyridyl group, as seen in 25 ($IC_{50} = 0.10\,nM$).

C. VX-745 Inhibitor Class

A third major class of p38 inhibitors (examples are 26 and 27) was first disclosed in a patent from Vertex Pharmaceuticals in 1998 (Bemis et al., 1998). At GlaxoSmithKline (GSK), our interest these compounds was stimulated by

TABLE 10.1 ***In Vitro*** **Data for Boehringer Ingelheim Diaryl Ureas**

#	R group	Fluorescence K_D, nM	Exchange curve K_D, nM	Thermal denaturation K_D, nM	Inhibition of TNFα, THP-1 cell IC_{50}, nM
20	—	350			5900
21	phenyl	3	21		180
22	1-naphthyl	1	1	11	370
23	A	88	310		6200
24	B		0.097	0.046	18
25	C			0.10	20

As noted in the table headings, p38 potency was determined for the unactivated enzyme using a variety of binding assays.

the SAR disclosed in the patent. This SAR suggested that the 4-fluorophenyl groups of 26 and 27 might bind in the gatekeeper pocket originally described for SB203580 (5). Furthermore, we envisioned a binding model in which the carbonyl oxygen atoms of 26 and 27 replaced the pyridyl nitrogen of 5 as the hydrogen bond acceptor interacting with the amide backbone of the kinase linker domain (Figure 10.7). These two features of the Vertex inhibitors appeared to fulfill key aspects of a structure-based design strategy which we had initiated at GSK to discover novel p38 inhibitor chemotypes.

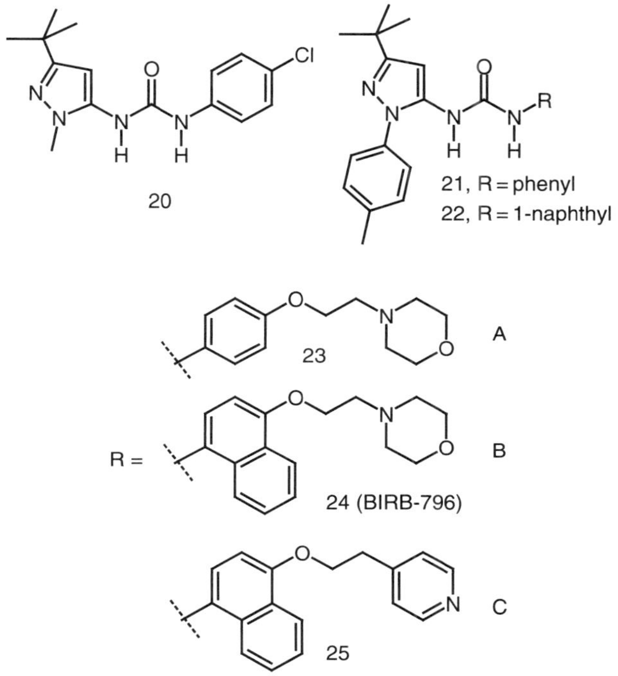

20

21, R = phenyl
22, R = 1-naphthyl

R =

23 A

24 (BIRB-796) B

25 C

FIGURE 10.6 Boehringer Ingelheim Diaryl Ureas: BIRB-796. For this model we assumed that the intramolecular hydrogen bond of the urea NH with the pyridine locked the conformation of 26 as illustrated. Compounds 26 and 27 are described in a Vertex patent application (Bemis et al., 1997).

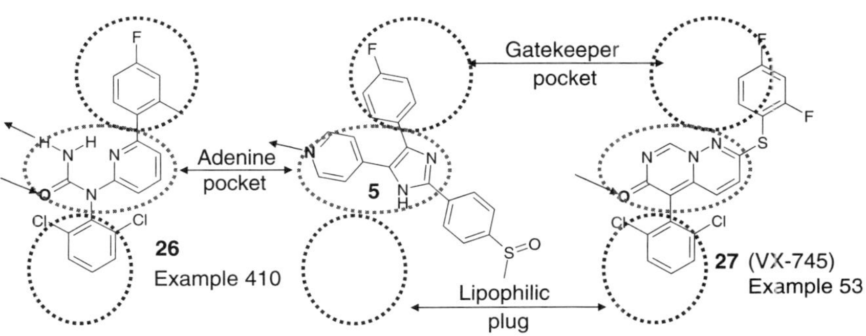

FIGURE 10.7 Binding hypothesis for Vertex inhibitors.

Specifically, for the design of a new p38 inhibitor class we desired to 1) retain a group that bound to the gatekeeper pocket in order to impart kinase selectivity and 2) maintain the adenine-mimetic backbone interaction to the kinase linker domain. In order to avoid the problems of P450 inhibition that had plagued the pyridinylimidazole template, we hoped to employ an amide carbonyl, such as that found in staurosporine, as the backbone binding adenine-mimetic. The required gatekeeper and adenine binding pocket fragments taken from the two inhibitor classes are illustrated Figure 10.8. Of secondary consideration was retention of an aryl ring to fill the lipophilic plug at the front of the ATP binding pocket.

The execution of our design is illustrated in Figure 10.9. Key variables that were examined included the size and nature of the amide ring (five-, six-, or seven-membered; additional heteroatoms), the point of attachment for the aryl ring binding in the gatekeeper pocket, and the nature of the linker connecting the two fragments. Molecular models were constructed, docked into the apo-p38 structure, and scored manually. Additionally, X-ray structures in p38α were obtained for examples of the two structural classes illustrated in Figure 10.7, thereby confirming the proposed binding mode. The top-rated scaffolds from this assessment were structures having a six-

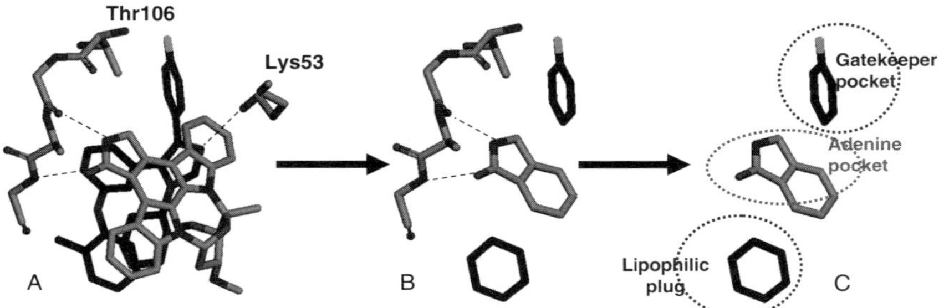

FIGURE 10.8 **Design of new template.** Panel A, Overlay of X-ray structures of SB 267030 (18) in p38 and staurosporine in PKA; Pane B, retained inhibitor fragments are the oxindole portion of staurosporine and two aryl rings of 18; Panel C, illustrates inhibitor fragment positions in the pharmacophore model.

X = NH, $NHCH_2$, $NHCH_2CH_2$
Y = O, S, N, or a bond
Z = N, CH_2

28 a–f

28	R1	R2	IC50, nM
a	Ph	Ph	91
b	Ph	Bn	78
c	Ph	CH_2CH_2Ph	4400
d	Ph	i-Pr	320
e	OPh	Ph	2,000
f	OPh	H	>17,000

29 a–d

29	R3	R4	IC50, nM
a	Ph	Ph	>17,000
b	NPh	Ph	1,660
c	SPh	Ph	2,820
d	NHC(O)Ph	Ph	>17,000

FIGURE 10.9 **Design of new inhibitor template.** Left panel, templates explored computationally by connection of gatekeeper pocket-bound phenyl ring from a pyridinylimidazole with the linker region-binding cyclic amide fragment of staurosporine Right panel, examples of quinazolinones prepared to optimize linker and site of attachment.

membered ring amide or urea. The choice of Z = NH versus CH2 was made to maintain the planarity of the fused ring system. Synthetically Z = NH was attractive in that it eliminated a potential chiral center and afforded a simple way to vary the substituent at R2/R4. Finally, the close mimicry of the quinazolone to that of the Vertex scaffold was a strong point in its favor. We could not discern the preferred point of attachment (five versus six) or nature of the linker Y, and as such we prepared a set of 22 compounds varying both factors, as well as the nature of the R2/R4 substituent (see Figure 10.9).

From these analogues a clear preference emerged favoring five- over six-substitution (28 versus 29) of the quinazolone and a direct attachment of R1 over the O-linked compound. In contrast to our initial design assumptions based upon the pyridinylimidazoles, a group occupying the lipophilic plug (28, R2 phenyl or benzyl) proved to be crucial for activity. An X-ray structure of 28a in p38 confirmed the expected binding mode in which R1 occupies the gatekeeper binding pocket, and the amide NH of Met-109 formed a hydrogen bond with the carbonyl oxygen of the inhibitor. Also noteworthy was the near orthogonal dihedral angles of both the R1 and R2 aryl groups of 28a to the plane of the quinazolinone, a feature that closely matched the placement of the analogous aryl groups in the urea 26. This data convinced us that further optimization of the 1,5-diaryl scaffold of 28a was warranted.

Prior to initiating this effort, we wanted to explore further changes in the scaffold to improve synthetic accessibility of analogues. We also desired a scaffold that would provide a facile method to introduce a third group (R3) which could be used as a handle to adjust physiochemical properties to provide "drug-like" molecules. These goals led us to consider the introduction of nitrogen atoms at positions six and/or eight of the scaffold, which would facilitate the introduction of both R1 and R2, and in addition would allow for the introduction of an R3 group (Figure 10.10). Molecular modeling indicated that these newly introduced nitrogen atoms at six and eight would be similarly positioned to those of the imidazole in the pyridinylimidazoles, and the nitrogen replacing C-6 might therefore hydrogen bond to Lys-53 (in the structure of 28a this atom is four angstroms from the amine of Lys-53). Moreover the vector of the R3 group of 30 would place this group in the region of the sugar and phosphate binding pockets, a region known to be tolerant of functional groups that could be used to improve solubility.

For our initial targets we chose to vary R3 and prepare R1 = 2,6-difluorophenyl and R2 = 2-Me-4-F-phenyl. These aryl substituents were chosen to enforce the orthogonal diaryl dihedral angles revealed in the X-ray structure and to maintain the well-documented preference for a 4-fluorophenyl

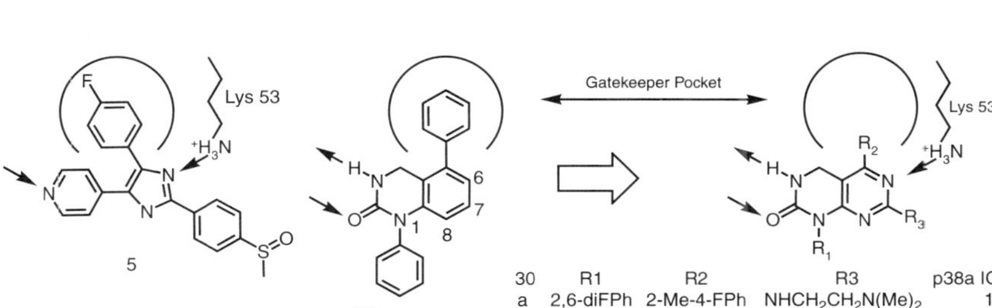

30	R1	R2	R3	p38a IC50, nM
a	2,6-diFPh	2-Me-4-FPh	$NHCH_2CH_2N(Me)_2$	16
b	2,6-diFPh	2-Me-4-FPh	SMe	26
c	Ph	Ph	SMe	630

FIGURE 10.10 Further design considerations.

moiety in the gatekeeper binding pocket. Furthermore, this substitution pattern was favored in the compounds illustrated in the Vertex patent and was supported by the limited SAR that we had obtained for the quinazolinones (28 and 29). The importance of the aryl substitution pattern was confirmed for the intermediate thiomethyl (30b) for which a ~24 fold enhancement in p38 potency was observed versus 30c (R1 = R2 = Ph). Oxidation of 30b to the sulfone followed by displacement with *N,N*-dimethylethylenediamine afforded 30a, a potent p38 inhibitor ($IC_{50} = 16\,nM$). This compound inhibited TNF production *in vitro* (LPS-stimulated human whole blood; $IC_{50} = 0.30\,\mu M$) and *in vivo* (murine LPS-stimulated TNF, $ED_{50} = 0.18\,mg/kg$). Further optimization of this scaffold led to the discovery of more potent inhibitors.

In close parallel to the GSK research, Fitzgerald et al. (2003) at Merck have reported the design and synthesis of quinazolinone and pyridol-pyrimidine p38 inhibitor scaffolds inspired by VX-745. A unique feature of the Merck work is the detailed X-ray crystallographic analysis of the binding of these compounds. As expected, these structures revealed the cyclic urea carbonyl to be hydrogen bonded to the backbone amide NH of Met-109, the same interaction observed between the N-1 of adenine and the 4-pyridyl nitrogen of the pyridinylimidazoles. Unexpectedly, the remaining lone pair of the carbonyl oxygen was engaged in a backbone amide hydrogen bond to the adjacent residue, Gly-110. This interaction required the amide to flip 180 degrees from the position in which it is usually observed. Because this amide flip was energetically most favorable for glycine, the Merck group speculated that kinases not having glycine at this position would be less sensitive to inhibition by VX-745 and related scaffolds. To test this hypothesis, the G110A and G110D site-specific mutants of p38 were prepared and were found to be less potently inhibited by VX-745 (43 and 150 fold, respectively), whereas SB 203580 was found to be a slightly more potent inhibitor of these mutants.

D. Kinase Selectivity

The obvious shortcomings of the early pyridinylimidazole inhibitors—poor pharmacokinetics and potent P450 inhibition—were readily addressed. What was less apparent, but of equal concern, was the kinase selectivity of these compounds. In the mid-1990s, the selectivity of kinase inhibitors was typically determined by testing a few phylogenetically related kinases plus examples of major kinase classes that were commercially available. Later, when a larger panel of 28 protein kinases was used to test the available kinase tool inhibitors, this initial approach to selectivity profiling was shown to be wholly inadequate, as many of the compounds proved to be potent inhibitors of a broad spectrum of kinases (Davies et al., 2000). In contrast, SB 203580 (5) and related inhibitors demonstrated better, but not absolute target selectivity. The kinases most sensitive to inhibition by 5 were those possessing a sterically small residue at the gatekeeper position, such as that present in p38α (Thr-106). More recently, an affinity chromatography approach using an immobilized derivative of 5 has been used in a proteome-wide survey for pyridinylimidazole binding proteins expressed in HeLa cells (Godl et al., 2003). This technique captured ~30 proteins, most of which were either protein kinases or highly expressed proteins. The captured protein kinases were JNK1 and 2, JAK1, RICK, GAK, PKNβ, CK1, and p38. Subsequent kinase activity assays using free 5 confirmed that RICK, GAK, and CK1 were potently inhibited, whereas at physiologically relevant ATP concentrations, the JNK enzymes were weakly inhibited. Both GAK and RICK possess a threonine at the gatekeeper position. This observation provides further support for the importance of a small gatekeeper residue in determining sensitivity to the pyridinylimidazoles.

A complementary and potentially more versatile approach to screening the human kinome has been described by researchers at Ambit Biosciences (Fabian et al., 2005). The Ambit technology relies upon phage display of kinase domains fused to T7 bacteriophage and a small set of ATP-competitive immobilized probe molecules, which are then allowed to compete for displacement of an inhibitor. The initially validated assay set contained 113 kinase domains, out of the ~518 protein kinases present in the human kinome. A clear advantage of this approach is that once the set of immobilized probe molecules is defined, any ATP-competitive inhibitor can be screened as the unmodified free ligand. The Ambit technology has been applied to representative members of the p38 inhibitor classes discussed in this chapter (Table 10.2).

TABLE 10.2 P38 Kinase Selectivity for Representative Compounds

Gene Symbol	SB 203580			BIRB-796		VX-745
	Davies	Godl	Fabian	Kuma	Fabian	Fabian
ABL1			>10		1.5	0.87
ABL2			>10		>10	0.22
c-Raf	2					
CSNK1E		0.124	0.22		>10	>10
EPHA3			>10		0.58	>10
EPHA5			>10		1.3	1.9
EPHA6			>10		0.43	>10
EPHA7			>10		0.22	>10
EPHA8			>10		0.14	3.1
EPHB4			>10		>10	1
FGR			>10		>10	1.2
FLT3			>10		1.3	>10
FRK			4.4		0.36	3.3
GAK		0.135	0.039		>10	>10
HCK			>10		>10	10
JNK1	>10		1.2		>10	>10
JNK2		11	0.095	50%	0.0056	>10
JNK3			0.045		0.062	>10
LCK	68%		>10	82%	1.1	1.8
MKNK2			>10		1.1	>10
p38α	0.05	0.038	0.017	0.08	0.0002	0.0032
p38β	0.5		0.25	0.38	0.22	0.16
p38γ	>10		1.7	0.92	0.014	>10
PDGFRβ			>10		1	0.87
RIPK2		0.018	2.1		>10	>10
SLK			4.6		0.086	>10
SRC			>10		>10	0.98
STK10			>10		0.0071	>10
STK36			0.86		>10	>10
TEK			>10		0.011	>10
TNIK			1.5		0.0099	>10
TTK			>10		1.2	>10

The reported IC50s are in μM using the assays outlined in the following references: Davies et al., 2000; Godl et al., 2003; Fabian et al., 2005, and Kuma et al., 2005. Kinases that had IC_{50} values >2 μM for all three compounds are not listed.

In common with the affinity chromatography method, this technique revealed the binding of 5 to GAK, RICK, and CK1. Importantly, the Ambit technology appears to have improved sensitivity, detecting interactions of 5 with several kinases not seen with the affinity chromatography method. Considering that both methodologies measure a competition in binding for a small-molecule ATP binding site ligand, and not the inhibition of substrate phosphorylation, the success of these techniques is remarkable.

Contributing to the desire to develop new inhibitor templates of increased selectivity is the concern that off-target kinase inhibition might lead to an unacceptable side effect profile. The initial publications describing the BIRB-796 (24) (Pargillis et al., 2002) and VX-745-like inhibitor classes (Fitzgerald et al., 2003) suggested improved selectivity of these compounds versus SB-203580. When evaluated in the broader Ambit panel of kinases, no clear selectivity advantage emerges for these new templates. Instead, what is clear is that each template has a different inhibition profile (see Table 10.2). These biochemical techniques do not address the importance of these binding results in a cellular environment and, as such, they can only be considered a screen for potential off-target interactions. In this regard, a recently described chemical genomics approach to identify the intracellular targets of kinase inhibitors provides a look forward to the type of new methodologies that may allow us to better determine inhibitor selectivity in the context of the intact cell (Kung et al., 2005).

III. CLINICAL STATUS

In those cases where clinical data have been reported, p38 inhibitors have generally demonstrated good oral bioavailability, excellent pharmacodynamic profiles, and no significant adverse events in single dose studies. In repeat dose studies, however, liver-enzyme elevations have been noted in a significant portion of the treated cohorts with three different inhibitors. For AMG-548, which was derived from the pyridinylimidazole inhibitor class, increased hepatic transaminase levels were found not to be dose or exposure dependent, and increases in bilirubin or alkaline phosphatase were not observed (Domiguez, et al., 2005). For BIRB-796 (24), increases in liver enzymes were seen at the highest doses in repeat dose studies. The compound did complete Phase II studies in rheumatoid arthritis (RA), Crohn's disease, and psoriasis. VX-745 (27), an example of a third structural class, completed a 12-week Phase II clinical trial in RA. Efficacy in RA was demonstrated in this trial against an ACR20 endpoint, but development was not continued because of CNS concerns discovered in preclinical studies. The most frequent adverse event was elevation of liver transaminases starting in week four. In place of 27, Vertex has advanced VX-702, a p38 inhibitor that does not penetrate the CNS, for evaluation in patients with acute coronary syndrome that have undergone bypass surgery. Encouragingly, five days of dosing VX-702 resulted in reduced serum levels of C-reactive protein, a marker of inflammation that is an accepted risk factor in coronary injury, and this reduction was maintained for 4 weeks following dosing. An additional compound, SCIO 469, is in Phase II studies for the treatment of pain in dental patients, multiple myeloma, and rheumatoid arthritis. This compound has demonstrated a significant increase in the medium time to use of rescue medication (ibuprofen) when compared

TABLE 10.3 P38 Inhibitors in Clinical Studies

Compound Name	Company	Indication	Status
PS 504466	Pharmacopeia/ Bristol-Myers Squibb	Rheumatoid arthritis (RA)	Phase I
Pfizer—unknown	Pfizer	Rheumatoid arthritis	Phase I
TAK-715	Takeda	RA	Phase II
AVE-9940	Sanofi-Aventis	RA, bone disorders	Phase II
VX-702	Vertex	Acute coronary syndrome	Phase II
SCIO 469	Scios, Johnson & Johnson	RA, myeloma, pain	Phase II
VX-745 (**27**)	Vertex	RA	Discontinued
SB-242235 (**17**)	GlaxoSmithKline	RA	Discontinued
RWJ-67657 #	Johnson & Johnson	RA	Discontinued
BIRD 796 (**24**)	Boehringer Ingelheim	RA, psoriasis	Discontinued
AMG-548*	Amgen	RA	Discontinued

Except as noted the information on clinical status was obtained from the commercial databases, R&D Focus or Pharmaprojects; # Fijen et al., 2001; * Dominguez et al., 2005.

with placebo. In conclusion, indications of clinical efficacy in diverse inflammatory conditions and acute pain have been reported for three p38 inhibitors. These results provide the first clinical proof of concept for p38 inhibitors.

IV. CONCLUSION

With the availability of high-density binding arrays (Ambit screen), it is now appreciated that selectivity is a double-edged sword. For some kinase drug targets, selectivity is critical, and yet for others, cross reactivity between close members of the kinase family is desirable (Fabian et al., 2005). For p38α, the preclinical data demonstrating an irreplaceable role for this kinase in inflammatory cytokine signaling and synthesis provide a strong rationale to support the thesis that selective inhibition of p38α should be sufficient to have a profound impact on both acute and chronic inflammatory disease. However, p38α, which is present in all cells and activated by a diverse set of signals in a context-dependent manner, may also play a central role in cellular processes required for normal cell function. In this regard the p38 inhibitors currently in the clinic may provide answers to questions of both efficacy and safety.

Central to the validity of any mechanistic conclusions that may be drawn from the clinical evaluation of p38 inhibitors, or of kinase inhibitors in general, is the quality of the tools used in the evaluation. One key aspect of this evaluation for kinase inhibitors is the ability to determine pharmacodynamic (PD) markers related to mechanism of action, and to delineate the precise pathways in which the target kinase operates. This is particularly crucial in understanding on-target versus off-target effects. For example, a reliable and predictive PD marker would greatly facilitate the transition from preclinical studies to clinical trials. Many PD markers rely on gene expression and/or phosphoprotein profiling. The former approach has been used extensively either through taqman or gene chip technologies. Phosphoprotein profiling remains a work in progress, as there are many technical hurdles to overcome. Currently, this method requires antibodies to specific phosphoepitopes; however, a nonantibody strategy using LC-MS is rapidly being advanced. The ability to integrate data obtained from various platforms is crucial in linking inhibition of the target, p38α, to cellular activity and effects on signaling pathways, and to the observed *in vivo* pharmacology.

Following the discovery of the first examples of highly "selective" ATP-competitive inhibitors of protein kinases, our structural understanding of protein kinases and their interaction with inhibitors has grown immensely. Likewise, so has our ability to develop and deploy kinase assays that now promise to cover the entire human kinome. In the case of p38 kinase this increasing level of sophistication has led to the discovery of inhibitor templates that take advantage of two unusual features present in p38α, the small gatekeeper residue Thr-106 and Gly-110 in the kinase linker domain. These advances promise to provide inhibitors of increasing selectivity. The coupling of these advances in selectivity with the improved *in vivo* profiling techniques discussed above should provide for a more precise determination of proof of concept, as well as the potential to determine the mechanism of potential undesirable toxicity.

A decade of intense focus on p38 kinase inhibition has led to an increase in potential therapeutic opportunities. The initial focus on p38 inhibitors as potential disease-modifying agents to treat rheumatoid arthritis has expanded to the exploration of psoriasis, Crohn's disease, inflammatory bowel disease, chronic obstructive pulmonary disease, cardiovascular disease, stroke, cardiac hypertrophy, Alzheimer's disease, vascular injury, and others. Furthermore the expansion of our knowledge of both upstream activators and downstream effectors of p38α signaling has opened up new drug targets for examination. One such promising new target is MAPKAPK-2, whose central role in p38-induced inflammatory cytokine synthesis has been validated in a knock-out mouse (Kotlyarov et al., 1999). Undoubtedly as our understanding of p38 signaling pathways increases, new opportunities will arise moving us closer to the ultimate goal of developing new medicines to treat human disease.

REFERENCES

Adams, J. L., A. M. Badger, et al. (2001). p38 MAP kinase. molecular target for the inhibition of pro-inflammatory cytokines. *Prog. Med. Chem.* 38: 1–60.

Adams, J. L., J. C. Boehm, et al. (1998). Pyrimidinylimidazole inhibitors of CSBP/p38 kinase demonstrating decreased inhibition of hepatic cytochrome P450 enzymes. *Bioorg. Med. Chem. Lett.* 8: 3111–3116.

Bemis, G. W., F. G. Salituro, et al. (1997). *PCT Int. Appl.* WO 9827098.

Cheung, P. C. F., D. G. Campbell, et al. (2003). Feedback control of the protein kinase TAK1 by SAPK2a/p38. *EMBO J.* 22: 5793–5805.

Cuenda, A., J. Rouse, et al. (1995). SB 203580 is a specific inhibitor of a MAP kinase homologue which is stimulated by cellular stresses and interleukin-1. *FEBS Lett.* 364: 229–233.

Davies, S. P., H. Reddy, et al. (2000). Specificity and mechanism of action of some commonly used protein kinase inhibitors. *Biochem. J.* 351: 95–105.

Deak, M., A. Clifton, et al. (1998). Mitogen and stress-activated protein kinase-1 (MSK1) is directly activated by MAPK and SAPK2/p38, and may mediate activation of CREB. *EMBO J.* 17: 4426–4441.

Dean, J., G. Sully, et al. (2004). The involvement of AU-rich element-binding proteins in p38 mitogen-activated protein kinase pathway-mediated mRNA stabilisation. *Cell. Sign.* 16: 1113–1121.

de Laszlo, S. E., D. Visco, et al. (1998). Pyrroles and other heterocycles as inhibitors of p38 kinase. *Bioorg. Med. Chem. Lett.* 8: 2689–2694.

Dominguez, D., D. A. Powers, and N. Tamayo (2005). p38 MAP kinase inhibitors: Many are made, but few are chosen. *Curr. Opin. Drug Discov. Devel.* 4: 421–430.

Dong, C., R. Davis, and R. Flavell (2002). MAP kinases in the immune response. *Annu. Rev. Immunol.* 20: 55–72.

Dumas, J., R. Sibley, et al. (2000). Discovery of a new class of p38 kinase inhibitors. *Bioorg. Med. Chem. Lett.* 10: 2047–2050.

Fabian, M. A., W. H., III, Biggs, et al. (2005). A small molecule-kinase interaction map for clinical kinase inhibitors. *Nat. Biotechnol.* 23: 329–336.

Fijen, J. W., J. G. Zijlstra, et al. (2001) Suppression of the clinical and cytokine response to endotoxin by RWJ-67657, a p38 mitogen-activated protein-kinase inhibitor, in healthy human volunteers. *Clin. Exp. Immunol.* 124: 16–20.

Fitzgerald, C. E., S. B. Patel, et al. (2003). Structural basis for p38alpha MAP kinase quinazolinone and pyridol-pyrimidine inhibitor specificity. *Nat. Struct. Biol.* 10: 764–769.

Foster, M. L., F. Halley, and J. E. Souness (2000). Potential of p38 inhibitors in the treatment of rheumatoid arthritis. *Drug News Perspect.* 13: 488–497.

Freshney, N., L. Rawlinson, et al, (1994). Interleukin-1 activates a novel protein kinase cascade that results in the phosphorylation of Hsp27. *Cell* 78: 1039-1049.

Frevel, M., T. Bakheet, et al. (2003). P38 mitogen-activated protein kinase-dependent and independent signaling of mRNA stability of AU-rich element-containing transcripts. *Mol. Cell Biol.* 23: 425–436.

Gallagher,T.F., G. L. Seibel, et al. (1997). Regulation of stress-induced cytokine production by pyridinylimidazoles; inhibition of CSBP kinase. *Bioorg. Med. Chem.* 5: 49–64.

Ge, B., H. Gram, et al. (2002). MAPKK-independent activation of p38alpha mediated by TAB1-dependent autophosphorylation of p38alpha. *Science* 295: 1291–1294.

Godl, K., J. Wissing, et al. (2003). An efficient proteomics method to identify the cellular targets of protein kinase inhibitors. *Proc. Natl. Acad. Sci. U. S. A.* 100: 15,434–15,439.

Gum, R. J., M. M. McLaughlin, et al. (1998). Acquisition of sensitivity of stress-activated protein kinases to the p38 inhibitor, SB 203580, by alteration of one or more amino acids within the ATP binding pocket. *J. Biol. Chem.* 273: 15,605–15,610.

Han, J., J. Lee, et al. (1994). A MAP kinase targeted by endotoxin and hyperosmolarity in mammalian cells. *Science* 265: 808–811.

Jackson, P. F., J. L. Bullington (2002) Pyridinylimidazole based p38 MAP kinase inhibitors. *Curr. Top. Med. Chem.* 2: 1011–1020.

Jiang, Y., C. Chen, et al. (1996). Characterization of the structure and function of a new mitogen-activated protein kinase (p38b). *J. Biol. Chem.* 271: 17,920-17,926.

Jiang, Y., H. Gram, et al. (1997). Characterization of the structure and function of the fourth member of p38 group mitogen-activated protein kinases, p38dalta. *J. Biol. Chem.* 272: 301, 220–301, 228.

Kotlyarov, A., A. Neininger, et al. (1999). MAPKAP kinase 2 is essential for LPS-induced TNF-alpha biosynthesis. *Nat. Cell Biol.* 1, 94–97.

Kotlyarov, A., Y. Yannoni, et al. (2002). Distinct cellular functions of MK2. *Mol. Cell. Biol.* 22: 4827–4835.

Kracht, M. and J. Saklatvala (2002). Transcriptional and post-transcriptional control of gene expression in inflammation. *Cytokine* 20: 91–106.

Kumar, S., J. Boehm, and J. Lee (2003). P38 MAP kinases: Key signalling molecules as therapeutic targets for inflammatory diseases. *Nat. Rev.* 2: 717–726.

Kumar, S., P. McDonnell, et al. (1997). Novel homologues of CSBP/p38 MAP kinase: Activation, substrate specificity and sensitivity to inhibition by pyridinyl imidazoles. *Biochem. Biophys. Res. Commun.* 235: 533–538.

Kuma, Y., G. Sabio, et al. (2005). BIRB796 inhibits all p38 MAPK isoforms *in vitro* and *in vivo*. *J. Biol. Chem.* 280: 19,472–19,479.

Kung, C., D. M. Kenski, et al. (2005). Chemical genomic profiling to identify intracellular targets of a multiplex kinase inhibitor. *Proc. Natl. Acad. Sci. U. S. A.* 102: 3587–3592.

Lasa, M., K. Mahtani, et al. (2000). Regulation of cyclooxygenase 2 mRNA stability by the mitogen-activated protein kinase p38 signaling cascade. *Mol. Cell. Biol.* 20: 4265–4274.

Lee, J. C., D. E. Griswold, et al. (1988). Inhibition of monocyte IL-1 production by the anti-inflammatory compound, SK&F 86002. *Int. J. Immunopharmacol.* 10: 835–843.

Lee, J. C., J. T. Laydon, et al. (1994). A protein kinase involved in the regulation of inflammatory cytokine biosynthesis. *Nature* 372: 739–746.

Liu,Y., K. Shah, et al. (1998). A molecular gate which controls unnatural ATP analogue recognition by the tyrosine kinase v-Src. *Bioorg. Med. Chem.* 6: 1219–1226.

Liverton, N. J., J. W. Butcher, et al. (1999). Design and synthesis of potent, selective, and orally bioavailable tetrasubstituted imidazole inhibitors of p38 mitogen-activated protein kinase. *J. Med. Chem.* 42: 2180–2190.

Mahtani, K., M. Brook, et al, (2001). Mitogen-activated protein kinase p38 controls the expression and posttranslational modification of tristetraprolin, a regulator of tumor necrosis factor alpha mRNA stability. *Mol. Cell Biol.* 21: 6461–6469.

McLaughlin, M., S. Kumar, et al, (1996). Identification of mitogen-activated protein (MAP) kinase-activated protein kinase-3, a novel substrate of CSBP p38 MAP kinase. *J. Biol. Chem.* 271, 8488–8492.

Pargellis, C., L. Tong, et al. (2002). Inhibition of p38 MAP kinase by utilizing a novel allosteric binding site. *Nat. Struct. Biol.* 9: 268–272.

Regan, J., S. Breitfelder, et al. (2002). Pyrazole urea-based inhibitors of p38 MAP kinase. from lead compound to clinical candidate. *J. Med. Chem.* 45: 2994–3008.

Reunanen, N., S. Li, et al. (2002). Activation of p38 alpha MAPK enhances collagenase-1 (matrix metalloproteinase (MMP)-1) and stromelysin-1 (MMP-3) expression by mRNA stabilization. *J. Biol. Chem.* 277: 32,360–32,368.

Rouse, J., P. Cohen, et al, (1994). A novel kinase cascade triggered by stress and heat shock that stimulates MAPKAP kinase-2 and phosphorylation of the small heat shock proteins. *Cell* 78: 1027–1037.

Saklatvala, J., (2004). The p38 MAP kinase pathway as a therapeutic target in inflammatory disease. *Curr. Opin. Pharma.* 4: 372–377.

Shaw, G. and R. Kamen (1986). A conserved AU sequence from the 3′ untranslated region of GM-CSF mRNA mediates selective mRNA degradation *Cell* 46: 659–667.

Tchen, C., M. Brook, et al. (2004). The stability of tristetraprolin mRNA is regulated by mitogen-activated protein kinase p38 and by tristetraprolin ttself. *J. Biol. Chem.* 279: 32,393–32,400.

Tong, L., S. Pav, et al. (1997). A highly specific inhibitor of human p38 MAP kinase binds in the ATP pocket. *Nat. Struct. Biol.* 4, 311–316.

Traxler, P. and P. Furet (1999). Strategies toward the design of novel and selective protein tyrosine kinase inhibitors. *Pharmacol. Ther.* 82: 195–206.

Waskiewicz, A., A. Flynn, et al. (1997). Mitogen-activated protein kinases activate the serine/threonine kinases Mnk1 and Mnk2. *EMBO J.* 16: 1909–1920.

Waterfield, M., Jin W., et al. (2004) I B kinase is an essential component of the Tpl2 signaling pathway. *Mol. Cell. Biol.* 24: 6040–6048.

Winzen, R., M. Kracht, et al. (1999). The p38 MAP kinase pathway signals for cytokine-induced mRNA stabilization via MAP kinase-activated protein kinase 2 and an AU-rich region-targeted mechanism. *EMBO J.* 18: 4969–4980.

Zarubin, T. and J. Han (2005). Activation and signaling of the p38 MAP kinase pathway. *Cell Res.* 15: 11–18.

11

IKK-2/NF-κB–DEPENDENT TRANSCRIPTION

F. CHRISTOPHER ZUSI,* WILLIAM J. PITTS,† AND JAMES R. BURKE†

**Ph.D., Bristol-Myers Squibb Co., Pharmaceutical Research Institute, Wallingford, Connecticut*
†Ph.D., Bristol-Myers Squibb Co., Princeton, New Jersey

Inhibitors of IKK-2, which inhibit signaling in the NF-κB pathway (initiated by TNF-α binding to its receptor), represent potential new drug candidates, as shown by the biology/biochemistry of the target and the *in vivo* activity of inhibitors in a variety of disease-relevant models. In this chapter, the roles of NF-κB and IKK-2 in various disease processes are reviewed, followed by a discussion of the enzymology of IKK-2 and the data supporting its role. A summary of literature reports concerning BMS-345541, an IKK-2 inhibitor with activity in multiple disease models, follows, along with briefer reports on the published activity of other IKK-2 inhibitors. Next, a general discussion of the possible consequences of inhibiting this enzyme is given, followed by some general conclusions and outlook for clinical success.

I. INTRODUCTION

During the past 30 years, enormous progress has been made in elucidating the biochemical signaling pathways associated with the inflammatory/immune system. This has led and still leads toward the development of new and improved therapeutics for the clinic. There are now three biological agents approved in the United States for the treatment of rheumatoid arthritis which block the activity of tumor necrosis factor alpha (TNFα): etanercept, infliximab, and adalimumab. These agents are also being investigated clinically for a variety of other conditions. Downstream cellular signaling events triggered by TNFα thus become targets for additional pharmacological intervention, to potentially improve efficacy and side-effect profiles. One key mediator in the TNFα response is the transcription factor, nuclear factor binding to kappa-B (NF-κB).

Target Validation in Drug Discovery

The NF-κB pathway is central to the expression of a variety of genes crucial to inflammatory and immune responses and to cell survival and proliferation. Examples of these factors include the cytokines TNFα, IL-6, IL-8, and IL-1β, the adhesion molecules E-selectin, ICAM-1, and VCAM-1, and the enzymes nitric oxide synthase and COX-2 (Baeuerle et al., 1997; Siebenlist et al., 1994). Such a central signaling cascade has stimulated considerable long-term interest from the pharmaceutical industry as a source of potential therapies for such diseases as asthma, arthritis, transplant rejection, inflammatory bowel disease, etc. One attractive target in the NF-κB pathway is the kinase, I-kappa-B kinase-2 (IKK-2). This chapter describes currently-studied IKK-2 inhibitors along with their in vivo activities, which validate the pathway as a clinical target.

II. BIOLOGICAL ROLES OF NF-κB

A. General Immunology

TNFα was isolated and identified in the mid-1970s as a powerful antitumor factor. It was successfully cloned and sequenced almost 10 years later (Pennica et al., 1984), but subsequent attempts to generate small-molecule agonists/antagonists met with little success. The first reports of activity of anti-TNFα antibodies in rheumatoid arthritis (Elliott et al., 1994) spurred further research, and it gradually became apparent that many inflammatory processes were linked to levels of TNFα activity. One central mediator in the cascade is NF-κB.

NF-κB is a protein transcription factor which normally resides in the cytoplasm of unstimulated cells as an inactive complex with a member of the IκB inhibitory protein family. This class of proteins includes IκBα, IκBβs, and IκBε, which all contain ankyrin repeats necessary to form a complex with NF-κB (Whiteside and Israel, 1997). There are at least two pathways for NF-κB activation; the "canonical" activation pathway is shown in simplified form in Figure 11.1.

Upon binding of TNFα to its membrane receptor (1), the signal is transmitted to the inner membrane (2), and the IKK complex (see section IV.B) is activated (3). IKK-2 (also known as IKK-β) phosphorylates the IκB (4), which is bound to NF-κB, shown as the prototypical heterodimer of p50 and p65. The phosphorylated IκB is then ubiquitinated (5), leading to its degradation by the proteasome (6). The liberated NF-κB, with a nuclear localization signal region exposed by the loss of the IκB, is imported into the cell nucleus (7), where it binds to DNA and initiates the transcription (8) of the pro-inflammatory signal molecules listed above.

Because of the importance of NFκB in regulating inflammation, it is expected that inhibitors of the pathway will have potent and broad-spectrum anti-inflammatory activity in a number of human disorders. Consistent with this, glucocorticoids are effective anti-inflammatory agents whose activity is mediated primarily through the inhibition of NF-κB–mediated transcription (Saklatvala, 2002). Because the use of glucocorticoids is limited by toxicities unrelated to their effects on NF-κB, agents which more selectively affect NF-κB ought to have glucocorticoid-like therapeutic effects without glucocorticoid receptor-mediated side-effects.

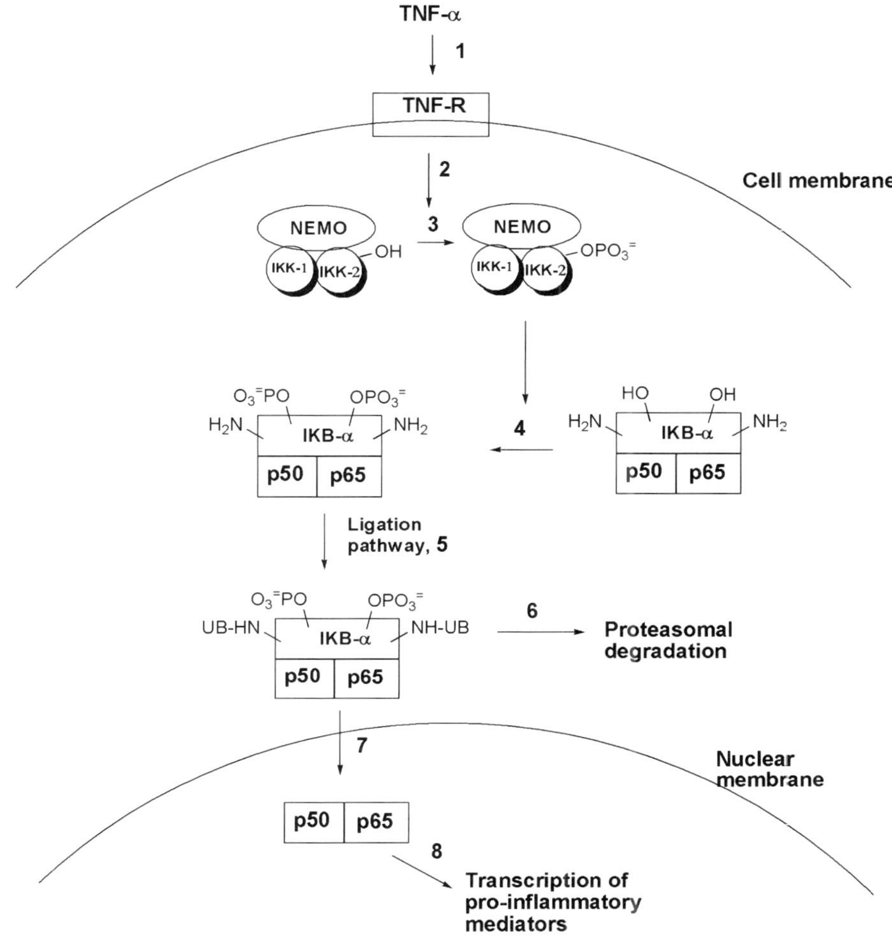

FIGURE 11.1 Canonical NF-κB activation pathway.

B. Inflammation/Immunity

Rheumatoid arthritis (RA) is characterized by inflammation of the joints with associated destruction of cartilage and bone. The transcription of cytokines thought to play a critical role in its pathobiology is NF-κB dependent, as is the transcription of a number of other mediators of the inflammatory and destructive pathways. For example, matrix metalloproteinases (MMPs) that catalyze the destruction of collagen within the joint are transcriptionally regulated by NF-κB, whereas their endogenous inhibitor, TIMP-1, is not (Bondeson et al., 1999). Consistent with a role for IKK-2 in rheumatoid arthritis, intraarticular administration of an adenoviral construct expressing a dominant-negative IKK-2 to rats significantly ameliorated the severity of an adjuvant-induced arthritis (Tak et al., 2001).

Osteoarthritis is not thought to have a significant inflammatory component, but the destructive mechanisms are analogous to those in rheumatoid arthritis, with MMPs playing a major role. In osteoarthritis, MMP-13 is thought to participate in the joint destruction, and the transcription of this key MMP, like that of MMP-1 and MMP-3, is regulated by NFκB (Mengshol et al., 2000). COX-2 expression is also NF-κB dependent, so that a pathway inhibitor should inhibit both MMP-driven joint destruction and prostaglandin-mediated pain.

Inflammatory bowel diseases, such as ulcerative colitis and Crohn's disease, are characterized by chronic relapsing inflammation. Therapy for these disorders typically includes sulfasalazine and short-term glucocorticoid use, and although these agents are moderately efficacious for inflammatory bowel diseases, improved therapies are still needed. Clinical reports of successful treatment using anti-TNFα antibodies (e.g., infliximab) indicate that cytokines are also involved here (Keating et al., 2002). The use of a peptide inhibitor of NF-κB nuclear localization has been shown to be effective in a murine model of IBD (Fujihara et al., 2000). Other pathway inhibitors, therefore, may prove to be very useful in the treatment of inflammatory bowel disease.

With pulmonary inflammatory disorders such as asthma and chronic obstructive pulmonary disease (COPD), the argument for the use of NF-κB pathway inhibitors is also compelling. Asthma is characterized by a chronic underlying inflammation, with involvement of T_H2 lymphocytes and eosinophils. In addition to many of the chemokines, cytokines, and cell adhesion-molecules mentioned previously, eotaxin, granulocyte macrophage colony stimulating factor (GM-CSF), and RANTES also play an important role in disease, mediating the chemoattraction, activation, and longevity of eosinophils and T_H2 lymphocytes. These all depend on NF-κB for transcription. The production of the key T_H2 cytokine IL-5 is also dependent on NF-κB, as is the transcription of chemokines MIP-1α and MIP-1β, which have been implicated in T cell recruitment to sites of inflammation (Yang et al., 1998). The receptor for eotaxin, CCR3, is also transcriptionally regulated by IKK/NF-κB (Huber et al., 2002). Glucocorticoids (systemic and inhaled), acting primarily through suppression of NF-κB, as mentioned above, are the best current treatment for controlling the underlying asthmatic inflammation. Potentially serious side effects thought to result from the transactivation of glucocorticoid receptor–dependent genes quite often limit the dose and duration of therapy, particularly in children.

COPD is now recognized to involve the chronic inflammation of small airways and lung parenchyma, with neutrophils and macrophages playing a major role. Mediators of the inflammatory and destructive mechanisms in COPD include IL-8, which is chemotactic for neutrophils, and TNFα. Both are regulated by NF-κB/IKK (Barnes, 2002; Churg et al., 2002). In addition, MMPs play an important role in the degradation of components of the lung parenchyma (i.e., emphysema). Of relevance to the chronic bronchitis aspect of COPD, mucin expression may also be directly regulated by NF-κB (Lee et al., 2002). Interestingly, disease severity in patients with COPD is associated with increased epithelial levels of NF-κB (Distefano et al., 2002).

Other inflammatory and autoimmune disorders (e.g., atopic dermatitis, psoriasis, lupus, multiple sclerosis) may also be effectively treated by an NF-κB pathway inhibitor. In addition, it is now appreciated that atherosclerosis is an inflammatory disease of the vasculature, with chemokines/cytokines such as MCP-1 and TNFα playing an important role (Cyrus et al., 2002; Ross, 1999). Like TNFα, the expression of MCP-1 is NF-κB–dependent.

C. Transplant Rejection

Graft rejection following solid organ transplantation is a complex process involving numerous immune mediators. Important roles for both T cells and cytokines have been demonstrated in multiple animal models of graft rejection, as well as in human transplant patients. The current standard of care is the prophylactic treatment of graft rejection by inhibiting the activation of T cells with immunosuppressive drugs, such as cyclosporine and mycophenolate mefetil, as well as the inhibition of cytokine production and activity by glucocorticoids and the anti-IL2R antibodies. IL-2 and its receptor, IL-2R, are both transcriptionally regulated by IKK/NF-κB. In addition, IKK-2 is essential for effective dendritic cell antigen presentation (Andreakos et al., 2003). Targeting the IKK/ NF-κB pathway, therefore, may prove to be useful in the prevention of transplant rejection.

D. Cancer

NF-κB plays a key role in many aspects of tumor development and progression. Aberrant sustained activation of NF-κB has been reported in many tumors and has been implicated in tumorigenesis (Baldwin, 2001; Luque and Gelinas, 1997; Rayet et al., 1999), with the increase of NF-κB activation correlating with increased IKK activity (Romieu-Mourez et al., 2001; Yang and Richmond, 2001). Moreover, other tumors show NF-κB activity being induced following chemotherapy, possibly contributing to resistance to therapy (Wang et al., 1996).

There are many mechanisms by which NF-κB may affect carcinogenesis. First, NF-κB regulates the transcription of antiapoptotic factors such as XIAP, IEX-1, and the Bcl-2 homologues A1/Bfl-1, Nr13, and BCL-XL (Wu et al., 1998; Zong et al., 1999; Stehlik et al., 1998). Direct regulators of cell-cycle progression, such as cyclin D1 and the growth-promoting genes c-myc and c-myb, are also dependent on NF-κB (Duyao et al., 1992; Hinz et al., 1999; Toth et al., 1995). Angiogenesis, key to providing blood supply to growing tumors, may also be affected by NF-κB through regulation of the angiogenic factor VEGF (Chilov et al., 1997).

In agreement with a role in carcinogenesis, recent reports have highlighted the potential of blocking IKK/NF-κB as promising therapy for the treatment of various cancers. For instance, expression of a super-repressor IκBα suppressed the tumorigenicity of a nonmetastatic human pancreatic cancer cell line, PANC-1, in an orthotopic nude-mouse model (Fujioka et al., 2003). In addition, mouse embryonic fibroblasts derived from mice lacking both IKK-1 and IKK-2 showed increased cell death and p53 induction in response to doxorubicin (Tergaonkar et al., 2002), and inactivation of the IKK complex in Hs578T cells via expression of a dominant negative NEMO reduced soft agar colony growth (Romieu-Mourez et al., 2001). In another report, adenoviral expression of a dominant negative IKK-2 in A549 human-lung carcinoma cells resulted in the cells being much more sensitive to apoptotic cell death following TNFα treatment (Sanlioglu et al., 2001). Moreover, the proteasome inhibitor bortezomib (Millennium), which acts in part by blocking NF-κB, was recently approved for the treatment of relapsed and refractory multiple myeloma, (Adams, 2001; Schenkein, 2002).

In non-Hodgkin's lymphoma, the use of either a super-repressor IκBα or dominant negative form of IKK-2 was toxic to activated B cell-like diffuse large B cell lymphoma cells but not in germinal center B-like diffuse large B-cell lymphoma cells showing that NF-κB activity is required for survival of the activated B-cell-like sub-group (Davis et al., 2001). Because lymphoma in the activated B-cell-like sub-group is associated with a particularly poor 5-year survival rate, the possibility of effective treatment with IKK-2 inhibitors is intriguing.

In human melanoma cells, IKK is constitutively active, which leads to NF-κB activation and the resulting aberrant over-expression of chemokines such as CXCL1 and CXCL8 (Yang and Richmond, 2001). These chemokines have been implicated in melanocyte transformation and melanoma tumor progression both *in vivo* and *ex vivo*.

E. Ischemia/Reperfusion Injury

In cerebral ischemia, NF-κB regulates the transcription of a number of genes involved in the pathogenesis of the resulting damage. These include inducible nitric oxide synthase, IL-1α, TNFα, ICAM-1, and IL-6. After middle cerebral artery occlusion and reperfusion in mice, NF-κB was shown to be activated, and the use of p50 knock-out mice demonstrated that abrogation of NF-κB significantly reduced the ischemic damage (Schneider et al., 1999). In a similar ischemic brain damage model, either neuronal-specific deletion of IKK-2 or expression of a dominant negative inhibitor of IKK-2 markedly reduced infarct size, whereas constitutive activation of IKK-2 enlarged infarct size, indicating a key function of IKK-2 in ischemic brain damage (Herrmann et al., 2005).

Reperfusion injury is also critical in myocardial infarction, and targeting NF-κB activation is a compelling target for this indication as well. The use of either NF-κB "decoy" oligonucleotides or proteasome inhibitors to block NF-κB has reduced myocardial reperfusion injury in rodents (Gao et al., 2002; Morishita et al., 1997; Pye et al., 2003).

Gut ischemia-reperfusion that is inflicted, for instance, by abdominal injuries, surgery, or bowel infarction can lead to multiple organ dysfunction syndrome (MODS) and acute respiratory distress syndrome (ARDS). This is mediated by various cytokines and inflammatory mediators that contribute to tissue injury and failure, and blockade of NF-κB may reduce MODS and ARDS severity and mortality. Consistent with this hypothesis, enterocyte-specific deletion of IKK-2 in mice completely prevents systemic inflammation and lung failure after gut ischemia-reperfusion (Chen et al., 2003). This demonstrates that local, IKK-dependent expression of pro-inflammatory mediators at the site of injury is critical in the resulting systemic organ failure and that IKK inhibitors may be beneficial in the treatment of acute MODS.

F. Diabetes

High doses of salicylates, such as aspirin, have been shown to prevent fat-induced insulin resistance in rats and improve glucose metabolism in patients with type 2 diabetes (Hundal et al., 2002; Kim et al., 2001). Because salicylates are (weak) inhibitors of IKK, their action may be attributed to inhibition of the NF-κB pathway. Consistent with this, heterozygous reduction of IKK-2 protected against development of insulin resistance during high-fat feeding and in obese $Lep^{ob/ob}$ mice (Yuan et al., 2001). This is especially compelling because these heterozygotes have only a 50% reduction in IKK-2. It has been hypothesized that NF-κB–dependent chronic inflammation may represent a triggering factor in insulin resistance (Pickup et al., 1998), and indeed transgenic mice lacking IKK-2 in myeloid cells were protected from insulin resistance in a model of obesity-induced insulin resistance (Arkan, 2005). Interestingly, IKK-2 has been implicated in the TNFα-induced phosphorylation of insulin receptor substrate 1 (IRS-1), co-immunoprecipitates with IRS-1 from intact cells, and was shown to directly phosphorylate IRS-1 *in vitro* at sites homologous to the IκBα phosphorylation site (Gao et al., 2002).

G. Chronic Heart Failure

Chronic heart failure (CHF) is a syndrome in which the heart is unable to pump sufficient blood to support other organs and can occur in response to chronic hypertension, ischemia, and infection. A dominant underlying mechanism in CHF involves a sustained cardiac hypertrophy, and NF-κB is essential for cardiomyocyte hypertrophy induced by G protein-coupled receptors (Gupta et al., 2002; Hirotani et al., 2002). Consistent with this, sustained activation of NF-κB has been observed in a rat model of CHF (Frantz et al., 2003). Phenylephrine, endothelin-1, and angiotensin II–induced cardiomyoctye hypertrophy was abrogated by expression of dominant-negative IKK-2 in cardiomyocytes (Purcell et al., 2001).

III. ENZYMOLOGY

A. Background

The involvement of NF-κB in many diseases, as discussed above, has prompted investigations into points of intervention into the signaling cascade. Kinases are a well-represented, "druggable" class of targets and would be an attractive way to modulate the process. The I-kappa-B kinase complex, lying downstream of the TNFα signal, represents a central "control point" for NF-κB modulation.

B. Multi-subunit IKK Complex

The IKKs were identified in 1997 (DiDonato et al., 1997; Mercurio et al., 1997; Woronicz et al., 1997) as components of a large "signalsome." The signalsome is a multi-meric complex (500 to 900 kDa) composed of two catalytic subunits, termed IKK-1 and IKK-2 (also known as IKK-α and IKK-β, respectively), as well as a regulatory subunit, termed NEMO (or IKKγ), which is involved in transducing upstream signals into activation of the catalytic subunits by mechanisms that are not fully understood. The multi-subunit complex is composed of heterodimers of IKK-1/IKK-2 held together by dimeric interactions between two NEMO molecules (Miller and Zandi, 2001), although complexes containing homodimeric IKK-1 or IKK-2 have also been isolated (Khoshnan et al., 1999). The catalytic subunits are highly homologous, each containing a kinase domain followed by leucine zipper and helix-loop-helix domains. Both catalytic subunits are capable of catalyzing the phosphorylation of IκBα (Burke et al., 1999).

The use of gene-targeting experiments has clearly shown that all known pro-inflammatory stimuli, including cytokines, viruses, and bacterial lipopolysaccharide (LPS) require the IKK-2 subunit for NF-κB activation (Ghosh et al., 2002). While the role of IKK-1 in NF-κB activation is still unclear, recent evidence suggests that IKK-1 may play a role only in response to certain stimuli (e.g., RANK-ligand and Blys/BAFF) and in select cells such as mammary epithelial cells and B lymphocytes (Cao et al., 2001). IKK-1 has also been implicated in NF-κB activation by non–IκBα-mediated effects, either phosphorylating the p65 subunit of NF-κB or histone H3 directly (Anest et al., 2003; Sizemore et al., 2002; Yamamoto et al., 2003). Consistent with the dominant role of IKK-2 in the signal-induced activation of NF-κB through IκBα phosphorylation, IKK-2 was found to be 650 times more efficient than IKK-1 in catalyzing the phosphorylation of IκBα (Burke et al., 1999). Intriguingly, two recent reports have implicated that IKK-1 plays a checkpoint role in the proper control of IKK-2 activity and is important in the resolution of inflammation (Lawrence et al., 2005; Li et al., 2005). These findings suggest that targeting IKK-2 instead of IKK-1 would be desirable in an anti-inflammatory agent.

IV. IKK INHIBITORS

A. Preliminary Leads

The isolation and identification of the IKKs led to considerable efforts to identify inhibitors of the enzymes. Salicylates and salicylate-derived agents—such as 5-aminosalicylic acid, sulfasalazine, and aspirin—as well as the food derivative curcumin have been shown to be weak inhibitors of IKK-2 (Jobin et al., 1999; Weber et al., 2000; Yan et al., 1999; Yin et al., 1998). Cyclopentenone prostaglandins have also been shown to be inhibitors of IKK-2, acting by covalently modifying a cysteine residue within the activation loop of the kinase (Rossi et al., 2000). A number of other natural products capable of alkylation reactions are also inhibitors.

Because of their relatively poor potency and lack of specificity, the above-mentioned molecules do not represent attractive leads for drug development. Because many pharmaceutical companies have drug-discovery efforts targeting kinases, however, a significant body of relevant experience, including compound collections, already exists. Starting principally from high-throughput screening of such collections, many groups have generated IKK-2 inhibitors. A number of disclosures have appeared in the patent literature, containing varying levels of detail, and there is a growing body of peer-reviewed publications.

Kinase inhibitor selectivity is an important consideration during structure optimization. There are more then 500 protein kinases in the human genome (Manning, 2002). An analysis of the current generation of clinically important kinase inhibitors demonstrates that compounds in development have wide variations in selectivity when profiled against approximately 120 kinases (Fabian, 2005). As previously discussed, IKK inhibitors have potential utility in a number of therapeutic applications. Their degree of selectivity may affect whether off-target toxicity is observed. Thus, off-target toxicity profiles might affect the choice of clinical indication(s) for any particular agent. *In vivo* activity is often viewed as the best therapeutic validation of the target; however multiple kinases are known to play a role in inflammation. Thus, observations of efficacy must be regarded carefully (depending on the suitability of the kinase selectivity profile for safety and tolerability in the intended clinical use). Assays for the entire 'kinome' are not currently available; therefore the ultimate kinase selectivity will be unknown to some extent for this generation of inhibitors. As there have been no reports to date of clinical activity for IKK-2 inhibitors; the next sections will describe the results of *in vivo* testing.

B. BMS-345541

In the course of a general program to find selective inhibitors of a variety of protein kinases thought to be disease-relevant, scientists at Bristol-Myers Squibb identified BMS-345541 as a potent and selective inhibitor of IKK-2 (Burke et al, 2003) (Figure 11.2). The IC_{50} for IKK-2 (0.3 μM) was approximately 1 log lower than for IKK-1 (4 μM), and there was no appreciable activity against a panel of other serine/threonine and tyrosine kinases. BMS-345541 displayed nonlinear inhibition kinetics but is not ATP-competitive (unusual for a classic kinase inhibitor). When dosed at noncytotoxic concentrations to THP-1 cells stimulated with LPS, the production of several pro-inflammatory cytokines was inhibited in the same concentration range as the inhibition of IKB phosphorylation. This is consistent with a common pathway for production of the cytokines and the central importance of IKK-2 to the activation.

The compound's inhibitory potency and kinase selectivity led to the selection of BMS-345541 as a tool compound to confirm the involvement of IKK-2/NFKB in various murine disease models and investigate potential therapeutic utility. Preliminary pharmacokinetic characterization in mice revealed that it was highly water-soluble, nearly 100% orally bioavailable, had an intravenous half-life of 2.2 hours, and reached micromolar levels in the plasma after an oral dose of 10 mpk. Thus BMS-345541 was tested in a number of *in vivo* pharmacodynamic and disease models.

Intraperitoneal administration of LPS to mice results in the release of TNFα in the serum (Ghezzi et al., 2000). Orally administered BMS-345541 dose-dependently inhibited serum TNFα production (Figure 11.3).

The 10 mg/kg dose produced approximately 50% inhibition, consistent with the exposure mentioned above and the THP cell potency. The highest dose (100 mg/kg) produced a nearly complete inhibition.

BMS-345541 was tested orally in the murine collagen-induced arthritis (CIA) model in both a prophylactic and a therapeutic mode (McIntyre et al., 2003). The mice are inoculated on day zero with bovine type II collagen and boosted on day 21. Disease onset occurs within 7 days of the second injection. In the prophylactic mode, compound dosing is initiated on day zero, while in the therapeutic mode, dosing is initiated at the first manifestation of disease. BMS-345541 at 10, 30, and 100 mg/kg daily displayed dose-dependent

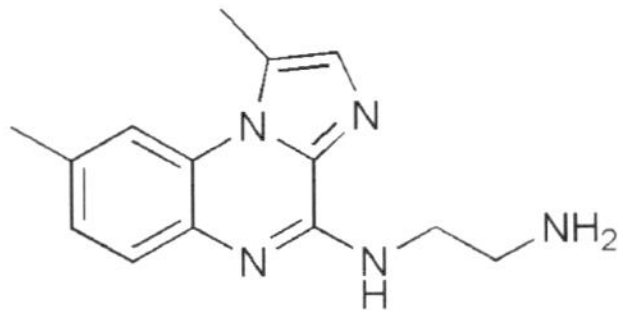

N N N N H NH2

BMS-345541

FIGURE 11.2 Structure of BMS-345541.

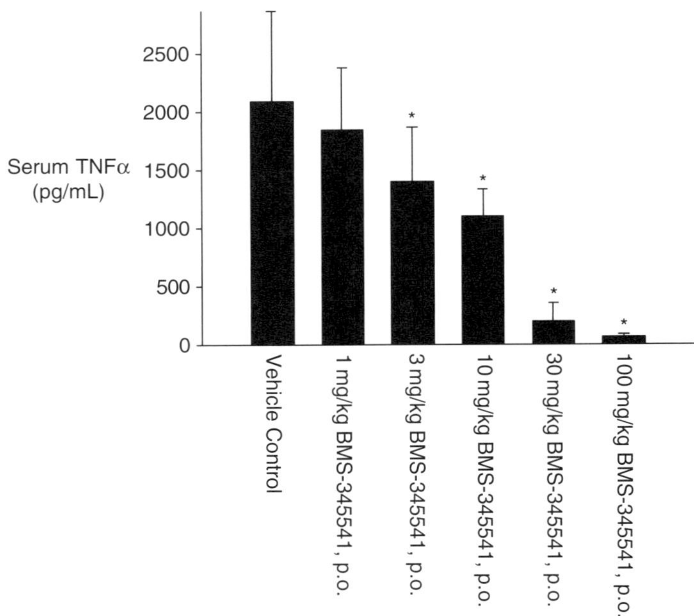

FIGURE 11.3 The effect of BMS-345541 on serum TNF concentrations induced by intraperitoneal injection of LPS. BMS-345541 was administered perorally 60 minutes prior to LPS challenge, and blood was drawn 90 minutes subsequent to challenge. Republished with permission from the *Journal of Biological Chemistry* (Burke et al., 2003), permission conveyed through Copyright Clearance Center, Inc.

inhibition of disease incidence and severity, with nearly complete suppression of all scores at the highest dose.

In the therapeutic mode (Figure 11.4), the 30 mg/kg dose prevented further disease progression, while 100 mg/kg resulted in improvement of existing symptoms.

The histological evaluations mirrored the clinical scores. Efficacy in the therapeutic dosing mode with an IKK-2 inhibitor is noteworthy because most agents, even methotrexate (the mainstay therapy for rheumatoid arthritis), are not effective in this model when administered after disease onset. Only glucocorticoids have previously been shown to be active therapeutically in this model, consistent with the idea that IKK-2 inhibitors will have glucocorticoid-like activity.

Cytokines are thought to play a critical role in the pathobiology of Crohn's disease, as discussed above. BMS-345541 was tested in the murine dextran sulfate sodium (DSS)-induced colitis model, compared to the positive control, sulfasalazine (MacMaster et al., 2003). Mice were given 6% DSS in drinking water for 1 week, with test compounds administered orally daily on days two to nine. On day 10, the animals were euthanized and the colons removed and graded for severity of crypt injury and inflammation. The results are shown in Table 11.1.

BMS-345541 at both doses significantly reduced the crypt injury and total inflammatory score, in contrast to sulfasalazine. Sufasalazine shows moderate clinically effectiveness in the treatment of inflammatory bowel disease. The superior results for the IKK inhibitor in this stringent model bode well for the eventual exploitation of this mechanism in the clinic.

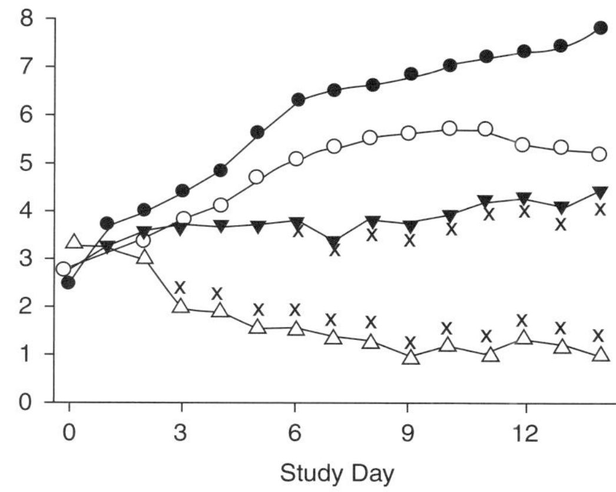

FIGURE 11.4 **Effect of BMS-345541 administered in the therapeutic dosing mode on disease severity in the murine model of collagen-induced arthritis.** BMS-345541 was administered by oral gavage once daily after disease onset had occurred. Data represent the average gross clinical scores for mice treated with 0 mg/kg (vehicle control) (filled circles), 10 mg/kg (open circles), 30 mg/kg (filled triangles), or 100 mg/kg (open triangles) BMS-345541. The x-axis denotes the number of days after initiation of treatment with BMS-345541 (i.e., after disease onset). $* = P < 0.05$ versus vehicle control. Reproduced from McIntyre et al., 2003, with permission from John Wiley & Sons, Inc.

TABLE 11.1 BMS-345541 Colitis Model results

Treatment	Clinical Score	% Reduction in injury score	% Reduction in inflammation score
Vehicle (water)	2.5 ± 0.2	0	0
BMS 345541, 100 mg/kg	1.0 ± 0.2*	33*	44*
BMS 345541, 30 mg/kg	1.0 ± 0.2*	43*	39*
Sulfasalazine, 100 mg/kg	2.6 ± 0.2	8	15

*$P < 0.05$ vs. vehicle control.

Anti-TNFα antibodies have been shown to be effective in prolonging graft survival in animal models. Therefore, BMS-345541 was also tested in a murine nonvascular cardiac transplant model (Townsend et al., 2004). Neonatal hearts were harvested from newborn mice and transplanted into the ears of adults, following a long-established model (Fulmer et al., 1963). The contractile activity of the transplanted organ was monitored daily by ECG, and the timing of the rejection was defined as the day contraction ceased.

By itself, BMS-345541 had no effect on graft survival. When co-administered with a suboptimal dose of either the co-stimulation inhibitor, CTLA4-Ig, or the immunosuppressive agent, cyclosporin A, a signficant extension of graft survival compared to either agent alone was seen. Median graft survival was 12 days for vehicle-treated or BMS-345541-treated mice. A suboptimal dose of CTLA4-Ig alone prolonged graft survival to 21 days, and the addition of 50 mg/kg BMS-345541 increased the median to 32 days. Similarly, a suboptimal dose of cyclosporin A (15 mg/kg) gave median graft survival of only 14 days, which was increased to 29 days by the addition of 50 mg/kg BMS-345541. The authors indicated that the likely mechanism of action was the modulation of the production of relevant cytokines, principally IL-2. These

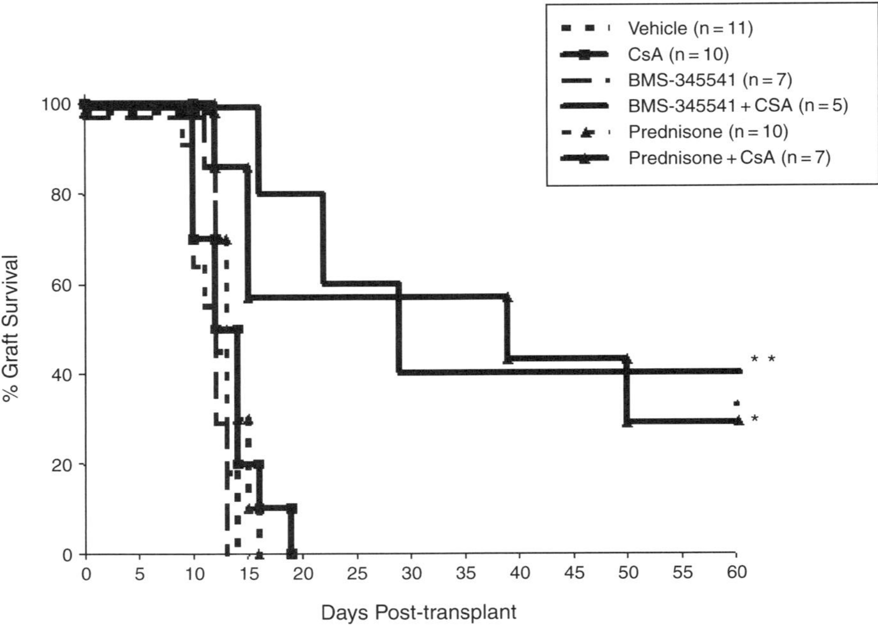

FIGURE 11.5 **Prolongation of graft survival. Neonatal C57BL/6 hearts were transplanted into the ears of adult BALB/c mice, and graft viability was monitored daily.** Mice that underwent transplantation were treated orally with water (Veh), BMS-345541 (50 mg/kg), or cyclosporin A (CsA) (15 mg/kg) at the time of transplant and daily thereafter for 60 days or until graft rejection occurred. Prednisone (3 mg/kg) was administered intramuscularly at the time of transplant and daily thereafter for 60 days or until graft rejection occurred. Data are presented as graft survival in days. ***P* less than 0.001 versus CsA and BMS-345541 alone. **P* less than 0.01 versus prednisone and BMS-345541 alone. Reprinted from Townsend *et al.*, 2003, with permission from Lippincott Williams & Wilkins.

effects of IKK inhibition in transplant rejection are similar to the synergistic effects of glucocorticoids with cyclosporin A, reiterating that IKK inhibitors may be effective as glucocorticoid replacements with potentially reduced side effects.

Consistent with the putative role of IKK in metastatic melanoma (Yang and Richmond, 2001), BMS-345541 was shown to reduce NF-κB activity and cell survival in melanoma cells through mitochondria-mediated apoptosis. In nude mice inoculated with various melanoma lines, BMS-345541 administered at doses ranging from 10 to 75 mg/kg once daily for 21 days showed dose-dependent inhibition of tumor growth, with the 75 mg/kg dose showing greater than 65% reduction in tumor growth (Yang et al., 2005). TUNEL staining of tumor sections from these mice showed BMS-345541 induced apoptosis in the tumors.

In a mouse model of stroke (cerebral ischemia) induced upon middle cerebral artery occlusion (MCAO), BMS-345541 administered by intracerebroventricular injection reduced the infarct volume by 60%, and the neuroprotective effect could be detected 2 weeks after MCAO (Herrmann et al., 2005). BMS-345541 was able to reduce infarct size when administered up to 4.5 hr after onset of MCAO. This is especially intriguing because a drug for stroke treatment would need to be effective hours after onset of ischemia since stroke patients receive medical attention after a time delay.

C. Other Compounds

Scientists at Millennium have identified a compound designated PS-1145 as a potent and selective IKK-2 inhibitor.

A 50-mg/kg dose in mice, 1 hour before LPS challenge, reduced the stimulated TNFα levels by 60% at the 5-hour time point (Castro et al., 2003). This compound has also been used to demonstrate the potential of a small-molecule inhibitor of IKK in cancer, first by showing inhibition of both IL-6 secretion from bone marrow stromal cells triggered by multiple myeloma-cell adhesion as well as the increased proliferation of adherent multiple myeloma cells. The compound also augmented TNFα-induced cell killing in multiple myeloma cells (Hideshima and Chauhan, 2002). In addition, PS-1145 was toxic to activated B cell–like diffuse large B cell lymphoma cells but not germinal center B-like diffuse large B-cell lymphoma cells, consistent with the requirement of IKK/NF-kB in the survival of the activated B cell-like subgroup (Lam et al., 2005). A second compound, MLN-120b (IKK-2 $IC_{50} = 0.062\ \mu M$), was reported to be efficacious in chronic models of inflammation (Nagashima et al., 2006). MLN-120b was efficacious in the rat adjuvant arthritis model (Wilkerson, 1994) with an ED_{50} of 11 mg/kg when dosed orally twice per day to provide 74% inhibition of paw swelling and protection against bone loss. MLN-120b was also efficacious in the rat collagen-induced arthritis model (Trentham et al., 1977) when dosed orally twice per day at 7 mg/kg.

GlaxoSmithKline scientists have reported the activity of TPCA-1 in the murine CIA model (Podolin et al., 2005).

In untreated mice, the time course of nuclear p65 activity in animals immunized on day zero and boosted on day 21 was closely correlated with the appearance of clinical symptoms. When the inhibitor was tested intraperitoneally at 3, 10, or 20 mg/kg b.i.d. from days 1 to 48, the clinical score was significantly reduced at the 10 and 20 mg/kg doses. If test article administration was delayed until the appearance of symptoms, only the 20 mg/kg dose showed

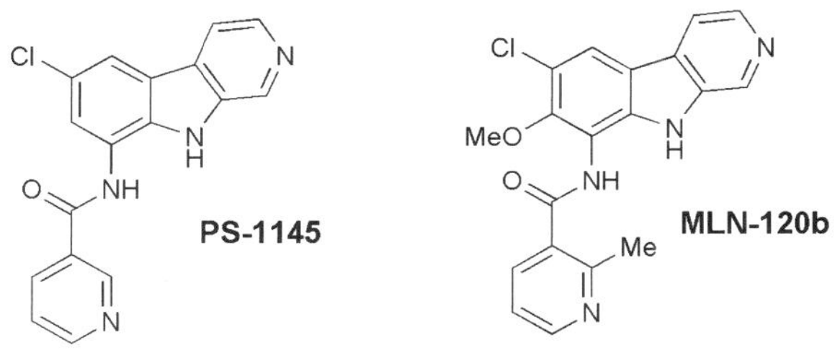

Cl NH O N H N N PS-1145

Cl MeO NH O N H N N Me MLN-120b

FIGURE 11.6 Millennium IKK-2 inhibitors.

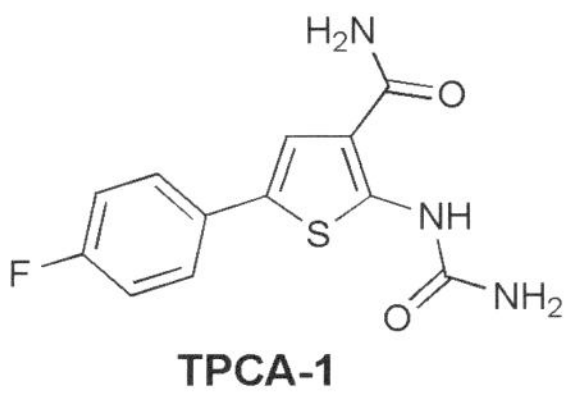

H_2N O S NH F O NH_2 TPCA-1

FIGURE 11.7 GlaxoSmithKline IKK-2 inhibitor.

activity. The tissue levels of cytokines and *ex vivo* antigen-produced T-cell proliferation decreased along with the clinical scores.

Scientists at the University of Tokyo have reported the salicylate-derived IMD-0354 and IMD-0560 as IKK-2 inhibitors, with multiple *in vivo* activities (Okazaki et al., 2005; Onai et al., 2004). No data have been presented showing direct inhibition of the isolated enzyme with these inhibitors (Figure 11.8).

In an ischemia-reperfusion model in rats, in which ischemia was produced by ligation of the left anterior descending coronary artery for 30 minutes followed by 24 hr of reperfusion, intraperitoneal injection of IMD-0354 at 5 or 10 mg/kg just prior to initiation of reperfusion reduced the size of the necrotic area by approximately 50% compared to the vehicle control. A reduction in inflammatory mediators was co-incident with the reduction in reperfusion damage. IMD-0560 was shown to be effective in a murine collagen-induced arthritis model at doses of both 1 and 3 mg/kg (IP), with significant effects on disease incidence and severity as well as pathological measurements of disease progression (Okazaki et al., 2005).

Effects against diet-induced insulin resistance with IMD-0354 was shown in KKA^y mice fed a high-fat diet (Kamon et al., 2004). When dosed i.p. daily for 14 days, IMD-0354 significantly decreased plasma glucose levels when insulin resitance testing was performed on day seven, and when glucose tolerance tests were performed on day 10, the plasma glucose levels were likewise lower in the treated group. These observations are consistent with IMD-0354 decreasing the insulin resistance. A likely mechanism is via the demonstrated elevation in plasma adiponectin levels.

The Tokyo Bayer group has reported that the compound ACHP, with an IKK-2 IC_{50} of 8.5 nM, has moderate aqueous solubility (120 μg/mL), good Caco permeability, and oral bioavailabilities in mice of 16% and in rats, 60% (Murata et al., 2004) (Figure 11.9).

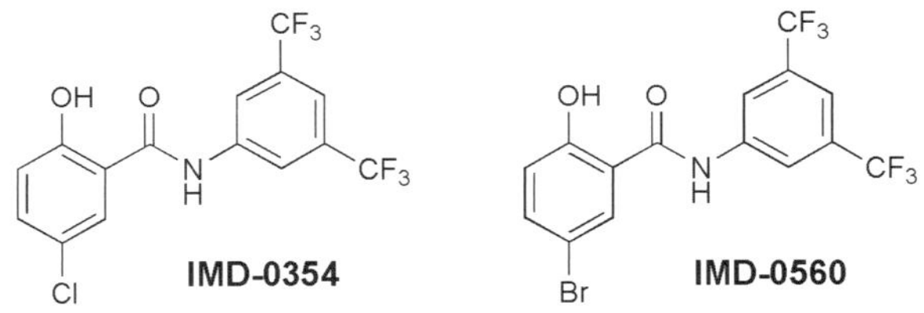

FIGURE 11.8 University of Tokyo IKK-2 inhibitors.

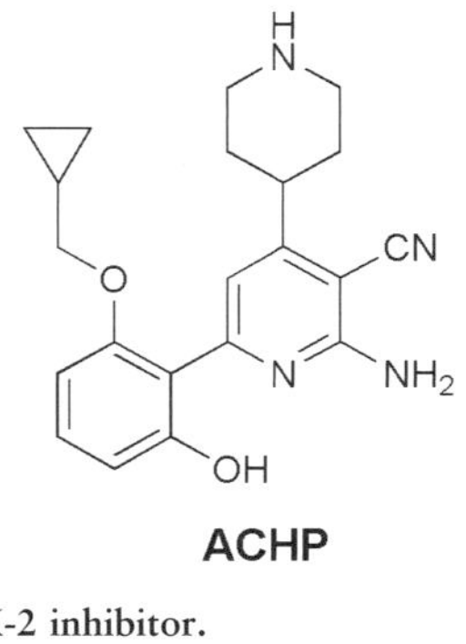

FIGURE 11.9 Tokyo Bayer IKK-2 inhibitor.

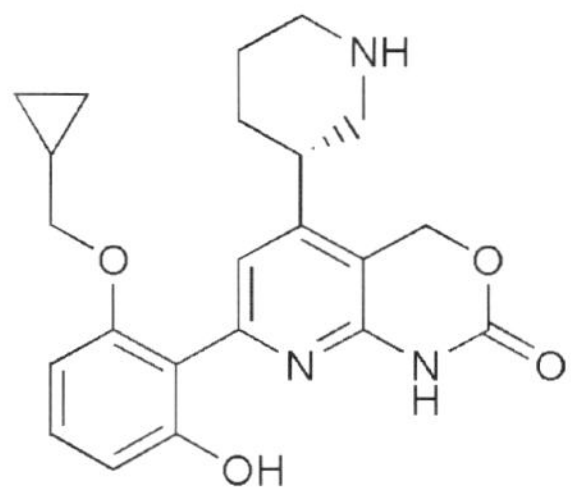

Compound A

FIGURE 11.10 Tokyo Bayer IKK-2 inhibitor.

It was tested in the arachidonic acid–induced murine ear inflammation model, which is related to COX-2 expression, in turn regulated by NF-κB activation. Oral dosing of 1 and 3 (but not 0.3) mg/kg 60 minutes prior to the topical application of the arachidonic acid significantly inhibited the ear swelling.

Related "Compound A" had an IKK-2 IC_{50} of 4 nM (Ziegelbaure et al., 2005) and was likewise active in both the arachidonic acid–induced and phorbol ester–induced ear models (Figure 11.10).

This compound was 36% and 69% orally bioavailable in mice and rats, respectively, with a Cmax of 76 nM. Its ED_{50} in the murine LPS-induced TNFα assay was 6.6 mg/kg. The racemic mixture of Compound A plus its enantiomer, when tested in a cockroach allergen airway model (Campbell et al., 1998) reduced the antigen-induced increase in airway resistance by over 50% at doses of 10 and 50 mg/kg. It also markedly reduced the influx of leukocytes in the ovalbumin-induced lung inflammation model. Reduction in the levels of a number of cytokines paralleled the activity.

Novartis has reported high-throughput screening Lead 1 (IKK-2 IC_{50} = 15 μM) and Lead 2 (IKK-2 IC_{50} = 30 μM). A focused medicinal chemistry effort (Waelchli et al., 2006) produced a hybrid series that was optimized to Compound 16 (IKK-2 IC_{50} = 0.04 μM). Compound 16, when dosed subcutaneously or orally to rats at 30 mg/kg 1 hour prior to challenge with LPS, inhibited plasma TNFα release (86% and 75%, respectively). Compound 16 was also active in a murine thioglycolate-induced peritonitis model (Cecconi et al., 1994). At 10 mg/kg, neutrophil migration into the peritoneal cavity was reduced by about 50%.

Aventis has reported that an IKK-2 inhibitor of undisclosed structure, AVE-1627, has an enzyme IC_{50} of 10 nM but a short $t_{1/2}$ in rats. (Tegeder et al., 2004). It was therefore dosed repeatedly intraperitoneally at 30 mg/kg to achieve a C_{max} in plasma of 2.6 uM. While having no effect on responses to applied heat or tactile stimuli in naïve rats, it reversed the hyperalgesia to both stimuli after injection of zymosan into the paws, as well as reducing the inflammatory edema.

V. DISCUSSION/MECHANISM-BASED TOXICITY

The foregoing discussion, while not exhaustive, clearly demonstrates that IKK-2 inhibitors have *in vivo* activity in a wide variety of disease models.

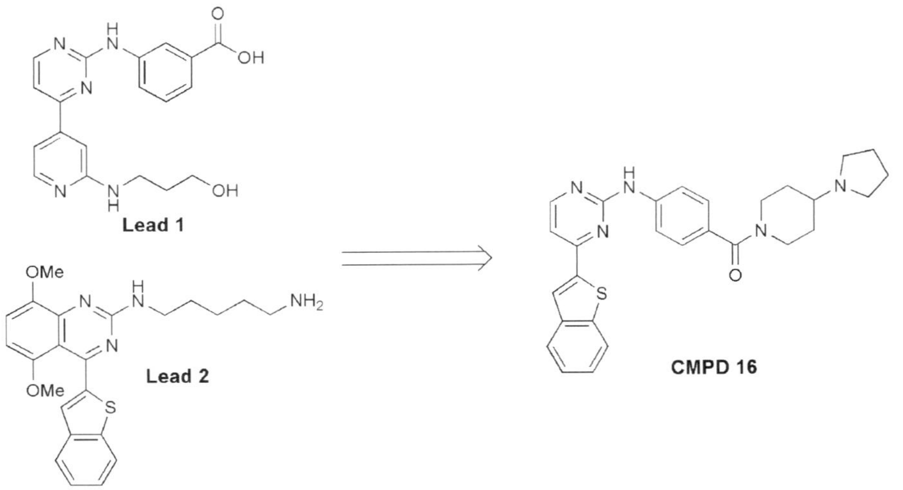

FIGURE 11.11 Novartis IKK-2 inhibitors.

This rather conclusively establishes the validity of inhibiting this signaling pathway and hence, the validity of NF-κB–mediated transcription as a therapeutic target. As mentioned above, it is thought that glucocorticoids exert their anti-inflammatory effect predominantly through suppression of NF-κB. This appears to result from both a direct interaction with ligand-bound glucocorticoid receptor as well as an indirect effect through glucocorticoid response element (GRE)-dependent transcription of IκBα (Auphan et al., 1995). While glucocorticoids represent some of the most potent and effective anti-inflammatory and immunosuppressive drugs available, their use is limited by GRE-dependent metabolic (e.g., altered glucose and lipid metabolism) and bone (e.g., osteoporosis) toxicities. IKK inhibitors, therefore, would be expected to have glucocorticoid-like activity without the GRE-dependent toxicities.

Inhibition of NF-κB is not without concern. It is well understood that members of the NF-κB family play a critical role in fetal development (Beg et. al., 1995; Iotsova et al., 1997), and IKK-2 knockouts are embryonic lethal. Therefore, IKK-2 inhibitors may be teratogens. But because heterozygotes develop normally even though NF-κB activation is inhibited by 70% to 80% (ZW Li et al., 1999), teratogenic effects might only result from continuous and complete inhibition of IKK-2.

The embryonic lethality observed in IKK-2 knockouts is a result of extensive liver damage from apoptosis. These mice could be rescued by the deletion of the gene encoding TNF receptor 1 (Q Li et al., 1999a). This is a similar phenotype to that observed with p50/p65 NF-κB knockouts and is consistent with the understood role of NF-κB in the development of the liver as well as other organs and tissues (Beg et al., 1995; Doi et al., 1999; Horwitz et al., 1997; Iotsova et al., 1997). While NF-κB certainly plays an important role in liver apoptosis during fetal development, there is considerable evidence that IKK-2/NF-κB inhibition will not predispose to mechanism-based hepatotoxicity in healthy adults. The first comes from a recent report investigating conditional transgenic mice with liver-specific expression of a super-repressor IκBα (Lavon et al., 2000). Mice with high expression of the transgene and complete inhibition of NF-κB in the liver were followed for as long as 15 months and were not distinguishable from normal mice whether in a barrier facility or a "dirty" environment in which parasitic and viral infections were common. The mice were unable to clear a *Listeria monocytogenes* infection, however, when specifically challenged.

Another report investigated the effect of NF-κB inhibition on alcohol-induced liver injury by using a similar super-repressor IκBα (Uesugi et al., 2001). Adenoviral expression of this super-repressor actually protected rats from alcohol-induced liver injury, consistent with reports that alcohol induces liver injury through the expression of cytokines that are NF-κB dependent. Thus, an IKK inhibitor may be protective in this context. Moreover, mice lacking IKK-2 in hepatocytes displayed significantly reduced liver necrosis and inflammation than wild-type mice (Luedde et al., 2005). An IKK-2 inhibitor, therefore, may be expected to protect from liver injury due to I/R without sensitizing them toward TNF-induced apoptosis. These findings argue that mechanism-based toxicity may not be problematic in adults with healthy livers and may actually protect against certain types of liver injury.

IKK-1 knockouts are also embryonic lethal, but the phenotype was distinct from that of IKK-2 knockouts. Instead of liver apoptosis, the most striking effect observed was a skin defect characterized by a massive epidermal hyperproliferation, which is the result of a defect in keratinocyte differentiation (Hu et al., 1999; Q Li et al., 1999b). This defect in differentiation in the knockouts is a result of lack of protein rather than a lack of catalytic activity. Transgenics in which the IKK-1 allele was replaced with a mutant, unactivatable IKK-1 have no skin or limb abnormalities, although mammary gland development during pregnancy is perturbed (Hu et al., 2001). Therefore, skin defects would not be expected from a small molecule inhibitor of IKK-1.

An important consideration relating to mechanism-based toxicity is whether the use of IKK-2 inhibitors will result in enhanced susceptibility to infection given the important role of NF-κB in innate and adaptive immunity. Anti-TNFα biologics are associated with increased reactivation of latent tuberculosis infection, although this appears to be manageable in the clinic. With IKK inhibitors, the challenge for therapy will be to only dampen the immune system in situations with aberrant activation rather than shutting it down completely. (Susceptibility to infection does not appear to be problematic with the use of glucocorticoids).

While the concept of treating cancers with inhibitors of IKK is intriguing, the potentially beneficial effects may be partially offset by the immunosuppressive effects of an IKK-2 inhibitor, which could be a disadvantage in cancer therapy. Emerging evidence suggests inflammation and cancer are, in some instances, intimately connected, and that the activation of NF-κB by the IKK-2–dependent pathway is a crucial mediator of inflammation-induced tumor growth and progression (Karin and Green, 2005). The potential of IKK inhibitors in cancer will become more clear as compounds are profiled in preclinical animals and, ultimately, in the clinic.

VI. CONCLUSIONS AND OUTLOOK

The recent work, both genetic and pharmacological, on the role of IKK-2 has shown it to be critical in NF-κB–dependent disorders. While the potential utility of IKK-2 inhibitors extends into human disorders as diverse as diabetes, cancer, and stroke, the most compelling use of these inhibitors will most likely be in treating inflammatory and autoimmune disorders given the central role of IKK/NF-κB in the immune system. Many pharmaceutical companies, large and small, are actively searching for inhibitors of IKK-2. Several different chemotypes have shown activity in an impressive range of inflammatory and immunological models, providing outstanding support for the NF-κB pathway in general and IKK-2 in particular as therapeutically useful and "druggable" targets. Drug discovery scientists are now engaged in the difficult task of converting potent lead compounds into clinical candidates by optimizing penetration, metabolic stability, etc. and designing out off-target effects. The number of different starting points and sets of properties is reflected in the diverse nature of the administration regimens in the published *in vivo* studies. The eventual utility of this class of compounds will be dictated by the clinical therapeutic index. The promise (and expectations) are high that IKK-2 inhibitors will occupy an important place in the clinician's armamentarium.

RECOMMENDED OTHER READING

Delhalle, S., R. Blasius, et al. (2004). A beginner's guide to NF-κB signaling pathways *Ann. N. Y. Acad. Sci.* 1030: 1–13.

Characterization of signal transduction pathways leading to NF-κB activation and inhibition of several of the upstream steps

Haefner, B. (2005). The transcription factor NF-κB as drug target. *Prog. Med. Chem.* 43: 137–188.

Discusses role of NF-KB in development, host defense and disease pathogenesis, with an involved discussion of small-molecule NF-κB pathway inhibitors

Hanada, T. and A. Yoshimura (2002). Regulation of cytokine signaling and inflammation. *Cytokine Growth Fact. Rev.* 13: 413–421.

Pro- and anti-inflammatory cytokine network and animal disease model

Hatada, E. N., D. Krappmann, and C. Scheidereit (2000). NF-κB and the innate immune response. *Curr. Opin. Immunol.* 12: 52–58.

Review of NF-κB pathway in the innate immune response, focusing on IKK's

Hayden, M. S. and S. Ghosh (2004). Signaling to NF-κB. *Genes Dev.* 18: 2195–2224.

Detailed review of NF-κB induction pathways and their regulation

Hehlgans, T. and K. Pfeffer (2005). The intriguing biology of the tumour necrosis factor/tumour necrosis factor receptor superfamily: Players, rules, and the games. *Immunol.* 115: 1–20.

Immunobiology of TNF and TNFR superfamily, covers signaling pathway affecting cell proliferation, survival, differentiation, apoptosis, immune organ development, and host defense

Karin, M., Y. Yamamoto, and Q. M. Wang (2004). The IKK NF-κB system: a treasure trove for drug development. *Nature Rev. Drug Disc.* 3: 17–26.

Review of recent progress on drug candidates affecting the pathway with discussion of potential applications in cancer and inflammatory/immune disorders

Nagashima, K., V. G. Sasseville, et al., (2006). Rapid TNFR1-dependent lymphocyte depletion *in vivo* with a selective chemical inhibitor of ικκβ. *Blood* 107: 4266–4273.

Newton, R. C. and C. P. Decicco (1999). Therapeutic potential and strategies for inhibiting Tumor Necrosis Factor α. *J. Med. Chem.* 42: 2295–2314.

Older but valuable review of TNFα and approaches to inhibiting various steps in the pathway

Pande, V. and M. J. Ramos (2005). NF-κB in human disease: Current inhibitors and prospects for *de novo* structure based design of inhibitors. *Curr. Med. Chem.* 12: 357–374.

Review focusing on NF-κB involvement in human disease and different classes of recently discovered inhibitors

Viatour, P., M.-P. Merville, et al. (2004). Phosphorylation of NF-κB and IκB proteins: Implications in cancer inflammation. *Trends Biochem. Sci.* 30: 43–52.

Review emphasizing roles of phosphorylated NFκB and IκB proteins; covers kinases in addition to IKK-2

Wu, J. T. and M. D. Kral (2005). The NF-κB/ IκB signaling system: A molecular target in breast cancer therapy. *J. Surg. Res.* 123: 158–169.

Relationship between uncontrolled/aberrant NF-κB activity as a regulator of mammary carcinogenesis

Yamamoto, Y. and R. B. Gaynor (2003). IκB kinases: Key regulators of the NF-κB pathway. *Trends Biochem. Sci.* 29: 72–79.

Specific review of IκB kinases. Role in NFκB activation and downstream effects

REFERENCES

Adams. J. (2001). Proteasome inhibition in cancer: Development of PS-341. *Semin. Oncol.* 28: 613–619.

Andreakos, E., C. Smith, et al. (2003). IκB kinase 2 but not NF-κB-inducing kinase is essential for effective DC antigen presentation in the allogeneic mixed lymphocyte reaction. *Blood* 101: 983–991.

Anest, V., J. L. Hanson, et al. (2003). A nucleosomal function for IκB kinase-α in NF-κB-dependent gene expression. *Nature* 423: 659–663.

Arkan, M. C., A. L. Hevener, et al. (2005). IKK-β links inflammation to obesity-induced insulin resistance. *Nat. Med.* 11: 191–198.

Auphan, N., J. A. DiDonato, et al. (1995). Immunosuppression by glucocorticoids: Inhibition of NF-κB activity through induction of IκB synthesis. *Science* 270: 286–290.

Baeuerle, P. A. and D. Baltimore (1997). NF-κB: Ten years after. *Cell* 87: 13–20.

Baldwin, A. S. (2001). Control of oncogenesis and cancer therapy resistance by the transcription factor NF-κB. *J. Clin. Invest.* 107: 241–246.

Barnes, P. J. (2002). New treatments for COPD. *Nat. Rev. Drug Discov.* 1: 437–446.

Beg, A. A., W. C. Sha, et al. (1995). Embryonic lethality and liver degeneration in mice lacking the RelA component of NF-κB. *Nature* 376: 167–170.

Bondeson, J., B. Foxwell, et al. (1999). Defining therapeutic targets by using adenovirus: Blocking NF-κB inhibits both inflammatory and destructive mechanisms in rheumatoid synovium but spares anti-inflammatory mediators. *Proc. Natl. Acad. Sci. U. S. A.* 96: 5668–5673.

Burke, J.R., M. K. Wood, et al. (1999). Peptides corresponding to the N and C termini of IκB-α, -β, and -ε as probes of the two catalytic subunits of IκB kinase, IKK-1 and IKK-2. *J. Biol. Chem.* 274: 36,146–36, 152.

Burke, J. R., M. A. Pattoli, et al (2003). BMS-345541 is a highly selective inhibitor of IκB kinase, binds at an allosteric site of the enzyme, and blocks NFκB-dependent transcription in mice. *J. Biol. Chem.* 278: 1450–1456.

Campbell, E.M., S. L. Kunkel, et al.(1998). Temporal role of chemonkines in a murine model of cockroach allergen-induced airway hyperreactivity and eosinophilia. *J. Immunol.* 161: 7047–7053.

Cao, Y., G. Bonizzi, G., et al. (2001). IKKα provides an essential link between RANK signaling and cyclin D1 expression during mammary gland development. *Cell* 107: 763–775.

Castro, A. C., L. C. Dang, et al. (2003). Novel IKK Inhibitors: β-Carbolines. *Bioorg. Med. Chem. Lett.* 13: 2419–2422.

Cecconi, O., R. M. Nelson, et al. (1994). Inositol polyanions. Noncarbohydrate inhibitors of L- and P-selectin that block inflammation. *J. Biol. Chem.* 269: 15,060–15,066.

Chen, L. W., L. Egan, et al. (2003). The two faces of IKK and NF-κB inhibition: Prevention of systemic inflammation but increased local injury following intestinal ischemia-reperfusion. *Nat. Med.* 9: 575–581.

Chilov, D., Kukk, E., et al. (1997). Genomic organization of human and mouse genes for vascular endothelial growth factor C. *J. Biol. Chem.* 272: 25,176–25,183.

Churg, A., J. Dai, et al. (2002). Tumor necrosis factor-α is central to acute cigarette smoke-induced inflammation and connective tissue breakdown. *Am. J. Respir. Crit. Care Med.* 166: 849–854.

Cyrus, T., S. Sung, et al. (2002). Effect of low-dose aspirin on vascular inflammation, plaque stability, and atherogenesis in low-density lipoprotein receptor-deficient mice. *Circulation* 106: 1282–1287.

Davis, R. E., K. D. Brown, et al. (2001). Constitutive nuclear factor κB activity is required for survival of activated B-cell-like diffuse large B cell lymphoma cells. *J. Exp. Med.* 194: 1861–1874.

DiDonato, J. A., M. Hayakawa, et al. (1997). A cytokine-responsive IκB kinase that activates the transcription factor NF-κB. *Nature* 388: 548–554.

DiStefano, A., G. Caramori, et al. (2002). Increased expression of nuclear factor-κB in bronchial biopsies from smokers and patients with COPD. *Eur. Respir. J.* 20: 556–563.

Doi, T. S., M. W. Marino, et al. (1999). Absence of tumor necrosis factor rescues RelA-deficient mice from embryonic lethality. *Proc. Natl. Acad. Sci. U. S. A.* 16: 2994–2999.

Duyao, M. P., D. J. Kessler, et al. (1992). Transactivation of the c-myc promoter by human T cell leukemia virus type 1 tax is mediated by NF-κB. *J. Biol. Chem.* 267: 16,288–16,291.

Elliott, M. J., R. N. Maini, et al. (1994). Randomised double-blind comparison of chimeric monoclonal antibody to tumor necrosis factor α (cA2) versus placebo in rheumatoid arthritis. *Lancet* 344: 1105–1110.

Fabian, M. A., W. H., III, Biffs, et al. (2005). A small molecule-kinase interaction map for clinical kinase inhibitors. *Nat. Biotechnol.* 23: 329–336.

Frantz, S., D. Fraccarollo, et al. (2003). Sustained activation of nuclear factor kappa B and activator protein 1 in chronic heart failure. *Cardiovasc. Res.* 57: 749–756.

Fujihara, S. M., J. S. Cleaveland, et al. (2000). A D-amino acid peptide inhibitor of NF-κB nuclear localization is efficacious in models of inflammatory disease. *J. Immunol.* 165: 1004–1012.

Fujioka, S., G. M. Sclabas, et al. (2003). Inhibition of constitutive NF-κB activity by IκBαM suppresses tumorigenesis. *Oncogene* 22: 1365–1370.

Fulmer, R. I., A. T. Cramer, et al. (1963). Transplantation of cardiac tissue into the mouse ear. *Am. J. Anat.* 113: 273.

Gao, Y., S. Lecker, et al. (2000). Inhibition of ubiquitin-proteasome pathway-mediated IκB alpha degradation by a naturally occurring antibacterial peptide. *J. Clin. Invest.* 106: 439–448.

Gao, Z., D. Hwang, et al. (2002). Serine phosphorylation of insulin receptor substrate 1 by inhibitor kappa B kinase complex. *J. Biol. Chem.* 277: 48,115–48,121.

Ghezzi, P., S. Sacco, et al. (2000). LPS induces IL-6 in the brain and in serum largely through TNF production. *Cytokine* 12: 1205–1210.

Ghosh, S. and M. Karin (2002). Missing pieces in the NF-κB puzzle. *Cell* 109: S81–S96.

Gupta, S., N. H. Purcell, et al. (2002). Activation of nuclear factor-κB is necessary for myotrophin-induced cardiac hypertrophy. *J. Cell Biol.* 159: 1019–1028.

Herrmann, O., B. Baumann, et al. (2005). IKK mediates ischemia-induced neuronal death. *Nat. Med.* 11: 1322–1329.

Hideshima, T. and D. Chauhan, (2002). NF-κB as a therapeutic target in multiple myeloma. *J. Biol. Chem.* 277: 16,639–16,647.

Hinz, M., D. Krappmann, et al. (1999). NF-κB function in growth control: regulation of cyclin D1 expression and G0/G1-to-S-phase transition. *Mol. Cell. Biol.* 19: 2690–2698.

Hirotani, S., K. Otsu, et al. (2002). Involvement of nuclear factor-κB and apoptosis signal-regulating kinase 1 in G-protein-coupled receptor agonist-induced cardiomyocyte hypertrophy. *Circulation* 105: 509–515.

Horwitz, B. H., M. L. Scott, et al. (1997). Failure of lymphopoiesis after adoptive transfer of NF-κB-deficient fetal liver cells. *Immunity* 6: 765–772.

Hu, Y., V. Baud (1999). Abnormal morphogenesis but intact IKK activation in mice lacking the IKKα subunit of IκB kinase. *Science* 284: 316–320.

Hu, Y., V. Baud, et al. (2001). IKKα controls formation of the epidermis independently of NF-κB. *Nature* 410: 710–714.

Huber, M. A., A. Denk, et al. (2002). The IKK-2/I-kappa-Balpha/NF-kappa-b pathway plays a key role in the regulation of CCR3 and eotaxin-1 in fibroblasts: A critical link to dermatitis in I-kappa-B-alpha-deficient mice. *J. Biol. Chem.* 277: 1268–1275.

Hundal, R. S., K. F. Petersen, et al. (2002). Mechanism by which high-dose aspirin improves glucose metabolism in type 2 diabetes. *J. Clin. Invest.* 109: 1321–1326.

Iotsova, V., J. Caamano, et al. (1997). Osteopetrosis in mice lacking NF-κB1 and NF-κB2. *Nat. Med.* 3:1285–1289.

Jobin, C., C. A. Bradham, et al. (1999). Curcumin blocks cytokine-mediated NF-κB activation and proinflammatory gene expression by inhibiting inhibitory factor IκB kinase activity. *J. Immunol.* 163: 3474–3483.

Kamon, J., T. Yamauchi, et al. (2004). A Novel IKKβ inhibitor stimulates adiponectin levels and ameliorates obesity-linked insulin resistance. *Biochem. Biophys. Res. Commun.* 323: 242–248.

Karin, M. and F. R. Greten (2005). NF-κB: linking inflammation and immunity to cancer development and progression. *Nat. Rev. Immunol.* 5: 749–759.

Keating, G. M. and C. M. Perry (2002). Infliximab: An updated review of its use in Crohn's disease and rheumatoid arthritis. *Bio. Drugs* 16:111–148.

Khoshnan, A., S. J. Kempiak, et al. (1999). Primary human CD4+ T cells contain heterogeneous IκB kinase complexes: Role in activation of the IL-2 promoter. *J. Immunol.* 163: 5444–5452.

Kim, J. K., Y. J. Kim, et al (2001). Prevention of fat-induced insulin resistance by salicylate. *J. Clin. Invest.* 108: 437–446.

Lam, L. T., R. E. Davis, et al. (2005). Small molecule inhibitors of IκB kinase are selectively toxic for subgroups of diffuse large B-cell lymphoma defined by gene expression profiling. *Clin Cancer Res* 11: 28–40.

Lavon, I., I. Goldberg, et al. (2000). High susceptibility to bacterial infection, but no liver dysfunction, in mice compromised for hepatocyte NF-κB activation. *Nat. Med.* (2000) 6: 573–577.

Lawrence T., M. Bebien, et al. (2005). IKKα limits macrophage NF-κB activation and contributes to the resolution of inflammation. *Nature* 434: 1138–1143.

Lee, H. W., D. H. Ahn, et al. (2002). Phorbol 12-myristate 13-acetate up-regulates the transcription of MUC2 intestinal mucin via Ras, ERK, and NF-κB. *J. Biol. Chem.* 277: 32,624–332,621.

Li, Q., D. Van Antwerp, et al. (1999a). Severe liver degeneration in mice lacking the IκB kinase 2 gene. *Science* 284: 321–325.

Li, Q., Q. Lu, et al. (1999b). IKK1-deficient mice exhibit abnormal development of skin and skeleton. *Genes Dev.* 13: 1322–1328.

Li, Z. W., W. Chu, et al. (1999). The IKKβ subunit of IκB kinase (IKK) is essential for nuclear factor κB activation and prevention of apoptosis. *J. Exp. Med.* 189: 1839–1845.

Li Q, Q. Lu, et al. (2005). Enhanced NF-κB activation and cellular function in macrophages lacking IκB kinase (IKK-1). *Proc. Natl. Acad. Sci. U. S. A.* 102: 12,425–12,430.

Luedde T, U. Assmus, et al. (2005). Deletion of IKK-2 in hepatocytes does not sensitize these cells to TNF-induced apoptosis but protects from ischemia/reperfusion injury. *J. Clin. Invest.* 115: 849–859.

Luque, I. and C. Gelinas (1997). Rel/ NF-κB and IκB factors in oncogenesis. *Semin. Cancer Biol.* 8: 103–111.

MacMaster, J. F., D. M. Dambach, et al. (2003). An inhibitor of IκB kinase, BMS-345541, blocks endothelial cell adhesion molecule expression and reduces the severity of dextran sulfate sodium-induced colitis in mice. *Inflamm. Res.* 52: 508–511.

Manning, G., D. B. Whyte, et al. (2002). The protein kinase complement of the human genome. *Science* 298: 1912–1916.

McIntyre, K. W., D. J. Shuster, et al. (2003). A highly selective inhibitor of IκB kinase, BMS-345541, blocks both joint inflammation and destruction in collagen-induced arthritis in mice. *Arthritis Rheum.* 48: 2652–2659.

Mengshol, J. A., M. P. Vincenti, et al. (2000). Interleukin-1 induction of collagenase 3 (matrix metalloproteinase 13) gene expression in chondrocytes requires p38, c-Jun N-terminal kinase, and nuclear factor κB: Differential regulation of collagenase 1 and collagenase 3. *Arthritis Rheum.* 43: 801–811.

Mercurio, F., H. Zhu, et al. (1997). IKK-1 and IKK-2: Cytokine-activated IκB kinases essential for NF-κB activation. *Science* 278: 860–866.

Miller, B. S. and E. Zandi (2001). Complete reconstitution of human IκB kinase (IKK) complex in yeast: Assessment of its stoichiometry and the role of IKKγ on the complex activity in the absence of stimulation. *J. Biol. Chem.* 276: 36, 320–36, 32–6.

Morishita, R., T. Sugimoto, et al. (1997). *In vivo* transfection of *cis* element "decoy" against nuclear factor-κB binding site prevents myocardial infarction. *Nat. Med.* 3: 894–899.

Murata, T., M. Shimada, et al. (2004). Synthesis and structure-activity relationships of novel IKK-β inhibitors. Part 3: Orally active anti-inflammatory agents. *Bioorg. Med. Chem. Lett.* 14: 4019–4022.

Okazaki, Y, T. Sawada, et al. (2005). Effect of nuclear factor-κB inhibition on rheumatoid fibroblast-like synoviocytes and collagen induced arthritis. *J. Rheumatol.* 32: 1440–1447.

Onai, Y., J. Suzuki, et al. (2004). Inhibition of IκB phosphorylation in cardiomyocytes attentuates myocardial ischemia/reperfusion injury. *Cardiovasc Res* 63: 51–59.

Pennica, D., G. E. Nedwin, et al. (1984). Human tumor necrosis factor: Precursor structure, expression, and homology to lymphotoxin. *Nature* 312: 724–729.

Pickup, J. C. and M. A. Crook (1998). Is type II diabetes mellitus a disease of the innate immune system? *Diabetologia* 41:1241–1248.

Podolin, P. L., J. F. Callahan, et al. (2005). Attenuation of murine collagen-induced arthritis by a novel, potent, selective small molecule inhibitor of IκB Kinase 2, TPCA-1 (2-[(aminocarbonyl)amino]-5-(4-fluorophenyl)-3-thiophenecarboxamide), occurs via reduction of proinflammatory cytokines and antigen-Induced T cell proliferation. *J. Pharmacol. Exp. Therapeut.* 312: 373–381.

Purcell, N. H., G. Tang, et al. (2001). Activation of NF-κB is required for hypertrophic growth of primary rat neonatal ventricular cardiomyocytes. *Proc. Natl. Acad. Sci. U. S. A.* 98: 6668–6673.

Pye, J., F. Ardeshirpour, et al. (2003). Proteasome inhibition ablates activation of NF-κB in myocardial reperfusion and reduces reperfusion injury. *Am. J. Physiol. Heart Circ. Physiol.* 284: H919–926.

Rayet, B. and C. Gelinas (1999). Aberrant rel/NF-κB genes and activity in human cancer. *Oncogene* 18: 6938–6947.

Romieu-Mourez, R., E. Landesman-Bollag, et al. (2001). Roles of IKK kinases and protein kinase CK2 in activation of nuclear factor-κB in breast cancer. *Cancer Res.* 61: 3810–3818.

Ross, R. (1999). Atherosclerosis: An inflammatory disease. *N. Engl. J. Med.* 340: 115–126.

Rossi, A., Kapahi, P., et al. (2000). Anti-inflammatory cyclopentenone prostaglandins are direct inhibitors of IκB kinase. *Nature* 403:103–108.

Saklatvala, J. (2002). Glucocorticoids: Do we know how they work? *Arthritis Res.* 4:146–150.

Sanlioglu, S., G. Luleci, and K. W. Thomas (2001). Simultaneous inhibition of Rac1 and IKK pathways sensitizes lung cancer cells to TNFα-mediated apoptosis. *Cancer Gene Ther.* 8: 897–905.

Schenkein. D. (2002). Proteasome inhibitors in the treatment of B-cell malignancies. *Clin. Lymphoma* 3: 49–55.

Schneider, A., A. Martin-Villalba, et al. (1999). NF-κB is activated and promotes cell death in focal cerebral ischemia. *Nat. Med.* 5: 554–559.

Siebenlist, U., G. Franzoso, and K. Brown (1994). Structure, regulation and function of NF-κB. *Annu. Rev. Cell Biol.* 10: 405–455.

Sizemore, N., N. Lerner, et al. (2002). Distinct roles of the IκB kinase alpha and beta subunits in liberating nuclear factor kappa B (NF-κB) from IκB and in phosphorylating the p65 subunit of NF-κB. *J. Biol. Chem.* 277: 3863–3869.

Stehlik, C., R. de Martin, et al. (1998). Nuclear factor (NF)-gκ B-regulated X-chromosome-linked iap gene expression protects endothelial cells from tumor necrosis factor alpha-induced apoptosis. *J. Exp. Med.* 188: 211–216.

Tak, P. P., D. M. Gerlag, et al. (2001). Inhibitor of nuclear factor κB kinase β is a key regulator of synovial inflammation. *Arthritis Rheum.* 44: 1897–1907.

Tegeder, I., E. Niederberger, et al. (2004). Specific inhibition of IκB kinase reduces hyperalgesia in inflammatory and neuropathic pain models in rats. *J. Neurosci.* 24: 1637–1645.

Tergaonkar, V., M. Pando, et al. (2002) p53 stabilization is decreased upon NFκB activation: A role for NFκB acquisition of resistance to chemotherapy. *Cancer Cell* 1: 493-503.

Toth, C. R., R. F. Hostutler, et al. (1995). Members of the nuclear factor kappa B family transactivate the murine c-myb gene. *J. Biol. Chem.* 270: 7661–7671.

Townsend, R. M., J. Postelnek, et al. (2004) A highly selective inhibitor of IκB kinase, BMS-345541, augments graft survival mediated by suboptimal immunosuppression in a murine model of cardiac graft rejection. *Transplantation* 77: 1090–1100.

Trentham, D. E., A. S. Townes, and A. H. Kang (1977). Autoimmunity to type II collagen: An experimental model of arthritis. *J. Exp. Med.* 146: 857–868.

Uesugi, T., M. Froh, et al. (2001). Delivery of IκB superrepressor gene with adenovirus reduces early alcohol-induced liver injury in rats. *Hepatology* 34: 1149–1157.

Waelchli, R., B. Bollbuck, et al. (2006). Design and preparation of 2-benzmido-pyrimidines as inhibitors of IKK. *Biorg. Med. Chem. Lett.* 16: 108–112.

Wang, C. Y., M. W. Mayo, and A. S., Jr., Baldwin (1996). TNF- and cancer therapy-induced apoptosis: Potentiation by inhibition of NF-κB. *Science* 274: 784–787.

Weber, C. K., S. Liptay, et al. (2000). Suppression of NF-κB activity by sulfasalazine is mediated by direct inhibition of IκB kinases α and β. *Gastroenterology* 119: 1209–1218.

Whiteside, S. T. and A. Israel (1997). IκB proteins: Structure, function and regulation. *Sem. Cancer Biol.* 8: 75–82.

Wilkerson, W. W., W. Galbraith, et al. (1994) Antiinflammatory 4, 5-diarylpyrroles: Synthesis and QSAR *J. Med. Chem.* 37: 988–998.

Woronicz, J. D., X. Gao, et al. (1997). IκB kinase-β: NF-κB activation and complex formation with IκB kinase-α and NIK. *Science* 278: 866–869.

Wu, M. X., Z. Ao, et al. (1998). IEX -1L, an apoptosis inhibitor involved in NF-κB-mediated cell survival. *Science* 281: 998–1001.

Yamamoto, Y., U. N. Verma, et al. (2003). Histone H3 phosphorylation by IKK-α is critical for cytokine-induced gene expression. *Nature* 423: 655–659.

Yan, F. and D. B. Polk (1999). Aminosalicylic acid inhibits IκB kinase phosphorylation of IκBα in mouse intestinal epithelial cells. *J. Biol. Chem.* 274: 36,631–36,636.

Yang, J. and A. Richmond (2001). Constitutive IκB kinase activity correlates with nuclear factor-κB activation in human melanoma cells. *Cancer Res.* 61: 4901–4909.

Yang, J., K. I. Amiri, et al. (2005). BMS-345541 targets IκB kinase and induces apoptosis in melanoma: Involvement of nuclear factor-κB and mitochondria pathways. *Clin Cancer Res.*, in press.

Yang, L., L. Cohn, et al. (1998). Essential role of nuclear factor κB in the induction of eosinophilia in allergic airway inflammation. *J. Exp. Med.* 188: 1739–1750.

Yin, M. J., Y. Yamamoto, and R. B. Gaynor (1998). The anti-inflammatory agents aspirin and salicylate inhibit the activity of IκB kinase-β. *Nature* 396: 77–80.

Yuan, M., N. Konstantopoulos, et al. (2001). Reversal of obesity- and diet-induced insulin resistance with salicylates or targeted disruption of IKKbeta. *Science* 293: 1673–1677.

Ziegelbaure, K., F. Gantner, et al. (2005). A selective novel low-molecular-weight inhibitor of IKB kinase-β (IKK-β) prevents pulmonary inflammation and shows broad anti-inflammatory activity. *Brit. J. Pharmacol.* 145: 178–192.

Zong, W. X., L. C. Edelstein, et al. (1999). The prosurvival Bcl-2 homolog Bfl-1/A1 is a direct transcriptional target of NF-κB that blocks TNFα-induced apoptosis. *Genes Dev.* 13: 382–387.

12

TNF SIGNALING PATHWAY INHIBITORS FOR INFLAMMATION-CCR2 ANTAGONISTS

KRIS VADDI

DVM, Ph.D., Incyte Corporation, Experimental Station, Wilmington, Delaware

A new era in the treatment of chronic inflammatory diseases has begun with the demonstration of profound clinical efficacy by TNFα blockers, Enbrel and Remicade, in rheumatoid arthritis. These therapies validated the mechanism of cytokine suppression as an effective therapeutic modality to treat chronic inflammation and stimulated great interest in the pursuit of small-molecule–based drugs that target cytokines. Because cytokines are often produced by infiltrating inflammatory cells, inhibiting influx of such cells is another approach to modulate cytokines. This approach was first attempted by blocking adhesion of leukocytes to vascular endothelium, the first event in their tissue migration. Recent clinical data with natalizumab, a neutralizing antibody to VLA-4, further validated that such an approach is effective. Adhesion molecules, however, are often shared by multiple subsets of infiltrating leukocytes, and therefore blocking adhesion does not allow for the modulation of a specific leukocyte subset. With the identification of chemokine receptors, the distribution of which is restricted to specific types of leukocytes, selective modulation of inflammatory cell subsets seemed possible. CCR2 as one such receptor, predominantly expressed on circulating monocytes, is an exciting target for small-molecule–based therapies to treat inflammatory conditions. Because macrophages are considered to be a major source of cytokines in inflammatory lesions, the benefits noted in clinical studies using anti-TNFα therapies are expected to be achieved by inhibition of tissue influx of macrophages. This review identifies several pathological conditions in which macrophages are implicated in the etiology of disease states and proposes CCR2 as a potential approach to modulate such diseases.

I. INTRODUCTION

Acute inflammation that manifests clinically as swelling, redness, and pain can be effectively managed by a number of available therapeutics such as analgesics and nonsteroidal anti-inflammatory drugs (NSAIDs), but inflammation

Target Validation in Drug Discovery

associated chronic diseases has proven to be a challenging area for drug discovery. The majority, if not all, of the chronic inflammatory diseases are autoimmune in nature and hence suppression of immune responses has been the traditional approach to treat chronic inflammation. Immune suppression is quite effective in the short term in controlling inflammation, but the side effects associated with frankly immunosuppressive therapies, such as steroids, limit their utility for chronic use (Tyndall and Gratwhol, 2000). The search for alternative mechanisms of suppressing local inflammation without global immunosuppression led to the investigation of approaches that neutralize or inhibit the major mediators of inflammation, known as cytokines. These mediators include interleukin-1 (IL-1), tumor necrosis factor alpha (TNFα), interferon gamma (IFNγ), interleukin-6 (IL-6), interleukin -12 (IL-12), interleukin -18 (IL-18), lymphotoxin-alpha (LTα), etc. (Ghezzi, 1997). While the therapeutic utility of inhibiting some of the cytokines has proven to cause unacceptable toxicities or to be not effective, therapies designed to inhibit TNFα have shown excellent clinical responses in a number of chronic diseases, including inflammatory bowl disease, psoriasis, and a variety of arthritides, including rheumatoid arthritis and ankylosing spondylitis (Smolen and Steiner, 2003).

Clinical validation of TNFα as a therapeutic target in chronic inflammation resulted in an intense effort to identify alternative approaches to modulate TNFα production using small-molecule–based therapies (Palladino, et al., 2003). Most of these approaches target inhibition of TNFα production or TNFα signaling by: 1) inhibition of the primary TNFα processing enzyme known as TACE (tumor necrosis factor α converting enzyme), which releases soluble TNFα from a membrane-associated form (Moss et al., 1997); 2) modulation of TNFα at the transcriptional level by inhibiting p38 kinase (Kumar et al., 2003) and PDE4 (Seldon et al., 1995); 3) suppression of TNFα production at the translational level by blocking mitogen-activated protein kinase 2 (MK2) (Neininger et al., 2002); and 4) modulation of TNFα signaling by inhibiting NFκB (Hass et al., 1990) (Figure 12.1). None of the above-mentioned mechanisms are very specific to TNFα and hence are likely to be associated with side effects, potentially limiting their therapeutic utility.

An entirely different approach to modulate local TNFα production in the inflammatory lesions is by stopping the influx of cells that produce TNFα. Numerous clinical and animal studies support the idea that activated

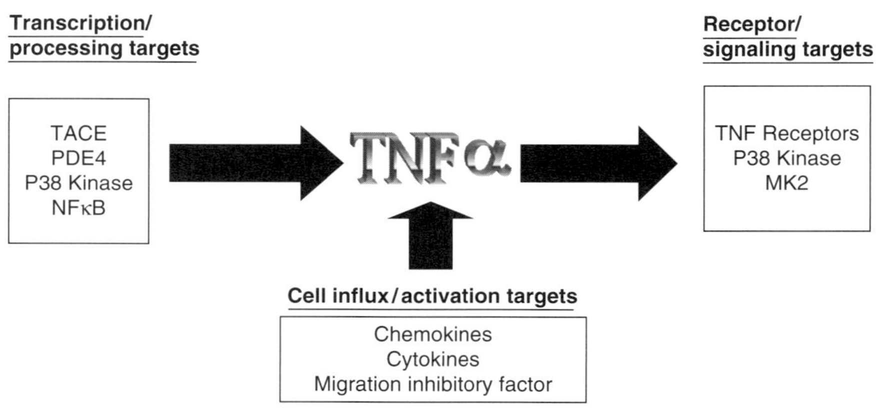

FIGURE 12.1 Therapeutic targets in TNF signaling pathway.

macrophages fit the description of a target cell type for this approach. (Fujiwara and Kobayashi, 2005). Early clinical validation of the approach to suppress tissue TNFα level by modulating the inflammatory cell influx comes from correlative studies that demonstrated that TNFα in chronic inflammatory lesions is primarily produced by infiltrating cells of hematopoietic origin rather than local tissue (Husby and Williams, 1988). In the case of rheumatoid arthritis, an excellent correlation was also observed between the tissue burden of infiltrating cells to either disease progression (Kraan et al., 2000), or responsiveness to therapies such as disease-modifying anti-rheumatic drugs (DMARDs) (Smith et al., 2001) and anti-TNFα therapies (Smeets et al., 2003). In addition, these studies have conclusively demonstrated the macrophage as a major pathogenic cell type because of its pleiotropic effects that perpetuate a chronic inflammatory process. These effects include the release of cytokines such as TNFa and IL-1 (Husby and Williams, 1988), chemokines including MCP-1, MIP-1α, RANTES etc. (Hamilton et al., 2002), tissue degrading metalloproteinases (MMPs) (Shapiro, 1999), and reactive oxygen intermediates (Forman and Torres, 2001), all of which have been implicated in inflammatory tissue destruction. One of the earliest documented approaches to selectively reduce the tissue load of inflammatory macrophages in rheumatoid arthritis was by local injection of liposomal clodronate (Van Lent et al., 1998). This approach results in the death of local tissue macrophages that selectively take up liposomal clodronate. Preclinical and pilot clinical studies showed early promise of this approach (Barrera et al., 2000), but the impractical nature of local injections in multiple joints did not allow larger studies to determine the true therapeutic utility of depletion of tissue inflammatory macrophages.

The quest for the identification of mechanisms to suppress tissue macrophage burden that are amenable to small-molecule therapies was greatly aided by the identification of a chemokine receptor expressed predominantly on monocytes known as "CC chemokine receptor-2 (CCR2)" (Charo et al., 1994). Following binding to its ligands, monocyte chemoattract proteins (MCP-1, -2, -3, -5), this receptor mediates influx of circulating monocytes into inflamed tissues. Mice with targeted deletion of CCR2 (Charo, 1999) provided novel insights into: 1) understanding how macrophages are called into inflammatory lesions, 2) the role of macrophage in tissue destruction, 3) the role of macrophages in host defense against various infectious agents, and 4) designing approaches to modulate macrophage influx. This review first describes the scientific evidence that implicates macrophages in rheumatoid arthritis, a disease condition in which anti-TNFα therapies have proven to be effective. Secondly, the potential role of the macrophage in other pathological conditions including atherosclerosis, multiple sclerosis, and obesity-associated insulin resistance will be discussed. Finally, an account of animal model studies utilizing CCR2 knockout mice or a selective small-molecule antagonist of CCR2 that support this novel approach to treat above mentioned diseases will be presented.

II. MACROPHAGES IN RHEUMATOID ARTHRITIS

With advances in arthroscopic techniques, our ability to study different cell populations in inflamed joint tissues has greatly improved (Youssef et al., 1998). Immunohistochemical studies using rheumatoid synovial biopsies indicated that >80% of the synovial lining cells are macrophage-like (Cutolo et al., 1993). Macrophages in the RA synovium are highly activated, as evidence by their morphology, surface class II HLA antigen expression, and synthesis of cytokines such as IL-1, TNFα, IL-6, granulocyte-macrophage colony-stimulating factor (GM-CSF), macrophage colony-stimulating factor (MCSF), and transforming growth-factor beta (TGFβ). In addition, macrophages in the joints differentiate into osteoclasts, the cell type known to be responsible for bone erosions in rheumatoid synovium (Gravallese, 2002).

To examine the relative contribution of individual synovial cell populations to polyarticular destruction in rheumatoid arthritis, Mulherin et al. (1996) performed serial measurements of disease activity and articular destruction on athroscopic tissue samples obtained prospectively in 28 RA patients followed for a mean of 5.8 years. This group determined the correlation of different synovial cell populations quantified using immunohistochemical techniques to radiological disease course. Results from this study indicated that synovial lining layer cell depth and sublining layer macrophage, but not lymphocyte, cell counts correlated significantly with radiological course. To evaluate changes in synovial tissue parameters that occur in conjunction with clinical response to an effective therapy, Gerlag et al. (2004) performed synovial biopsies on 21 RA patients receiving either oral prednisolone ($n = 10$) or placebo ($n = 11$) before and after 14 days of treatment. Immunohistochemical analysis of the cell infiltrate demonstrated that patients receiving active treatment had significantly fewer macrophages, but not other cell types, after therapy compared with those receiving placebo. A larger retrospective study was performed by Harringman et al. (2005) in 88 patients who participated in various clinical trials in which all patients underwent serial arthroscopy before initiation of treatment and after different time intervals. Immunohistochemical and digital image analysis of arthroscopic samples to detect changes in CD68+ synovial sublining macrophages in relationship to changes in the 28 joint count Disease Activity Score (DAS28) demonstrated a significant correlation between the change in the number of macrophages and the change in DAS28 (Pearson correlation 0.874, $p < 0.01$).

Most striking changes in synovial macrophage content were demonstrated following TNFα blockade in a recent study by Smeets et al. (2003). Within 48 hours of the administration of TNFα blocking antibody, infliximab, a dramatic reduction in synovial lining macrophages was observed in this study. Of particular note was the absence of an increase in the number of apoptotic cells after infliximab treatment suggesting that the decreased cell infiltration primarily results from early inhibition of cell migration. Taken together, the observations described above strongly support the idea that suppressing macrophage influx into inflammatory tissues in rheumatoid arthritis possesses tremendous therapeutic potential. In addition to down modulating the local TNFα production, suppression of macrophage content of joints is also likely to reduce several processes that cause tissue destruction in the affected joints, suggestive of the potential for a true disease modifying effect.

III. MACROPHAGES IN MULTIPLE SCLEROSIS

Multiple sclerosis (MS) is a chronic inflammatory disease of the central nervous system (CNS). Neurological deficits in MS are due to reduced signal conduction by demyelinated axons (Bruck and Stadelmann, 2005).

Inflammatory cellular infiltrates around demyelinated axons are believed to result in axonal loss, the key pathogenic process in MS. To identify specific cell types responsible for axonal injury, Ferguson et al. (1997) undertook detailed quantitative studies on axonal injury and loss in MS, using amyloid precursor protein as a marker for acute axonal injury. These studies demonstrated that massive axonal damage occurs during the stage of active demyelination in fresh lesions and that acute axonal injury correlated with the degree of macrophage infiltration in the lesions. In fact, these studies demonstrated that only macrophages, and not any other cell type, are closely attached to damaged axons. Presence and antigenic composition of myelin debris in macrophages further support the contention that macrophages mediate massive acute axonal injury that occurs during a small time window of about 2 weeks after onset of demyelination (Kornek et al., 2000).

Lesions in MS appear to begin with margination and diapedesis of lymphocytes and macrophages forming perivascular cuffs around capillaries and venules. This is followed by diffuse parenchymal infiltration by inflammatory cells, edema, macrophage activity with stripping of myelin from axons, astrocytic hyperplasia, and the appearance of increased numbers of lipid-laden macrophages and demyelinated axons.

Immunohistochemical analysis demonstrates that acute plaques contain macrophages throughout the lesion, and these macrophages exhibit profiles of phospholipid-rich myelinic bodies (Adams et al., 1989). In contrast, inactive ongoing lesions only show a rim of macrophages at the edge of the lesion. These macrophages show profiles of large vacuoles, thought to represent the sudanophilic esterified cholesterol formed during the demyelination process. In addition to acting as the effector cells causing tissue destruction in MS, macrophages also orchestrate a global inflammatory response in the CNS by secreting several chemokines such as RANTES, MIP-1α, and MIP-3 that attract other cell types including CD8+ T-cells, which promote disease progression (Benveniste, 1997).

Serial gadolinium-enhanced magnetic resonance imaging (MRI) scans reveal MS as a continuously active disease process. Therefore, interfering with the influx of new macrophages is likely to affect disease progression. Most recent data from multiple clinical trials with anti–very-late antigen-4 (VLA-4) antibody, natalizumab provided the clinical validation for targeting inflammatory cell influx as a therapeutic modality in MS (Steinman, 2005). While this drug is removed from the market recently because of a fatal side effect known as progressive multifocal leukoencephalopathy (PML) (Langer-Gould, 2005), this approach validates the basic concept that inhibiting leukocyte migration into the CNS is effective in halting disease progression in MS. VLA-4 is an adhesion molecule on monocytes as well as T-lymphocytes that allows these cells to adhere to endothelial lining of the blood vessels (Hemler, et al., 1987). In the context of inflammatory disease affecting the CNS, VLA-4 is critical for homing of inflammatory cells into the lesions in the brain. Clinical studies with natalizumab demonstrated a striking reduction in the relapse rate in multiple clinical trials in multiple sclerosis (O'Connor, et al., 2004). The current hypothesis for the fatal side effect of VLA-4 antagonism is that this therapy compromises the ability of T-lymphocytes to keep in check a virus known as JC virus, commonly present in most humans (Langer-Gould, 2005). Hence, if monocytes can be selectively inhibited from entering into brain without affecting lymphocyte recirculation, in theory, the benefits of

VLA-4 antagonism could be preserved without the side effects of the therapies targeting VLA-4.

IV. MACROPHAGES IN ATHEROSCLEROSIS

Atherosclerosis is a chronic disease that causes artery walls to thicken and become less elastic. The resulting restriction of blood flow can lead to various problems such as heart attacks, renal failure, abdominal aneurysms, stroke, and other serious cardiovascular complications (Ohashi et al., 2004). Despite the broad spectrum of clinical presentation, most of the acute manifestations of atherosclerosis share a common pathogenetic feature, the rupture of atherosclerotic plaques. Plaque disruptions may vary greatly in extent from tiny fissures or erosions of the plaque surface to deep intimal tears that extend into the soft lipid core of the lesions. In all these instances, at least some degree of thrombus formation occurs (van der Wal and Becker, 1999). Plaque rupture has been extensively studied in coronary arteries, where occlusive thrombus often leads to fatal complications, which suggest that a subset of atherosclerotic plaques described as "unstable" seem particularly prone to physical disruption (Falk, 1989). These studies identified a new approach in treating atherosclerosis by stabilizing the plaques, that is, rendering the atheroma more resistant to rupture and less thrombogenic. It is now believed that inhibiting the inflammatory process in atherosclerosis is likely provide increase plaque stability for the reasons described below.

Atherosclerosis is well recognized as a form of chronic inflammation occurring within the arterial walls (Libby and Theroux, 2005). The earliest lesion in atherosclerosis, known as "fatty streak," begins by attachment of circulating monocytes to the blood vessel wall followed by their accumulation in subendothelial space. Once resident in the arterial intima, monocytes accumulate lipid from the uptake of retained and modified apoB lipoproteins. This accumulation is considered to be a central event in atherogenesis (Tall et al., 1986). Cholesterol accumulation in these cells is thought to be mediated primarily by uptake of fully oxidized LDL via scavenger receptors A (SR-A) and CD36 (Febbraio et al., 2001). More-advanced atherosclerotic lesions, called fibro-fatty plaques, contain large numbers of foam cells together with smooth muscle cells and CD4+ T cells (Lucas and Greaves, 2001). Progressive accumulation of lipid-laden macrophages and remodeling of vessels wall leads to narrowing of the arterial lumen, which eventually leads to ischemia and related cardiovascular complications.

In addition to contributing to the morphological lesion of atherosclerosis, macrophages also play a key role in determining the stability of plaques. Ruptured plaques usually have thin fibrous caps overlying a large thrombogenic lipid core that is rich in lipid-laden macrophages (Boyle, 2005). These lesional macrophages also acquire other functional properties, including production of the potent procoagulant tissue factor, apolipoprotein E, and an increasing list of cytokines that may participate importantly in autocrine and paracrine signaling among leukocytes and vascular endothelial and vascular smooth muscle cells (VSMCs) (Libby and Geng, 1996). The integrity of the fibrous cap, and thus its resistance to rupture, depends critically on the collagenous extracellular matrix of the plaque's fibrous cap, which in turn, is regulated by the MMP activity of local macrophages (Boyle, 2005). Because macrophages are innate immune effectors, that is, they are activated without antigenic specificity, they are capable of causing indiscriminate tissue damage, in contrast to the activation of lymphocytes which is tightly regulated. Atherosclerotic plaque milieu contains a number of macrophage activation stimuli associated with atherosclerotic risk factors including oxLDL, advanced glycosylation end products (AGEs) of diabetes, angiotensin II, and endothelin. In addition to collagen breakdown, activated macrophages also reduce smooth muscle content of the plaque, a critical component in stable plaques. It is believed that macrophages induce plaque vascular smooth muscle cell apoptosis by cell to cell proximity, Fas-L, and nitric oxide generation.

Based on the above considerations, it is evident that selective inhibition of macrophage influx in to atherosclerotic plaques may provide multiple therapeutic benefits. In fact, data from experimental models suggests that a number of drugs currently used in diabetes and dyslipidemias such as thiazolidinediones (TZDs) and statins provide significant therapeutic benefit by attenuating macrophage activation (Dandona and Aljada, 2002; Martínez-González et al., 2001). Mechanisms that target macrophages by alternative approaches are likely to further complement the current therapeutic options to treat atherosclerosis.

V. MACROPHAGES IN METABOLIC SYNDROME

Obesity and the associated metabolic pathologies are the most common and detrimental metabolic diseases affecting over 50% of the adult population. Recent and emerging data clearly implicates chronic low-grade inflammatory disease that accompanies obesity as a major mechanism of multiple complications of obesity, including increased risk of cardiovascular diseases and insulin resistance that often leads to type 2 diabetes (Berg and Scherer, 2005). The white adipose tissue (WAT) of obese subjects is characterized by increased production and secretion of a wide panel of inflammatory molecules, including TNFα (Hotamisligil et al., 1993), IL-6, TGFα, and plasminogen activator inhibitor 1 (Bastard and Jardel, 1999). The inhibition of signaling downstream of the insulin receptor appears to be the primary mechanism through which inflammatory cytokines lead to insulin resistance. For example, exposure of cells to TNFα stimulates inhibitory phosphorylation of serine residues of IRS-1 (Saltiel and Pessin, 2002). This phosphorylation reduces both tyrosine phosphorylation of IRS-1 in response to insulin and the ability of IRS-1 to associate with the insulin receptor, thereby inhibiting downstream signaling and insulin action (Hotamisligil et al., 1996).

An interesting feature of the inflammatory response that emerges in the presence of obesity is that it appears to be triggered, and to reside predominantly, in adipose tissue (Wellen and Hotamisligil, 2003). Studies using mice that were fed a high-fat diet during a 24-week period demonstrate that the dramatic upregulation of inflammation-related *genes* is mostly restricted to WAT; the mRNA expression of these *genes* was barely detectable and essentially unchanged in muscle and liver of the obese mice (Xu et al., 2003). In addition, there was little change of expression of these *genes* in lung and spleen in the obese state, further demonstrating that the observed inflammation was adipose-specific. As a result of these observations, much emphasis was placed on understanding molecular changes that happen within WAT during obesity. Transcriptional profiling studies conducted on WAT from lean and obese animals and humans have pointed to striking changes in a large repertoire of inflammatory *genes* in adipose tissue. Studies to identify the source of increased inflammatory mediators in WAT clearly implicate infiltrating macrophages and not the adipocyte as the predominant source (Cancello et al., 2005). This obesity-induced macrophage accumulation in WAT appears to be highly correlated with, if not causative of, insulin resistance in various mouse models. Expression levels of selected macrophage and inflammation *genes* were upregulated before the increase in circulating-insulin levels (Xu et al., 2003), suggesting that macrophage influx/activation occurs after the increase of adiposity but before insulin resistance. Based on the well-established capabilities of macrophages in other inflammatory conditions and the evolving picture of the role of inflammation in obesity-associated insulin resistance, a hypothetical model can be envisioned in which macrophages infiltrate into adipose tissue during obesity and produce factors that lead to changes in the adipocyte phenotype that ultimately result in insulin resistance (Figure 12.2). The exact nature of the causative factors released by macrophages remains to identified.

While it is still unclear why macrophages infiltrate adipose tissue during obesity, evidence from several animal and clinical studies support the contention that modulating WAT infiltration of macrophages may hold promise for the treatment of obesity-related metabolic disease due to the pleotropic nature of macrophages. Relevance of data generated in experimental animal models to human metabolic syndrome has been established through the demonstration of similar increases of macrophage content in WAT between animals and humans. Furthermore, MCP-1 has been shown to be significantly elevated in obese humans (Sartipy and Loskutoff, 2003). Treatment with PPARγ agonists (thiozolidinediones), or induction of weight loss in obese subjects results in significant reduction of adipose tissue inflammation and reduced expression of CCR2 ligand, MCP-1, thereby leading to improvement of metabolic consequences of obesity (Bruun et al., 2005; Clement et al., 2004). For example, weight loss following caloric restriction in humans decreases the expression of inflammatory markers in white adipose tissue of obese subjects

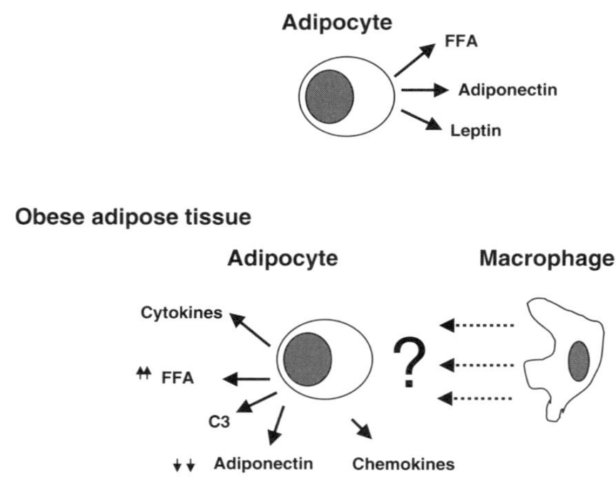

FIGURE 12.2 Altered phenotype of adipocytes in obesity.

and the vast majority of the changes in the gene transcripts that correlate with the reduction in the macrophage content of the WAT (Clement et al., 2004).

Modulating inflammation and, in particular, macrophage infiltration and activation in adipose tissue clearly offers an exciting new approach for the treatment of metabolic disease. Because of its potential to complement current therapeutic modalities—such as insulin sensitizers, which are not designed to correct inflammation—macrophage modulators could address one of the key pathophysiological mechanisms of this disease. Potential approaches to suppress WAT infiltration of macrophages include biologics—such as neutralizing antibodies to MCP-1, MIP-1α, and migration inhibitory factor (MIF-1)—which play a role in mediating tissue migration of macrophages; as well as small-molecule antagonists of chemokine receptors and inhibitors of cytokine generation.

VI. CCR2 ANTAGONISM AS A NEW APPROACH TO MODULATE INFLAMMATION

Data presented above strongly support the contention that the modulation of tissue influx of macrophages is likely to have therapeutic benefits in diseases that have proven to be amenable to anti-TNFα therapies. In addition, diseases like atherosclerosis, multiple sclerosis, and diabetes are additional therapeutic areas in which inhibition of the tissue influx of macrophages could be of potential benefit. The next several sections of this chapter will focus on blocking the chemokine receptor known as CCR2 as an approach to modulating the migration of circulation monocytes into inflammatory lesions. Data from CCR2 knockout mice as well as effects of a small molecule CCR2 antagonist, INCB3344, in various preclinical animal models will be presented.

CCR2 is a chemokine receptor predominantly expressed on monocytes that mediates their tissue influx in the context of immune-based inflammation (Power and Proudfoot, 2001). CCR2 is a G protein-coupled receptor, the ligands for which include the monocyte chemoattractant protein (MCP) family of chemokines (CCL2, CCL7, CCL8, etc.) (Boring et al., 1997). These ligands bind CCR2 with a high affinity (kDa of ~1 nM) and elicit a chemotactic signal that results in directed migration of the receptor-bearing cells. CCL2 has been shown to be present in high concentrations in various inflammatory lesions (Daly and Rollins, 2003), implicating this chemokine as a physiologically important chemotactic signal for monocytes. The critical role of CCL2-CCR2 interactions in the tissue influx of monocytes has been demonstrated most eloquently in studies of mice engineered to be deficient in either CCR2 or CCL2 (Kurihara et al., 1997). Mice deficient in the receptor CCR2 and the ligand CCL2, while phenotypically normal, show a selective defect in the migration of macrophages to sites of inflammation (Kuziel et al., 1997; Lu et al., 1998). In addition, a protein antagonist of CCR2 generated by N-terminal truncation of the first seven amino acids of CCL2, referred to as 7ND, has been used to determine the impact of intervention with a CCR2 antagonist in animal models of inflammatory disease (Kitamoto and Egashira, 2003), as described on page 232.

VII. CCR2 ANTAGONISM IN ARTHRITIS

Studies reported over the past several years have used various techniques to block the ligand: receptor interaction of MCP-1 and CCR2 to determine the impact of CCR2 antagonism on the development of arthritis in preclinical animal models. These include a protein antagonist known as 7ND (Kitamoto and Egashira, 2003), a DNA vaccine against MCP-1, the primary ligand for CCR2 (Youssef et al., 1998), antibody-based approaches (Lee et al., 2003), as well as receptor and ligand knockout studies. Most of these studies have clearly demonstrated a significant suppression of clinical and histological signs of inflammation in rat and mouse models utilizing complete Freund's adjuvant as an inducer of disease. A recent study by Quinones et al. (2004), however, observed an exacerbation of arthritis in the CCR2 knockouts as compared to wild-type littermates, which raised concerns about the mechanism as a potential approach to treat RA. Because macrophages are important in the clearance of mycobacterial antigens, which are typically used in the immunization protocols to induce arthritis, it is not clear if the effects noted by Quinones et al. (2005) are related to the blockade of macrophages into inflamed joints or related to a defect in the clearance of arthritogen in knockout mice. In contrast to the reported findings, studies performed by Brodmerkel et al. (manuscript in preparation, personal communication, 2005) using a small-molecule antagonist of CCR2, INCB3344, demonstrated suppression of both clinical and histopathological evidence of arthritis when administered in either preventive or therapeutic regimens. In contrast to the increased anti-collagen antibody titers observed in CCR2 knockouts, small-molecule antagonism of CCR2 did not result in increased titers, suggesting that there are some intrinsic differences with regard to the antibody generation between CCR2 knockout and wild-type mice.

Currently there are clinical studies ongoing with CCR2-blocking antibody (MLN1202; Millenium Pharmaceuticals, Cambridge, MA), as well as small-molecule antagonist of CCR2 (INCB3284, Incyte Corp., Wilmington, DE) in rheumatoid arthritis, the results of which will help understand potential benefit of this mechanism in rheumatoid arthritis. If these studies demonstrate an efficacy profile similar to that of anti-TNFα therapies, a natural extension of these studies would be to explore potential benefits in diseases such as Crohn's disease, ankylosing spondylitis, psoriatic arthritis, and psoriasis.

VIII. CCR2 AND MULTIPLE SCLEROSIS

As described previously, extensive literature on the pathogenesis of MS implicates infiltrating macrophages and lymphocytes as key determinants of disease progression, particularly in the relapsing and remitting form of the disease (Owens et al., 2002). Chemokines have been shown to play an important role in the activation and directional migration of cells to sites of CNS inflammation. Expression of the beta-chemokine receptors CCR2, CCR3, and CCR5 has been demonstrated in postmortem CNS samples from MS patients. In chronic active MS lesions, CCR2 is associated with foamy macrophages, activated microglia, as well as infiltrating lymphocytes. Ligands for CCR2 that include MCP-1 and MCP-3 are co-localized around vessels with the infiltrating leukocytes, but are also present in unaffected areas of cortex (Simpson, et al., 2000).

In an experimental model of MS, known as experimental autoimmune encephalomyelitis (EAE), monocyte recruitment to the CNS is a necessary step in the development of pathological inflammatory lesions (Mahad and Ransohoff, 2003). To test the contribution of CCR2 signaling in the initiation and progression of EAE, CCR2−/− and CCR2+/+ mice immunized with the encephalitogenic peptide MOGp35–55 were examined (Fife et al., 2000). Results from this study for the first time demonstrated a striking resistance to disease induction accompanied by markedly reduced numbers of infiltrating leukocytes into the CNS. To determine if the protection from disease in CCR2−/− mice is due to alterations in the MOGp35–55-specific immune responses, the phenotype of T cells generated by immunization with MOGp35–55 in CCR2-deficient and wild-type mice were examined. Draining LN cells from wild-type mice showed robust proliferative responses to the antigen, with stimulation indices approximately two times higher than in LN cells from the CCR2-deficient mice. IFN-γ, a proinflammatory Th1 cytokine known to be involved in the pathogenesis of EAE, was also decreased in CCR2-deficient mice compared with wild-type mice (Fife et al., 2000).

The critical role of CCR2 was also demonstrated in a relapsing remitting model of multiple sclerosis, in which a significant reduction in the number as well as the magnitude of relapses was observed in CCR2- as well as MCP-1 deficient mice (Izkison et al., 2000; Huang et al., 2001). These studies also addressed the question of whether the dependence of disease induction on the MCP-1/CCR2 pathway is because of its requirement for the generation of immune responses to the encephalitogen or because of the need for this receptor ligand pair for the CNS infiltration of effector macrophages. Adoptive transfer of MCP-1+/+ T-cells into MCP-1−/− negative mice failed to develop disease, while the transfer of T cells from MCP-1−/− mice to MCP-1+/+ mice developed normal disease, suggesting that lack of MCP-1 does not have an impact on the generation of normal immune responses. To determine the impact of MCP-1 gene deletion on humoral immune responses, sera from MCP-1+/+ and MCP-1−/− mice with EAE were analyzed for total IgG, IgG1, and IgG2a Abs against the immunizing MOG35–55 peptide. Wild-type and MCP-1–deficient mice produced similar amounts of total MOG-specific IgG, indicating no defect on humoral immune responses in MCP-1 deficient mice (Huang et al., 2001).

Based on the data from similar studies in knockout mice of other chemokine receptors present on monocytes such as CCR1 and CCR5, which demonstrated much less significant impact in the disease incidence and severity, it appears that CCR2 is the predominant receptor involved in the CNS infiltration of moncytes (Rottman et al., 2000; Tran et al., 2000). Ongoing clinical studies using CCR2-neutralizing antibody, MLN1202, as well as small-molecule antagonists of CCR2, will confirm if the promise of this mechanism holds true in human disease. Based on the demonstration of a lack of immunosuppression in CCR2 knockout mice (Kurihara et al., 1997), if CCR2 blockade is efficacious in MS, it offers a potentially safer alternative or complementary approach for the treatment of this disease.

IX. CCR2 AND ATHEROSCLEROSIS

Because of the established data implicating the macrophage as the predominant cell type in atherosclerotic plaques, gene expression of MCP-1 was examined in normal and diseased human arteries (Nelken et al., 1991). MCP-1 mRNA was detected in 16% of the cells in human carotid endarterectomy specimens, with highest expression seen in organizing thrombi (33%)

and in macrophage-rich areas bordering the necrotic lipid core (24%). Based on immunohistochemical staining of serial sections and on cell morphology, MCP-1 mRNA appeared to be expressed by VSMCs, mesenchymal appearing intimal cells, and macrophages. By contrast, few cells expressing MCP-1 mRNA were found in normal arteries (less than 0.1%). Several factors that are present in atherosclerotic lesions including thrombin, oxLDL and tissue factor have been shown to be strong stimuli for the generation of MCP-1 in human disease (Wenzel et al., 1995; Wang, 1997; Schecter et al., 1997).

Several groups have examined the role of MCP-1 and CCR2 in preclinical models of atherosclerosis utilizing knockout mice. The most commonly utilized rodent models of atherosclerosis include Apolipoprotein E (ApoE) and low-density lipoprotein (LDL) receptor knockout mice, which develop atherosclerotic plaques, primarily in aortic root, due to defective clearance of LDL particles (Jawein, et al., 2004). Abundant numbers of macrophages are present in the subendothelial space of coronary vessels in these mice, making them appropriate models to study the role of CCR2 on lesion development in atherosclerosis.

Studies using crosses of CCR2–/– and ApoE–/– or MCP-1–/– and LDL receptor –/– mice identified CCR2 and its ligand MCP-1 as genetic determinants of atherosclerosis and provided strong evidence for a direct effect of MCP-1 in macrophage recruitment and atherogenesis (Charo and Taubman, 2004). Despite having the same amount of total and fractionated serum cholesterol as LDL receptor –/– mice with wild-type MCP-1 alleles, LDL receptor/MCP-1–deficient mice had 83% less lipid deposition throughout their aortas. Consistent with MCP-1's monocyte chemoattractant properties, LDL receptor/MCP-1–/– mice also had fewer macrophages in their aortic walls (Gu et al., 1998). Similar results were also reported in CCR2/ApoE–/– mice (Boring et al., 1998). In addition, this group compared various types of cholesterols, including VLDL, LDL, IDL, HDL, and plasma triglyceride levels of CCR2–/– and wild-type mice and found no differences, thereby confirming that the reduced atherosclerotic plaque formation in CCR2–/– mice is due to reduced infiltration of macrophages into the subendothelial space.

Because plaque stability is a major cause of cardiovascular complications of atherosclerosis, it would be of interest to determine if CCR2 antagonism would be of therapeutic utility in improving plaque stability. Unlike human disease, murine atheromas do not rupture. A number of histological indices of plaque instability, such as decreased collagen content and decreased VSMC area of the plaque, do occur in preclinical models, which can be examined to determine the impact of a given therapeutic modality on plaque stability. Inoue et al. (2002) examined the effect of adenoviral gene delivery of a protein antagonist of CCR2, 7ND, on plaque composition in ApoE–/– mice following 8 weeks of therapy and noted a significant improvement in both collagen content and VSMC area of the plaques. Concomitant with these changes there was a significant reduction of mean lesional area, plaque macrophage content, and plaque lipid content, suggesting a favorable effect on both progression of atherosclerosis as well as improvement of plaque stability. Encouraged by these observations, Kitamoto et al., (2004) used the same method of blocking CCR2, that is, adenoviral delivery of 7ND, in a primate model of atherosclerosis. In this model, monkeys were fed a high-cholesterol diet (0.5% cholesterol and 6% corn oil) for 6 months and received balloon injury of the descending aorta and the right common iliac artery at 1 month after the initiation of the diet.

In this study, 7ND treatment was compared to a cholesterol lowering agent, pravastatin. Similar to the mouse study described in the report by Inove et al. (2002), 7ND treatment of primates resulted in a significant improvement in plaque stability in a manner similar to pravastatin. In addition, plaque regression was also noted as a result of treatment of either agent, supporting the idea that blocking CCR2 could be a potential approach for the treatment of atherosclerosis and related complications.

Among different therapeutic areas in which CCR2 antagonists could be efficacious, atherosclerosis appears to be the most logical disease to be responsive to this modality. Decades of research implicate macrophages in playing a causal role in this disease. Data generated in numerous animals models with multiple methods of interfering with the CCR2/MCP-1 pathway support a critical role for this mechanism in the pathology of atherosclerosis.

X. CCR2 AND DIABETES

Unlike the diseases described above, in which inflammation has been well established to play a major role in their pathogenesis, the potential importance of inflammation in the development of obesity and related complications is a recent discovery. Hence, there are fewer studies that have examined the CCR2 pathway in this disease. The earliest evidence that suggested a role for this pathway in diabetes is the finding that circulating MCP-1 concentrations are elevated in obese individuals (Christiansen et al., 2005). Based on the findings that a good correlation exists between adipose tissue macrophage content and obesity-associated metabolic dysregulation, Weisberg et al. (2005) examined the impact of CCR2 gene deletion as well as pharmacological inhibition of CCR2 on the metabolic phenotype in mice following a high-fat diet. In lean animals, no effect of CCR2 genotype on metabolic traits could be detected. In contrast, when mice are fed a high-fat diet, CCR2 gene deletion modulated feeding behavior, the development of obesity, and the development of obesity-associated adipose tissue inflammation, adipose tissue macrophage accumulation, adiponectin expression, hepatic steatosis and insulin sensitivity. In established obesity, short-term pharmacological inhibition of CCR2 with a specific antagonist, INCB3344, was able to affect the maintenance of obesity-associated phenotypes and specifically attenuate obesity-induced macrophage accumulation in adipose tissue and ameliorate insulin resistance.

While it is tempting to speculate that the positive effects of CCR2 antagonism in diet-induced obesity models are strictly due to suppression of macrophage influx into adipose, it is clear that additional studies will be required to fully understand the role of CCR2 in this disease.

XI. CONCLUDING REMARKS

TNFα-neutralizing therapies paved the way for serious clinical investigation of cytokine modulatory therapies in inflammatory diseases. While directly modulating TNFα activity is the most validated mechanism, currently available small-molecule based approaches to accomplish such an effect is complicated by potential non–mechanism-based toxicities. Clear demonstration of the importance of macrophage as a key producer of TNFα makes this suppressing the tissue influx of this cell type an alternative mechanism of TNFα modulation. Discovery of a GPCR target like CCR2 makes it pharmacologically feasible to modulate macrophage tissue migration by the use of small-molecule antagonists. Gene knockout studies using CCR2 and its ligand MCP-1 have demonstrated that CCR2 antagonsim is likely to provide therapeutic benefit in diseases that are not only responsive to TNFα-neutralizing agents but also several additional diseases that are known to be dependent on macrophages. Clinical studies using CCR2 antagonists are ongoing in rheumatoid arthritis, multiple sclerosis, atherosclerosis, and diabetes, and the data from these studies will ultimately prove if this mechanism is important in human disease.

RECOMMENDED RESOURCES

A. Chemokines and Cytokines

Series of short lectures on chemokines:
http://www.bio.davidson.edu/Courses/immunology/chemokinespeech/chemokinetalks.html
Nature special focus on chemokines:
http://www.nature.com/ni/focus/chemokines/index.html#rv
Cytokine & Chemokine family cDNA Database: http://cytokine.medic.kumamoto-u.ac.jp/
Encyclopedia of cytokines: http://www.copewithcytokines.de/
Cytokine information system: http://advice.i2r.a-star.edu.sg/cytokine/

B. TNF

Comprehensive molecular and biochemical information: http://www.genecards.org/cgi-bin/carddisp.pl?gene=TNF

C. CCR2

Comprehensive molecular and biochemical information: http://www.genecards.org/cgi-bin/carddisp.pl?gene=CCR2

D. General Information on Inflammatory Diseases

Rheumatoid Arthritis: http://www.rheumatology.org/; http://www.eular.org/
Multiple Sclerosis: http://www.nationalmssociety.org/; http://www.msfacts.org/
Atherosclerosis: http://www.americanheart.org/presenter.jhtml?identifier=4440; http://www.athero.org/
Diabetes/Metabolic Syndrome:
http://www.americanheart.org/presenter.jhtml?identifier=4756;
http://www.diabetes.org/home.jsp;

E. Pathophysiology of Inflammatory Diseases

Rheumatoid Arthritis: http://www.medscape.com/viewarticle/487710_3;
http://www.healthcentral.com/rheumatoid-arthritis/introduction-000048_2-145.html
Multiple Sclerosis: http://www.healthcentral.com/ency/408/000737.html;
http://www.emedicine.com/emerg/topic321.htm
Atherosclerosis: http://www.healthcentral.com/ency/408/000171.html;

http://www.emedicine.com/med/topic182.htm
Diabetes/Metabolic Syndrome:
http://www.cmeondiabetes.us/pub/metabolic.syndrome.php;

REFERENCES

Adams, C. W., R. N. Poston, and S. J. Buk (1989). Pathology, histochemistry and immunocytochemistry of lesions in acute multiple sclerosis. *J. Neurol. Sci.* 92(2-3): 291–306.

Barrera, P. and A. Blom, et al. (2000). Synovial macrophage depletion with clodronate-containing liposomes in rheumatoid arthritis. *Arthritis Rheum.* 43(9): 1951–1959.

Bastard, J. P., C. Jardel, et al. (1999). Evidence for a link between adipose tissue interleukin-6 content and serum C-reactive protein concentrations in obese subjects. *Circulation* 99(16): 2221–2222.

Benveniste, E. N. (1997). Role of macrophages/microglia in multiple sclerosis and experimental allergic encephalomyelitis. *J. Mol. Med.* 75(3): 165–173.

Berg, A. H. and P. E. Scherer (2005). Adipose tissue, inflammation, and cardiovascular disease. *Circ. Res.* 96(9): 939–949.

Boring, L., J. Gosling, et al. (1997). Impaired monocyte migration and reduced type 1 (Th1) cytokine responses in C-C chemokine receptor 2 knockout mice. *J. Clin. Invest.* 100(10): 2552–2561.

Boring, L. and J. Gosling, et al. (1998). Decreased lesion formation in CCR2-/- mice reveals a role for chemokines in the initiation of atherosclerosis. *Nature* 394(6696): 894–897.

Boyle, J. J. (2005). Macrophage activation in atherosclerosis: Pathogenesis and pharmacology of plaque rupture. *Curr. Vasc. Pharmacol.* 3(1): 63–68.

Brodmerkel, C. M., Huber, K., et al. (2005). Discovery and pharmocological characterization of a novel rodent-active CCR2 antagonist, INCB 3344. *J. Immunol.* 175(8): 5370–5378.

Bruck, W. and C. Stadelmann (2005). The spectrum of multiple sclerosis: New lessons from pathology. *Curr. Opin. Neurol.* 18(3): 221–224.

Bruun, J. M., A. S. Lihn, et al. (2005). Monocyte chemoattractant protein-1 release is higher in visceral than subcutaneous human adipose tissue (AT): Implication of macrophages resident in the AT. *J. Clin. Endocrinol. Metab.* 90(4): 2282–2289.

Cancello, R., C. Henegar, et al. (2005). Reduction of macrophage infiltration and chemoattractant gene expression changes in white adipose tissue of morbidly obese subjects after surgery-induced weight loss. Diabetes 54(8): 2277–2286.

Charo, I. F. (1999). CCR2: From cloning to the creation of knockout mice. *Chem. Immunol.* 72: 30–41.

Charo, I. F., S. J. Myers, et al. (1994). Molecular cloning and functional expression of two monocyte chemoattractant protein 1 receptors reveals alternative splicing of the carboxyl-terminal tails. *Proc. Natl. Acad. Sci. U. S. A.* 91(7): 2752–2756.

Charo, I. F. and M. B. Taubman (2004). Chemokines in the pathogenesis of vascular disease. *Circ. Res.* 95(9): 858–866.

Christiansen, T., B. Richelsen, and J. R. Bruun (2005). Monocyte chemoattractant protein-1 is produced in isolated adipocytes, associated with adiposity and reduced after weight loss in morbid obese subjects. *Int. J. Obes. (Lond).* 29(1): 146–150.

Clement, K., N. Viguerie, et al. (2004). Weight loss regulates inflammation-related genes in white adipose tissue of obese subjects. *FASEB J.* 18(14): 1657–1669.

Cutolo, M., A. Sulli, et al. (1993). Macrophages, synovial tissue and rheumatoid arthritis. *Clin. Exp. Rheumatol.* 11(3): 331–339.

Daly, C. and B. J. Rollins (2003). Monocyte chemoattractant protein-1 (CCL2) in inflammatory disease and adaptive immunity: Therapeutic opportunities and controversies. *Microcirculation* 10(3-4): 247–257.

Dandona, P. and A. Aljada (2002). A rational approach to pathogenesis and treatment of type 2 diabetes mellitus, insulin resistance, inflammation, and atherosclerosis. *Am. J. Cardiol.* 90(5A): 27G–33G.

Falk, K. (1989). Morphologic features of unstable atherothrombotic plaques underlying acute coronary syndromes. *Am. J. Cardiol.* 63(10): 114–120.

Febbraio, M., D. P. Hajjar, and R. L. Silverstein (2001). CD36: A class B scavenger receptor involved in angiogenesis, atherosclerosis, inflammation, and lipid metabolism. *J. Clin. Invest.* 108 (6): 785–791.

Ferguson, B., M. K. Matyszak, et al. (1997). Axonal damage in acute multiple sclerosis lesions. *Brain* 120(3): 393–399.

Fife, B. T., G. B. Huffnagle, et al. (2000). CC chemokine receptor 2 is critical for induction of experimental autoimmune encephalomyelitis. *J. Exp. Med.* 192(6): 899–905.

Forman, H. J. and M. Torres (2001). Redox signaling in macrophages. *Mol. Aspects Med.* 22(4-5): 189–216.

Fujiwara, N. and K. Kobayashi (2005). Macrophages in inflammation. *Curr. Drug Targets. Inflamm. Allergy* 4(3): 281–286.

Gerlag, D. M., J. J. Haringman, et al. (2004). Effects of oral prednisolone on biomarkers in synovial tissue and clinical improvement in rheumatoid arthritis. *Arthritis Rheum.* 50(12): 3783–3791.

Ghezzi, P. (1997). Cytokines: Pharmacologically-oriented introduction. *Eur. Cytokine Netw.* 8(3): 289–290.

Gravallese, E. M. (2002). Bone destruction in arthritis. *Ann. Rheum. Dis.* 61 (Suppl 2) 84–86.

Gu, L., Y. Okada, et al. (1998). Absence of monocyte chemoattractant protein-1 reduces atherosclerosis in low density lipoprotein receptor-deficient mice. *Mol. Cell* 2(2): 275–281.

Haas, J. G., P. A. Baeuerle, et al. (1990). Molecular mechanisms in down-regulation of tumor necrosis factor expression. *Proc. Natl. Acad. Sci. U. S. A.* 87(24): 9563–9567.

Hamilton, T. A., Y. Ohmori, and J. Tebo (2002). Regulation of chemokine expression by antiinflammatory cytokines. *Immunol. Res.* 25(3): 229–245.

Haringman, J. J., D. M. Gerlag, et al. (2005). Synovial tissue macrophages: A sensitive biomarker for response to treatment in patients with rheumatoid arthritis. *Ann. Rheum. Dis.* 64(6): 834–838.

Hemler, M. E., C. Huang, et al. (1987). Characterization of the cell surface heterodimer VLA-4 and related peptides. *J. Biol. Chem.* 262(24): 11,478–11,485.

Hotamisligil, G. S., P. Peraldi, et al. (1996). IRS-1-mediated inhibition of insulin receptor tyrosine kinase activity in TNFa-alpha- and obesity-induced insulin resistance. *Science* 271(5249): 665–668.

Hotamisligil, G. S., N. S. Shargill, and B. M. Spiegelman (1993). Adipose expression of tumor necrosis factor-alpha: Direct role in obesity-linked insulin resistance. Science 259(5091): 87–91.

Huang, D. R., J. Wang, et al. (2001). Absence of monocyte chemoattractant protein 1 in mice leads to decreased local macrophage recruitment and antigen-specific T helper cell type 1 immune response in experimental autoimmune encephalomyelitis. *J. Exp. Med.* 193(6): 713–726.

Husby, G. and R. C. Williams, Jr. (1988). Synovial localization of tumor necrosis factor in patients with rheumatoid arthritis. *J. Autoimmun.* 1(4): 363–371.

Inoue, S., K. Egashira, et al. (2002). Anti-monocyte chemoattractant protein-1 gene therapy limits progression and destabilization of established atherosclerosis in apolipoprotein E-knockout mice. *Circulation* 106(21): 2700–2706.

Izikson, L., R. S. Klein, et al. (2000). Resistance to experimental autoimmune encephalomyelitis in mice lacking the CC chemokine receptor (CCR)2. *J. Exp. Med.* 192(7): 1075–1080.

Jawien, J., P. Nastalek, and R. Korbut (2004). Mouse models of experimental atherosclerosis. *J. Physiol. Pharmacol.* 55(3): 503–517.

Kitamoto, S. and K. Egashira (2003). Anti-monocyte chemoattractant protein-1 gene therapy for cardiovascular diseases. *Expert Rev. Cardiovasc. Ther.* 1(3): 393–400.

Kitamoto, S., K. Nakano, et al. (2004). Cholesterol-lowering independent regression and stabilization of atherosclerotic lesions by pravastatin and by antimonocyte chemoattractant protein-1 therapy in nonhuman primates. *Arterioscler. Thromb. Vasc. Biol.* 2004. 24(8): 1522–1528.

Kornek, B., M. K. Storch, et al. (2000). Multiple sclerosis and chronic autoimmune encephalomyelitis: A comparative quantitative study of axonal injury in active, inactive and remyelinated lesions. *Am. J. Pathol.* 157(1): 267–276.

Kraan, M. C., J. J. Haringman, et al. (2000). Quantification of the cell infiltrate in synovial tissue by digital image analysis. *Rheumatology* (Oxford) 39(1): 43–49.

Kumar, S., J. Boehm, and J. C. Lee (2003). p38 MAP kinases: Key signalling molecules as therapeutic targets for inflammatory diseases. *Nat. Rev. Drug Discov.* 2(9): 717–726.

Kurihara, T., G. Warr, et al. (1997). Defects in macrophage recruitment and host defense in mice lacking the CCR2 chemokine receptor. *J. Exp. Med.* 186(10): 1757–1762.

Kuziel, W. A., S. J. Morgan, et al. (1997). Severe reduction in leukocyte adhesion and monocyte extravasation in mice deficient in CC chemokine receptor 2. *Proc. Natl. Acad. Sci. U. S. A.* 94(22): 12,053–12,058.

Langer-Gould, A. (2005). Progressive multifocal leukoencephalopathy in a patient treated with natalizumab. *N. Engl. J. Med.* 353(4): 375–381.

Lee, I., L. Wang, et al. (2003). Blocking the monocyte chemoattractant protein-1/CCR2 chemokine pathway induces permanent survival of islet allografts through a programmed death-1 ligand-1-dependent mechanism. *J. Immunol.* 171(12): 6929–6935.

Libby, P., Y. L. Geng, et al. (1996). Macrophages and atherosclerotic plaque stability. *Curr. Opin. Lipidol.* 7(5): 330–335.

Libby, P., and P. Theroux (2005). Pathophysiology of coronary artery disease. *Circ.* 111(25): 3481–3488.

Lu, B., B. J. Rutledge, et al. (1998). Abnormalities in monocyte recruitment and cytokine expression in monocyte chemoattractant protein-1-deficient mice. *J. Exp. Med.* 187(4): 601.

Lucas, A. D. and D. R. Greaves (2001). Atherosclerosis: Role of chemokines and macrophages. *Expert Rev. Mol. Med.* 2001: 1–18.

Mahad, D. J., and R. M. Ransohoff (2003). The role of MCP-1 (CCL2) and CCR2 in multiple sclerosis and experimental autoimmune encephalomyelitis (EAE). *Semin. Immunol.* 15(1): 23–32.

Martínez-González, J., J. Alfon, et al. (2001). HMG-CoA reductase inhibitors reduce vascular monocyte chemotactic protein-1 expression in early lesions from hypercholesterolemic swine independently of their effect on plasma cholesterol levels. *Atherosclerosis* 159(1): 27–33.

Moss, M. L., S. L. Jin, et al. (1997). Structural features and biochemical properties of TNFa-alpha converting enzyme (TACE). *J. Neuroimmunol.* 72(2): 127–129

Mulherin, D., O. Fitzgerald, and B. Bresnihan (1996). Synovial tissue macrophage populations and articular damage in rheumatoid arthritis. *Arthritis Rheum.* 39(1): 115–124.

Neininger, A., D. Kontoyiannis, et al. (2002). MK2 targets AU-rich elements and regulates biosynthesis of tumor necrosis factor and interleukin-6 independently at different post-transcriptional levels. *J. Biol. Chem.* 277(5): 3065–3068.

Nelken, N. A., S. R. Coughlin, et al. (1991). Monocyte chemoattractant protein-1 in human atheromatous plaques. *J. Clin. Invest.* 88(4): 1121–1127.

O'Connor, P. W., A. Goodman, et al. (2004). Randomized multicenter trial of natalizumab in acute MS relapses: Clinical and MRI effects. *Neurology* 62(11): 2038–2043.

Ohashi, R., H. Mu, et al. (2004). Atherosclerosis: Immunopathogenesis and immunotherapy. *Med. Sci. Monit.* 10(11): 55–60.

Owens, T. (2002). Identification of new therapeutic targets for prevention of CNS inflammation. *Expert Opin. Ther. Targets* 6(2): 203–215.

Palladino, M. A., F. R. Bahjat, et al. (2003). Anti-TNFa-alpha therapies: The next generation. *Nat. Rev. Drug Discov.* 2(9): 736–746.

Power, C. A. and A. E. Proudfoot (2001). The chemokine system: Novel broad-spectrum therapeutic targets. *Curr. Opin. Pharm.* 1(14): 417–424.

Quinones, M. P., S. K. Ahuja, et al. (2004). Experimental arthritis in CC chemokine receptor 2-null mice closely mimics severe human rheumatoid arthritis. *J. Clin. Invest.* 113(6): 856–866.

Quinones, M. P., C. A. Estrada, et al. (2005). The complex role of the chemokine receptor CCR2 in collagen-induced arthritis: Implications for therapeutic targeting of CCR2 in rheumatoid arthritis. *J. Mol. Med.* 83(9): 672–681.

Rottman, J. B., A. J. Slavin, et al. (2000). Leukocyte recruitment during onset of experimental allergic encephalomyelitis is CCR1 dependent. *Eur. J. Immunol.* 30(8): 2372–2377.

Saltiel A. R. and J. E. Pessin (2002). Insulin signaling pathways in time and space. *Trends Cell Biol.* 12(2): 65–71.

Sartipy, P. and D. J. Loskutoff (2003). Monocyte chemoattractant protein 1 in obesity and insulin resistance. *Proc. Natl. Acad. Sci. U. S. A.* 100(12): 7265–7270.

Schecter, A. D., B. J. Rollins, et al. (1997). Tissue factor is induced by monocyte chemoattractant protein-1 in human aortic smooth muscle and THP-1 cells. *J. Biol. Chem.* 272(45): 28,568–28,573.

Seldon, P. M., P. J. Barnes, et al. (1995). Suppression of lipopolysaccharide-induced tumor necrosis factor-alpha generation from human peripheral blood monocytes by inhibitors of phosphodiesterase 4: Interaction with stimulants of adenylyl cyclase. *Mol. Pharmacol.* 48(4): 747–757.

Shapiro, S. D. (1999). Diverse roles of macrophage matrix metalloproteinases in tissue destruction and tumor growth. *Thromb. Haemost.* 82(2): 846–849.

Simpson, J., P. Rezaie, et al. (2000). Expression of the beta-chemokine receptors CCR2, CCR3 and CCR5 in multiple sclerosis central nervous system tissue. *J. Neuroimmunol.* 108(1-2): 192–200.

Smeets, T. J., M. C. Kraan, et al. (2003). Tumor necrosis factor alpha blockade reduces the synovial cell infiltrate early after initiation of treatment, but apparently not by induction of apoptosis in synovial tissue. *Arthritis Rheum.* 48(8): 2155–2162.

Smith, M. D., M. C. Kraan, et al. (2001). Treatment-induced remission in rheumatoid arthritis patients is characterized by a reduction in macrophage content of synovial biopsies. *Rheumatology* (Oxford) 40(4): 367–374.

Smolen, J. S. and G. Steiner (2003). Therapeutic strategies for rheumatoid arthritis. *Nat. Rev. Drug Discov.* 2(6): 473–488.

Steinman, L. (2005). Blocking adhesion molecules as therapy for multiple sclerosis: Natalizumab. *Nat. Rev. Drug Discov.* 4(6): 510–518.

Tall, A., D. Sammett, and E. Granot (1986). Mechanisms of enhanced cholesteryl ester transfer from high density lipoproteins to apolipoprotein B-containing lipoproteins during alimentary lipemia. *J. Clin. Invest.* 77(4): 1163–1172.

Tran, E. H., W. A. Kuziel, and T. Owens (2000). Induction of experimental autoimmune encephalomyelitis in C57BL/6 mice deficient in either the chemokine macrophage inflammatory protein-1alpha or its CCR5 receptor. *Eur. J. Immunol.* 30(5): 1410–1415.

Tyndall, A. and A. Gratwohl (2000). Immune ablation and stem-cell therapy in autoimmune disease: Clinical experience. *Arthritis Res.* 2(4): 276–280.

van der Wal, A. C. and A. E. Becker (1999). Atherosclerotic plaque rupture: Pathologic basis of plaque stability and instability. *Cardiovasc. Res.* 41(2): 334–344.

Van Lent, P. L., A. E. Holthuysen, et al. (1998). Local removal of phagocytic synovial lining cells by clodronate-liposomes decreases cartilage destruction during collagen type II arthritis. *Ann. Rheum. Dis.* 57(7): 408–413.

Wang, G. P., Z. D. Deng, et al. (1997). Oxidized low density lipoprotein and very low density lipoprotein enhance expression of monocyte chemoattractant protein-1 in rabbit peritoneal exudate macrophages. *Atherosclerosis* 133(1): 31–36.

Weisberg, S. P., D. Hunter, et al. (2005). CCR2 modulates inflammatory and metabolic phenotypes in mice after high fat feeding. *J. Clin. Invest.* 116(1): 115–124.

Wellen, K. E., G. S. Hotamisligil (2003). Obesity-induced inflammatory changes in adipose tissue. *J. Clin. Invest.* 112(12): 1785–1788.

Wenzel, U. O., B. Fouqueray, et al. (1995). Thrombin regulates expression of monocyte chemoattractant protein-1 in vascular smooth muscle cells. *Circ. Res.* 77(3): 503–509.

Xu, H., G. T. Barnes, et al. (2003). Chronic inflammation in fat plays a crucial role in the development of obesity-related insulin resistance. *J. Clin. Invest.* 112(12): 1821–1830.

Youssef, P. P., T. J. Smeets, et al. (1998). Microscopic measurement of cellular infiltration in the rheumatoid arthritis synovial membrane: A comparison of semiquantitative and quantitative analysis. *Br. J. Rheumatol.* 37(9): 1003–1007.

Youssef, S., G. Wildbaum, et al. (1998). Long-lasting protective immunity to experimental autoimmune encephalomyelitis following vaccination with naked DNA encoding C-C chemokines. *J. Immunol.* 161(8): 3870–3879.

13 THE DISCOVERY OF ELTROMBOPAG, AN ORALLY BIOAVAILABLE TpoR AGONIST

KEVIN J. DUFFY AND CONNIE L. ERICKSON-MILLER

Ph.D., GlaxoSmithKline Pharmaceuticals, Collegeville, Pennsylvania

High-throughput screening (HTS) of small-molecule compound collections is a well-validated and efficient methodology for discovering new lead molecules for drug discovery. The successful application of such an approach to the discovery of hematopoeitic cytokine agonists, however, is a daunting task because of the complicated and extensive nature of a cytokine-cytokine receptor interaction and the myriad downstream signaling pathways involved in transcriptional activation and cellular survival. Nevertheless, HTS for agonists of thrombopoietin receptor (TpoR)-activated signaling pathways, utilizing hematopoetic cell lines co-transfected with the human TpoR, a luciferase gene, and several requisite signal-transduction responsive elements, successfully identified several small-molecule leads. These leads were shown to activate TpoR-mediated signaling pathways in cell lines and, ultimately, to stimulate the formation of megakaryocyte progenitor cells ex vivo in human-bone-marrow tissue samples. The data from these assays were critical in the optimization of the molecular leads and in the discovery of a pharmacophore for TpoR activation. Ultimately, this understanding of the structure-activity relationships (SAR) for TpoR activation culminated in the design of SB-497115 (Eltrombopag), a potent, orally bioavailable TpoR agonist. The progression of Eltrombopag into clinical trials provides the final validation of our approach, wherein a dose-dependant stimulation of platelet production is observed in humans.

Target Validation in Drug Discovery

I. THROMBOCYTOPENIA AND MEGAKARYOCYTOPOIESIS

A. Thrombopoietin

The circulatory cytokine thrombopoietin (Tpo) plays a critical role in the regulation of megakaryocytopoiesis and is the primary regulator of platelet production (Kuter and Begley, 2002). Although the existence of Tpo was demonstrated 40 years ago, purification of the protein was achieved only relatively recently.(Kaushansky et al., 1994). Human Tpo is a 332 amino acid protein with a 154 amino acid N-terminal region arranged in the classic cytokine motif of four anti-parallel alpha-helices. Tpo acts through binding and activating the cell-surface receptor, Tpo receptor (TpoR; *c-mpl*), which is expressed on hematopoietic stem cells, immature hematopoietic progenitor cells, and megakaryocytes. The binding of Tpo to TpoR initiates a cell-signaling cascade involving several pathways, including the activation of JAK-2, (Drachman et al., 1999) STAT-5 (Miyakawa et al., 1996) and MAPK, (Rojnuckarin et al., 1999) leading to the irrevocable commitment of the progenitor cell along the megakaryocytic lineage. Mature megakaryocytes then undergo cell fragmentation to produce the platelet bodies that are an essential component in the process of wound healing through their subsequent cross-linking with fibrin, and the eventual formation of a protective thrombus in compromised blood vessels. The TpoR is also expressed on platelets therefore providing a biological feedback mechanism to inversely regulate the concentration of free, circulating Tpo with respect to platelet count.

B. Thrombocytopenia

Thrombocytopenia is a serious clinical complication in patients receiving intensive chemotherapy. Likewise, a number of other conditions can also contribute to thrombocytopenia, such as immune thrombocytopenic purpura (ITP), liver dysfunctions, drug treatments (interferons, [Garcia-Suarez et al., 2000] heparin, and viral infections [HIV (Mannucci and Gringeri, 2000), hepatitis C (Espanol et al., 2000)]. ITP is an autoimmune disorder characterized by autoantibody-induced platelet destruction, which can lead to mucocutaneous bleeding in patients with counts of <30,000 platelets/μL. The mortality rate from hemorrhage in such patients is estimated to be between 1.6 and 3.9% per patient-year, although the frequency is influenced by patient age and the duration of the thrombocytopenia (Cohen et al., 2000). Low platelet counts can also prevent the initiation and maintenance of interferon therapy in patients with chronic liver disease, thereby compromising the patient's ability to benefit from the sustained virological response possible with therapy (Kercher et al., 2004).

Patients receiving intensive chemotherapy or myeloablative regimens, as treatments for a variety of cancers, require repeated transfusions of platelets to control the associated thrombocytopenia. This accounts for approximately 40% of the platelet transfusions performed in the U. S. at this time (Gupta et al., 2000; Kaushansky, 2000; McCullough, 2000), with such transfusions becoming increasingly costly (Guest et al., 1998; Meehan et al., 2000; Webb and Anderson, 1999). Despite their effectiveness, platelet transfusions can lead to numerous, serious complications, including alloimmunization (therefore rendering the patient refractory to subsequent transfusions), febrile and allergic responses, circulatory overload and bacterial or viral infections (rare today, but often fatal) (McCullough, 2000).

C. Protein and Small-Molecule Agonist Therapeutics

The only approved platelet growth factor currently available is the recombinant human protein therapeutic interleukin 11 (rhIL-11, Neumega™) (Adams and Brenner, 2000), which is approved for prevention of severe thrombocytopenia and for the reduction of the need for platelet transfusions following myelosuppressive chemotherapy in patients with nonmyeloid malignancies who are at high risk of severe thrombocytopenia. rhTpo and pegylated megakaryocyte growth and development factor (peg-MGDF, a pegylated, N-terminal truncated portion of thrombopoietin) are two protein agents evaluated in numerous clinical trials for the treatment of thrombocytopenia before their discontinuation (Kuter and Begley, 2002). Nomura et al. (2002) first demonstrated an increase in platelet counts induced by MGDF in a small group of patients with ITP. AMG-531, a peptide-Fc fusion protein with amino acids nonhomologous to rhTpo, but with similar functional activity is currently being tested in clinical trials—administered subcutaneously—for the treatment of ITP (Wang et al., 2004). Preliminary reports confirm Nomura's data, providing further proof of principle that TpoR agonists are effective in this disease (Kuter et al., 2004; Newland et al., 2004). Although hematopoietic growth factors such as rhEPO (Epogen™), and rhG-CSF (Neupogen™), have proven to be successful prescribed medications, there remains tremendous opportunity for nonpeptidic, orally administered small-molecule agents in this area. The obvious benefits of such an agent are lower cost of production, the ease of out-patient treatment, and lack of immunogenicity and injection-site reactions. The feasibility of such an approach was first demonstrated by the discovery of the nonpeptidic mimic of granulocyte-colony stimulating factor (G-CSF), SB-247464 (Tian et al., 1998) (Figure 13.1). Most recently, another small-molecule Tpo mimetic, JTZ-132, has been described in the literature, but clinical data has not been reported (Inagaki et al., 2004). In the interim, preclinical data for small-molecule agonists have been reported for other members of the growth factor receptor superfamily, for example, for the G-CSF receptor (Kusano et al., 2004; Tian et al., 1998), the insulin receptor (Air et al., 2002; Zhang et al., 1999), and Epo receptor (Qureshi et al., 1999).

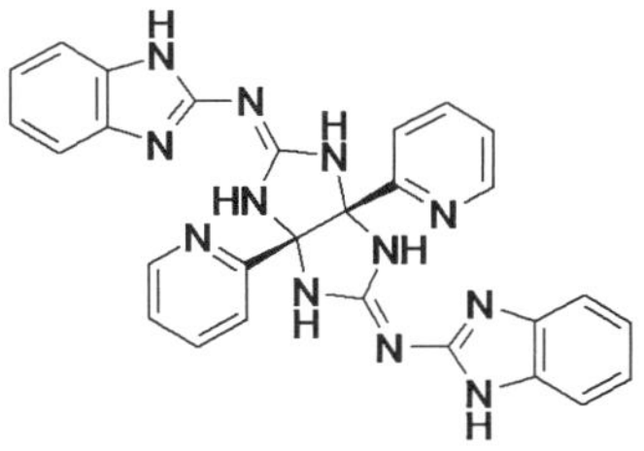

G-CSFR agonist SB-247464

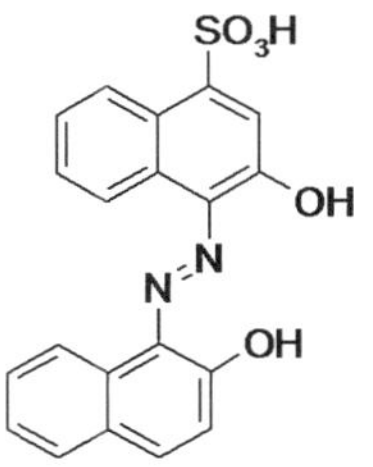

TpoR agonist SKF-56485

FIGURE 13.1 Small-molecule cytokine agonists identified by GSK HTS.

II. THE DISCOVERY OF SMALL-MOLECULE TpoR AGONISTS AT GLAXOSMITHKLINE

Typically, a cytokine-cytokine receptor interaction requires several thousands of square angstroms of surface-area contact to productively oligomerize the receptor and induce signal transduction (Aritomi et al., 1999; de Vos et al., 1992; Livnah et al., 1996). Because a typical orally active, drug-like molecule with a molecular weight of 500 Daltons possesses a total surface area on the order of two magnitudes smaller than that required to achieve this effect, it was clear from the outset that discovering a molecule that acted on the TpoR in a mode similar to that of Tpo itself would be very challenging, if not impossible. We therefore required a rapid screening methodology that would monitor Tpo agonist-like activity induced by the several hundred thousand small-molecules in our screening collection regardless of their mode of action (e.g., oligomerization of Tpo receptors or through a variety of possible downstream signaling mechanisms, such as phosphatase inhibition or kinase activation).

A. Development of a High-Throughput Screen (HTS) for TpoR Agonists

To this end, a high-throughput, cell-based assay was developed to detect compounds that activate hTpoR-linked signal transduction pathways. This screening assay utilizes a reporter gene encoding the protein luciferase, the expression of which is driven by a synthetic STAT-responsive promoter that is stably transfected into a Tpo-responsive cell line (Lamb et al., 1996; Lamb and Rosen 1997; Lamb et al., 1998). Tpo treatment of these cells results in the activation of STAT5, which binds to the synthetic promoter and increases luciferase expression. Thus, the murine hematopoietic progenitor cell line BAF-3 was co-transfected with cloned hTpoRcDNA and an IRF-1-luciferase reporter construct. Screening of a large number of stably transfected, drug-resistant clones resulted in the identification of a suitable Tpo-responsive cell line, designated BAF-3/hTpo-Rluc. A dose response in the BAF-3/hTpo-Rluc cell line with respect to Tpo consistently afforded a 15- to 20-fold maximal increase in luciferase production (typically at [Tpo] $\sim$0.1 ng/mL) with an $EC_{50} = 0.001-0.01$ ng/mL. This large window ensured the identification of even relatively weak activators of the Tpo signaling pathway with this assay.

B. HTS Results

HTS of the SmithKline Beecham and Ligand Pharmaceuticals proprietary compound collections resulted in the discovery of SKF-56485 (see Figure 13.1), with a reproducible $EC_{50} = 0.2\,\mu M$ and 38% efficacy (compared to that of maximal [Tpo]). SKF-56485, a deep purple diazonaphthalene, is one representative of a class of compounds historically used in the dyestuffs industry that is also known as Eriochrome® Blue Black R (C.I. 15705), often used as a colorimetric indicator for metal ions (Abollino et al., 1994).

Although SKF-56485 contains numerous unattractive and potentially reactive functionalities of concern for progression, the consistent activity of the compound as an agonist of Tpo-responsive intracellular signaling pathways prompted further study. The foremost metabolic concern was the possibility of *in vivo* azoreduction, particularly by intestinal bacteria, (Gingell et al., 1971) resulting in the destruction of the molecule prior to absorption. In addition to this potential loss of active drug substance, such a transformation would result in the production of lipophilic aminonaphthols, a compound class comprising known mutagens. Therefore, initial SAR studies focused on removing this undesirable functionality from SKF-56485.

C. The Evolution of Eltrombopag (SB-497115)

One approach to obviating any potential azoreduction liability of SKF-56485 was suggested by precedents from the dyestuffs literature itself. Orange food coloring tartrazine (Figure 13.2) is an azopyrazole that is known to be resistant to azoreduction both *in vivo* and *in vitro* (Ryan and Wright, 1962). It has been postulated that this resistance is due to its existence solely as the 5-oxo-pyrazol-4-ylidenehydrazine tautomer, which is not recognized by reductase enzymes (Conner et al., 1990). Therefore, replacement of the lower naphthol ring of SKF-56485 with a pyrazolin-5-one was undertaken with the expectation that potency would be retained with the potential for azoreduction reduced or, preferably, eliminated. Gratifyingly, such pyrazolinones are effective replacements for the naphthol group and further SAR studies resulted in the discovery of very potent and efficacious Tpo mimetic compounds, exemplified by SB-394725 (see Figure 13.2). SB-394725 demonstrates excellent activity in the BAF-3/hTpoR luciferase reporter assay, with an $EC_{50} \sim 0.03\,\mu M$ and 100% efficacy (compared to that of maximal [Tpo]) (Erickson-Miller et al., 2005).

Luciferase production was also induced by SB-394725 in cell lines in which the luciferase reporter gene was linked to the megakaryocyte specific gpIIb promoter. As in the assays using the IRF-1 promoter, $3\,\mu M$ SB-394725 reached 100% of the efficacy of Tpo, with an EC_{50} of $0.03\,\mu M$ in the 32D-mpl luciferase assay utilizing the gpIIb promoter (Figures 13.3A and B).

To confirm that the activity induced by these compounds was acting through the JAK/STAT signal transduction pathway, EMSA assays were done on the BAF3/TpoR screening cell line, as well as on N2C-Tpo, a Tpo-dependent human cell line that endogenously expresses TpoR. Figure 13.4A demonstrates that as little as $0.01\,\mu M$ SB-394725 activated STAT5; higher concentrations of compound shifted the IRF-1 probe in the EMSA in a dose dependent manner. The compound activated STATs with similar kinetics to Tpo; peak activity was demonstrated at 30 minutes and decreased to background levels at approximately 60 to 120 minutes (data not shown). Phosphorylation of

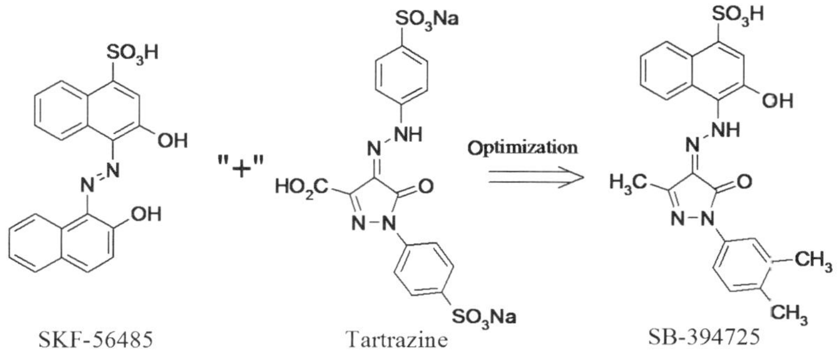

SKF-56485 "+" Tartrazine Optimization ⟹ SB-394725

FIGURE 13.2 Evolution of SB-394725 from SKF-56485.

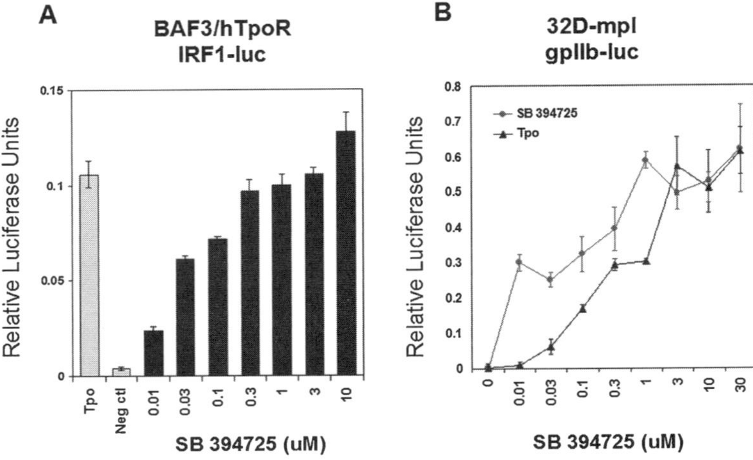

FIGURE 13.3 Luciferase production induced by SB-394725 in two Tpo-responsive cell lines.

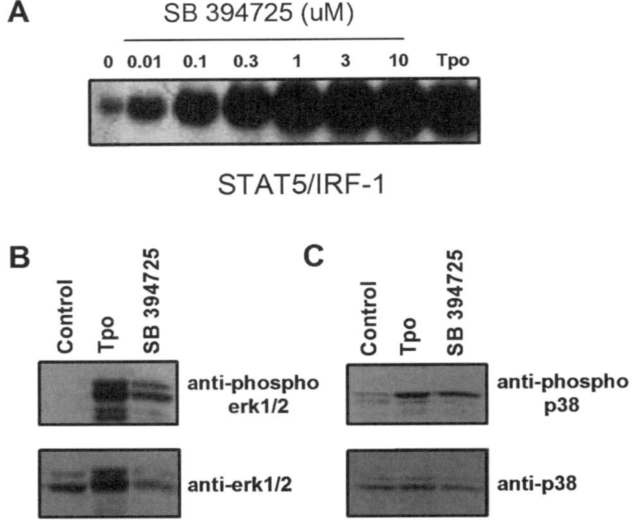

FIGURE 13.4 Detection of phosphorylated proteins by EMSA induced by SB-394725 in two Tpo-responsive cell lines.

the p42/44 (erk1/2) and p38 MAP kinases was also demonstrated in lysates of BAF3/TpoR cells following 30 minutes of treatment with rhTpo or 30 μM SB-394725 (Figures 13.4B and C).

Downstream of the activation of the STAT and MAPK pathways, the expression of certain early response genes was also demonstrable with SB-394725. *Fos*, *EGR-1*, and *CIS* message was detectable in the RNA of N2C-Tpo cells treated with either 100 ng/mL Tpo or 30 μM SB-394725 in as little as 30 minutes (Erickson-Miller et al., 2005). Because SB-394725 induces the signal transduction pathways expected of a TpoR agonist, further experiments measuring proliferation, STAT activation, and reporter gene expression in various transfected and nontransfected cell lines were performed to

TABLE 13.1 Selectivity of SB-394725 and eltrombopag for Tpo receptor expressing cells.

Cell Line	Cytokine Response	Assay Type	Positive Control	Tpo	SB-394725	Eltrombopag
BAF3/Tpo	Tpo/IL-3	Proliferation or STAT 5 EMSA	++	++	++	++
N2C-Tpo	Tpo	Proliferation or STAT5 EMSA	++	++	++	++
hMPL-HepG2	Tpo	STAT5 EMSA	++	++	+	ND
BAF3/G-CSF	G-CSF/IL-3	Proliferation or STAT5 EMSA	++	–	–	–
NFS-60	G-CSF	Proliferation	+++	–	–	ND
Kasumi 1	G-CSF	Proliferation	++	–	–	ND
TF-1	GM-CSF/Epo	Proliferation	+++	–	–	ND
UT7-Epo	Epo	Proliferation or STAT5 EMSA	+++	–	–	–
HEPG2	IFN-γ	STAT1 EMSA	+++	–	–	–

substantiate the selectivity of SB-394725 for the TpoR. Table 13.1 lists the cell lines tested to determine responses to either Tpo (100 ng/mL) or SB-394725 (0.001–100 μM). All of these lines respond as expected to control cytokines specific for each cell type and confirm the activity of SB-394725 only on cells expressing the TpoR, and not IFN-g, granulocyte colony stimulating factor (G-CSF), granulocyte macrophage colony stimulating factor (GM-CSF), Epo or IL-3 receptors.

Given this encouraging biological activity of SB-394725, further exploration into the chemical optimization of this TpoR agonist was undertaken. Not surprisingly, sulfonic acids such as SB-394725 have negligible exposure upon oral dosing in multiple species (e.g., Sprague-Dawley rats, Beagle dogs, Cynomolgous monkeys). While replacing the sulfonic acid of SB-394725 analogues with a carboxylic acid does result in orally bioavailable molecules (e.g., the oral bioavailability of the carboxylic acid analogue of compound SB-394725 is approximately 17% in male Sprague-Dawley rats), such carboxylic acid analogues are less potent than their parent sulfonic acids, as previously reported (Duffy et al., 2001).

In an attempt to understand this potency difference, an overlay of the three central heteroatoms of SB-394725 with potent thiosemicarbazone carboxylic acid SB-450572 was constructed (Figure 13.5). This latter class of Tpo agonists was also discovered through our HTS and was optimized contemporaneously with our pyrazolinones (Duffy et al., 2002). One major highlight of these two independent optimization efforts was the realization that, although chemically quite distinct, both classes share a similar pharmacophore for Tpo receptor agonism. This pharmacophore comprises a relatively flat topology with distal acidic and lipophilic functionalities along with a central core of two to three heteroatoms that can form a metal-chelating motif. Importantly, the most significant difference between the two classes of inhibitors is in the position of the acidic functionality on the proximal left-hand ring (as drawn in Figure 13.5). In the potent thiosemicarbazone acid SB-450572 (BAF-3/hTpoR

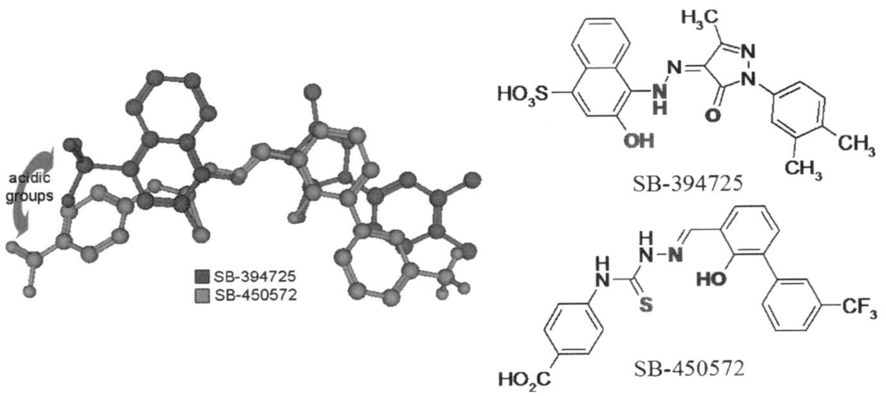

FIGURE 13.5 Molecular modeling overlay of pyrazolinone SB-394725 with thiosemicarbazone SB-450572. (Hydrogen atoms omitted for clarity). Red, green, and blue-colored atoms represent those pairs which are overlayed in the model.

luciferase assay $EC_{50} \sim 0.02\,\mu M$), the carboxylic acid is extended out much farther from the putative metal chelating-motif than it is in the pyrazol-4-ylidenehydrazine SB-349725.

Therefore, the extension of the acidic functionality in the pyrazol-4-ylidenehydrazine series was considered to access potent, nonsulfonic acid hydrazone derivatives with the potential for oral bioavailability. Molecular modeling was again used to evaluate potential molecules to probe this hypothesis, with a phenyl group emerging as a promising "spacer" to extend the position of the carboxylic acid functionality. All possible biphenyl substitution patterns were examined, with the 3,3-biphenyl substitution pattern emerging as the one most closely mimicking the thiosemicarbazone acidic group position. An overlay of this new, biphenyl compound, SB-497115, with the thiosemicarbazone SB-450572 is shown in Figure 13.6.

This overlay clearly shows that incorporation of the biphenyl extension results in a much closer superimposition of the acidic functionalities of these two agonists. Most gratifyingly, SB-497115 is indeed quite potent in our BAF-3/hTpoR luciferase assay under control of the IRF-1 promoter with an

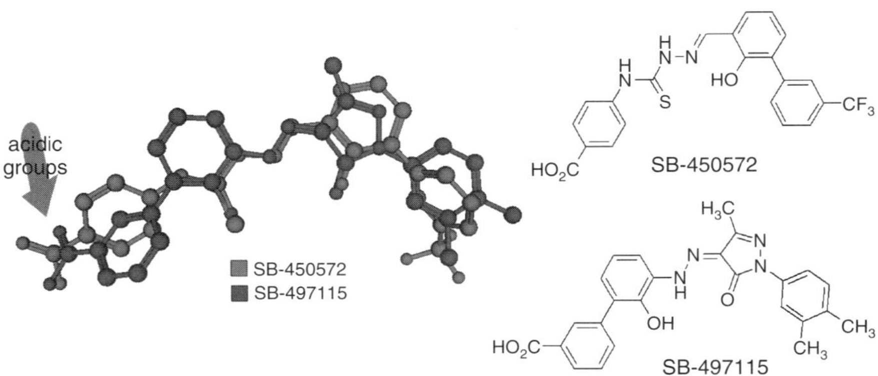

FIGURE 13.6 Overlay of thiosemicarbazone SB-450572 with biphenyl SB-497115. (Hydrogen atoms omitted for clarity). Red, green, and blue-colored atoms represent those pairs that are overlayed in the model.

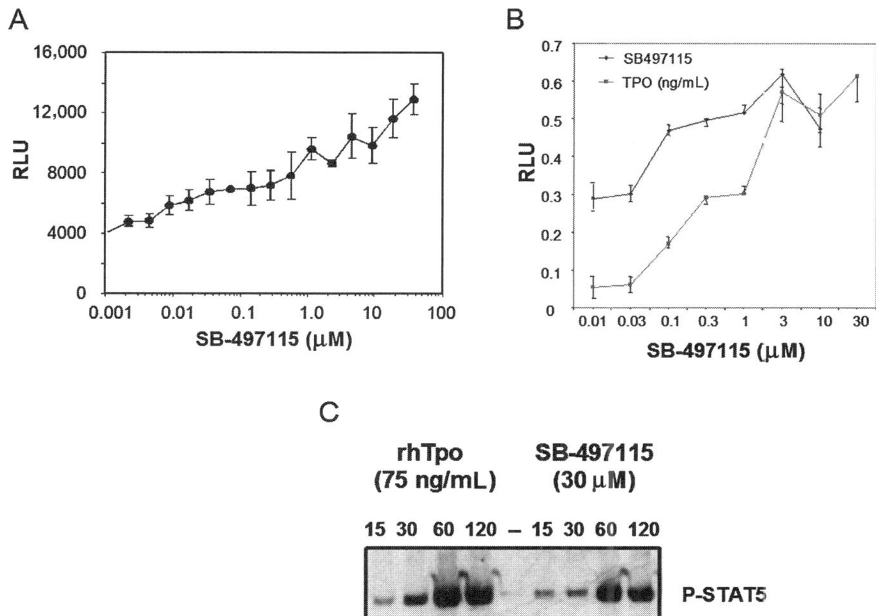

FIGURE 13.7 Luciferase production and kinetics of phosphorylated STAT5 induction by SB-497115 in Tpo-responsive cell lines.

$EC_{50} = 0.27\,\mu M$ and also induces luciferase production under the gpIIb promoter (Figures 13.7A and B, respectively). As with SB-394725, treatment of N2C-Tpo cells with 30 μM SB-497115 results in activation of STAT5 (Figure 13.7C) and p42/44 MAPK (data not shown).

Most importantly, SB-497115 is quite effective at stimulating the proliferation of the Tpo-dependent megakaryoblastic leukemia cell line, N2C/Tpo ($EC_{50} \sim 0.03\,\mu M$ – data not shown) and is equally efficacious at promoting the proliferation and differentiation of CD34 cells isolated from human bone marrow samples, as measured by the appearance of the CD41 marker for megakaryocytic differentiation ($EC_{50} \sim 0.10\,\mu M$) (Figure 13.8). The requirement of SB-497115 for TpoR was also demonstrated with various proliferation and activation assays as was demonstrated for SB-394725 (see Table 13.1).

SB-497115 possesses good oral bioavailability ($F > 30\%$) in all animals tested (rats, dogs, cynomolgus monkeys), with $T_{1/2}$ of 12, 14, and 7.7 hr, respectively. This supports the hypothesis that the poor oral bioavailability of the sulfonic acids such as SB-394725 is indeed because of the high polarity imparted by the sulfonic acid moiety.

Perhaps because of the potentially small interaction sites with large receptors, several nonpeptidic growth factor agonists have demonstrated specificity for particular animal species (Kusano et al., 2004; Tian et al., 1998). In the case of SB-497115, the molecule has activity only on the human and chimpanzee TpoRs (Figure 13.9).

Thus, in the absence of a small-animal model, a 5-day repeat dose study was conducted in chimpanzees administered either vehicle or SB-497115 (10 mg/kg/day) by oral gavage. Platelet counts increased in the

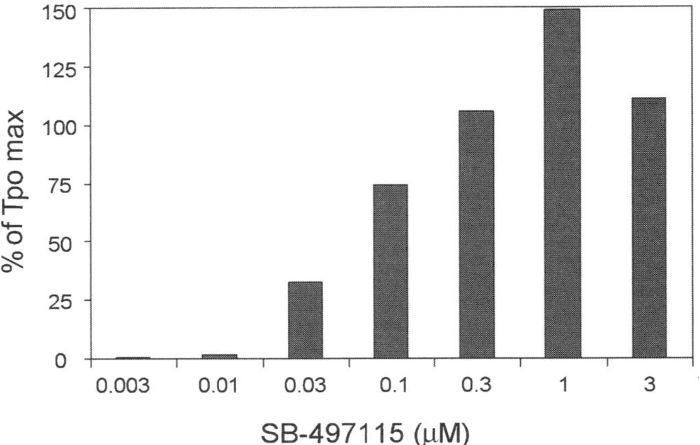

FIGURE 13.8 Differentiation assay for SB-497115. Megakaryocyte differentiation of CD34+ cells after 10 days in culture with SB-497115. The percent of CD41+ cells differentiating from CD34+ cells cultured for 10 days. CD34+ cells were purified and cultured for 10 days in SCF (100 ng/mL) + rhTpo (100 ng/mL) or SCF + SB-497115 (0.001–1 μM). The percent of CD41+ cells as measured by flow cytometry were determined. The data is represented as the percent of CD41+ cells stimulated by SB-497115 relative to that stimulated by 100 ng/mL rhTpo.

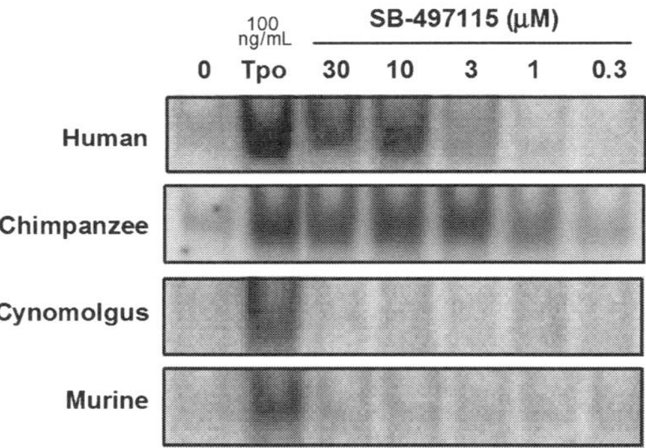

FIGURE 13.9 Comparison of STAT5 phosphorylation induced by SB-497115 in platelets isolated from several species.

SB-497115-treated chimps by up to two fold about day 11, and returned to baseline values by day 22 of the study (Figure 13.10).

D. Human Pharmacology of Eltrombopag (SB-497115)

In a Phase I study, eltrombopag was administered as oral capsules or tablets to healthy male subjects at doses of 5 to 75 mg once daily for 10 days. After 10 days of dosing, a dose dependent increase in platelet count was observed at the 30, 50, and 75 mg doses (Figure 13.11). The mean platelet count of six subjects administered 75 mg/day, the highest dose studied, began increasing on the fifth day of dosing and peaked on day 16 following initiation of drug (Figure 13.12). Other than platelets, there were no effects on other blood cell parameters, that is, hemoglobin, white blood cell and neutrophil counts were similar to

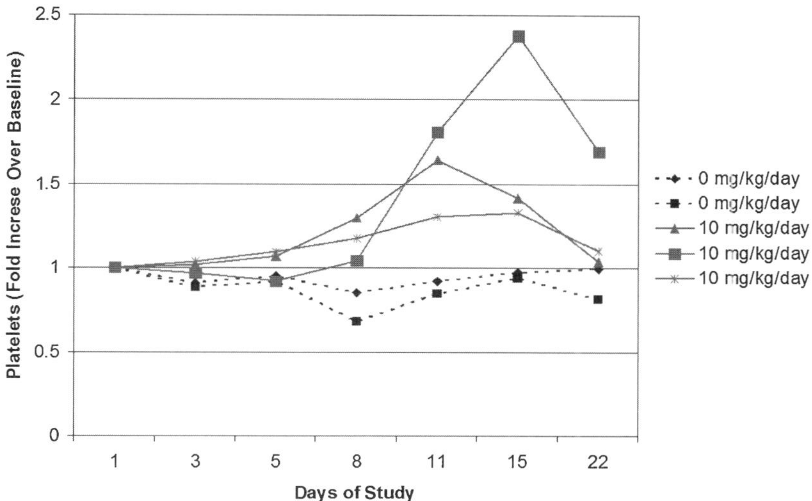

FIGURE 13.10 Increases in platelets of chimpanzees dosed once daily with 10 mg/kg/day for 5 days by oral gavage.

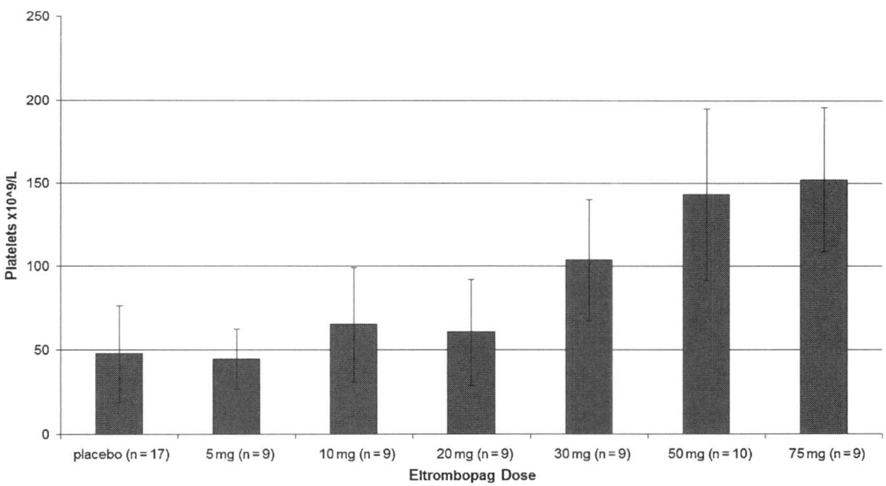

FIGURE 13.11 Dose-dependent effects of eltrombopag on platelets in normal human subjects.

placebo. Blood was drawn from subjects and platelets assayed for function by aggregometry, with no differences seen in *ex vivo* aggregation responses of platelets between eltrombopag- and placebo-treated subjects. Eltrombopag demonstrated only mild to moderate adverse events with no apparent dose relatedness, and no serious adverse events.

The pharmacokinetic profile of eltrombopag was dose-dependent and linear. In the cohort of subjects receiving 75 mg of eltrombopag, the mean Cmax was 7.3 μg/mL (15% CV), the mean half-life was 14.4 hr (49% CV), and the mean AUC was 79.0 μg/hr/mL (23% CV).

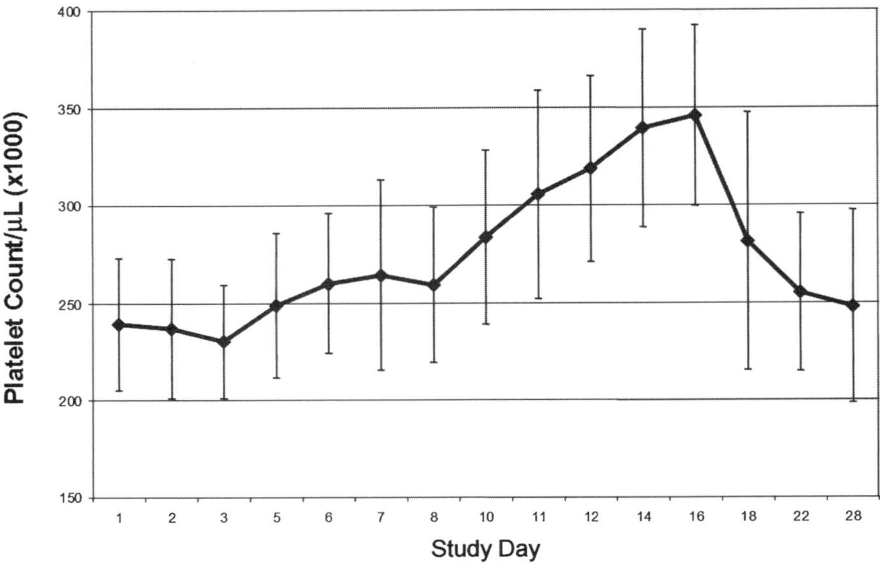

FIGURE 13.12 Kinetics of platelet response to eltrombopag in normal human subjects.

CONCLUSIONS

During the course of our studies on small-molecule TpoR agonists, we have defined a pharmacophore for TpoR activation exemplified by a flat molecular topology with distal acidic and lipophilic functionalities, and a central core of two to three heteroatoms that can form a metal-chelating motif. This pharmacophore was successfully utilized to merge the desired properties of two chemically disimilar classes of agonists, combining the potential for oral absorbtion of the thiosemicarbazone series with the excellent potency of the hydrazinonaphthalene series. The resulting biaryl (pyrazol-4-ylidene) hydrazine carboxylic acids demonstrate excellent potency at stimulating the proliferation of Tpo-responsive cell lines, and at promoting the differentiation of human bone marrow toward megakaryocyte progenitor cells. Most importantly, the biaryl (pyrazol-4-ylidene) hydrazines are well absorbed orally in all preclinical species tested (rat, dog, cynomolgous monkey). Specificity for chimpanzee and human TpoR precluded more typical preclinical efficacy studies in lower species such as rodents or dogs. These results, however, demonstrate that activity on the primary target cell, in this case human bone marrow, accurately predicted activity in humans. Eltrombopag is the first nonpeptidic, drug-like molecule to demonstrate *in vitro* megakaryocyte proliferation and differentiation, which translates into an increased platelet count when orally administered in normal subjects.

ACKNOWLEDGMENTS

This work was sponsored by GlaxoSmithKline. Kevin Duffy and Connie Erickson-Miller are employees and share-holders of GlaxoSmithKline. We thank the many members of the Tpo Receptor Agonist team at GSK and Ligand Pharmaceuticals.

REFERENCES

Abollino, O., C. Sarzanini, et al. (1994). Evaluation of stability constants of metal complexes with sulfonated azo-ligands. *Talanta* 41, 1107–1112.

Adams, V. R. and T. Brenner (2000). Oprelvekin (Neumega), first platelet growth factor for thrombocytopenia. *J. Am. Pharm. Assoc.* 39, 706–707.

Air, E. L., M. Z. Strowski, et al. (2002). Small molecule insulin mimetics reduce food intake and body weight and prevent development of obesity. *Nat. Med.* 8: 179–183.

Aritomi, M., N. Kunishima, et al. (1999). Atomic structure of the GCSF-receptor complex showing a new cytokine-receptor recognition scheme. *Nature* 401: 713–717.

Cohen, Y. C., B. Djulbegovic, et al. (2000). The bleeding risk and natural history of idiopathic thrombocytopenic purpura in patients with persistent low platelet counts. *Arch. Intern. Med.* 160: 1630–1638.

Conner, J. A., R. J. Kennedy, et al. (1990). Structural studies of arylazo and arylimino compounds: Nitrogen-15 NMR and x-ray crystallographic studies of azo-hydrazo tautomerism. *J. Chem. Soc.* Perkin Trans.2 203–207.

de Vos, A. M., M. Ultsch, and A. A. Kossiakoff (1992). Human growth hormone and extracellular domain of its receptor: Crystal structure of the complex. *Science* 255: 306–312.

Drachman, J. G., K. M. Millett, and K. Kaushansky (1999). Thrombopoietin signal transduction requires functional JAK2, not TYK2. *J. Biol. Chem.* 274: 13, 480–13, 484.

Duffy, K. J., M. G. Darcy, et al. (2001). Hydrazinonaphthalene and azonaphthalene thrombopoietin mimics are nonpeptidyl promoters of megakaryocytopoiesis. *J. Med. Chem.* 44: 3730–3745.

Duffy, K. J., A. N. Shaw, et al. (2002). Identification of a pharmacophore for thrombopoietic activity of small, non-peptidyl molecules. 1. Discovery and optimization of salicylaldehyde thiosemicarbazone thrombopoietin mimics. *J. Med. Chem.* 45: 3573–3575.

Erickson-Miller, C. L., E. Delorme, et al. (2005). Discovery and characterization of a selective, nonpeptidyl thrombopoietin receptor agonist. *Exp. Hematol.* 33: 85–93.

Espanol, I., A. Gallego, et al. (2000). Thrombocytopenia associated with liver cirrhosis and hepatitis C viral infection: Role of thrombopoietin. *Hepatogastroenterology* 47: 1404–1406.

Garcia-Suarez, J., C. Burgaleta, et al. (2000). HCV-associated thrombocytopenia: Clinical characteristics and platelet response after recombinant alpha2b-interferon therapy. *Br. J. Haematol.* 110: 98–103.

Gingell, R., J. W. Bridges, and R. T. Williams (1971). The role of the gut flora in the metabolism of prontosil and neoprontosil in the rat. Xenobiotica 1: 143–156.

Guest, J. F., V. Munro, and R. F. Cookson (1998). The annual cost of blood transfusions in the United Kingdom. *Cli. Lab. Haematol.* 20: 111–118.

Gupta, P., S. C. LeRoy, et al. (2000). Long-term blood product transfusion support for patients with myelodysplastic syndromes (MDS): Cost analysis and complications. *Leuk. Res.* 23: 953–959.

Inagaki, K., T. Oda, et al. (2004). Induction of megakaryocytopoiesis and thrombocytopoiesis by JTZ-132, a novel small molecule with thrombopoietin mimetic activities. *Blood* 104: 58–64.

Kaushansky, K., (2000). Use of thrombopoietic growth factors in acute leukemia. *Leukemia* 14: 505–508.

Kaushansky, K., S. Lok, et al. (1994). Promotion of megakaryocyte progenitor expansion and differentiation by the c-Mpl ligand thrombopoietin [see comments]. *Nature* 369: 568–571.

Kercher, K. W., A. M. Carbonell, et al. (2004). Laparoscopic splenectomy reverses thrombocytopenia in patients with hepatitis C cirrhosis and portal hypertension. *J. Gastrointest. Surg.* 8: 120–126.

Kusano, K., S. Ebara, et al. (2004). A potential therapeutic role for small nonpeptidyl compounds that mimic human granulocyte colony-stimulating factor. *Blood* 103: 836–842.

Kuter, D. J. and C. G. Begley (2002). Recombinant human thrombopoietin: Basic biology and evaluation of clinical studies. *Blood* 100: 3457–3469.

Kuter, D., J. B. Bussel, et al. (2004). A phase 2 placebo controlled study evaluating the platelet response and safety of weekly dosing with a novel thrombopoietic protein (AMG531) in thrombocytopenic adult patients (pts) with immune thrombocytopenic purpura (ITP). *ASH Annual Meeting Abstracts* 104, 511.

Lamb, P. and J. Rosen (1997). Drug discovery using receptors that modulate gene expression. *J. Recept. Signal Transduction Res.* 17: 531–543.

Lamb, P., H. M. Seidel, et al. (1996). The role of JAKs and STATs in transcriptional regulation by cytokines. *Annu. Rep. Med. Chem.* 31: 269–276.

Lamb, P., P. Tapley, and J. Rosen (1998). Biochemical approaches to discovering modulators of the JAK-STAT pathway. *Drug Disc. Today* 3: 122–130.

Livnah, O., E. A. Stura, et al. (1996). Functional mimicry of a protein hormone by a peptide agonist: The EPO receptor complex at 2.8 A. *Science* 273: 464–471.

Mannucci, P. M. and G. Gringeri (2000). HIV-related thrombocytopenias. *Ann. Ital. Med. Int.* 15: 20–27.

McCullough, J., (2000). Current issues with platelet transfusion in patients with cancer. *Semin. Hematol.* 37: 3–10.

Meehan, K. R., C. O. Matias, et al. (2000). Platelet transfusions: Utilization and associated costs in a tertiary care hospital. *Am. J. Hematol.* 64: 251–256.

Miyakawa, Y., A. Oda, et al. (1996). Thrombopoietin induces tyrosine phosphorylation of Stat3 and Stat5 in human blood platelets. *Blood* 87: 439–446.

Newland, A., M. T. Caulier, et al. (2004). An open-label, unit dose-finding study evaluating the safety and platelet response of a novel thrombopoietic protein (AMG531) in thrombocytopenic adult patients (Pts) with immune thrombocytopenic purpura (ITP). *ASH Annual Meeting Abstracts* 104: 2058.

Nomura, S., K. Dan, et al. (2002). Effects of pegylated recombinant human megakaryocyte growth and development factor in patients with idiopathic thrombocytopenic purpura. *Blood* 100: 728–730.

Qureshi, S. A., R. M. Kim, et al. (1999). Mimicry of erythropoietin by a nonpeptide molecule. *Proc. Natl. Acad. Sci. U. S. A.* 96: 12,156–12,161.

Rojnuckarin, P., J. G. Drachman, and K. Kaushansky (1999). Thrombopoietin-induced activation of the mitogen-activated protein kinase (MAPK) pathway in normal megakaryocytes: Role in endomitosis. *Blood* 94: 1273–1282.

Ryan, A. J. and S.E. Wright (1962). Biliary excretion of some azo dyes related to tartrazine. *Nature* (London) 195: 1009.

Tian, S. S., P. Lamb, et al. (1998). A small, nonpeptidyl mimic of granulocyte-colony-stimulating factor. *Science* (Washington, D.C.) 281: 257–259.

Wang, B., J. L. Nichol, and J. T. Sullivan (2004). Pharmacodynamics and pharmacokinetics of AMG 531, a novel thrombopoietin receptor ligand. *Clin. Pharmacol. Ther.* 76: 628–638.

Webb, I. J. and K. C. Anderson (1999). Risks, costs, and alternatives to platelet transfusions. *Leuk. Lymphoma* 34: 71–84.

Zhang, B., G. Salituro, et al. (1999). Discovery of a small molecule insulin mimetic with antidiabetic activity in mice. *Science* 284: 974–977.

14 ORALLY BIOAVAILABLE GLYCOPROTEIN IIb/IIIa ANTAGONISTS: A NEGATIVE CASE STUDY

DIETMAR A. SEIFFERT* **AND JEFFREY T. BILLHEIMER†**

**M.D., Bristol-Myers Squibb Co., Pennington, New Jersey*
†Ph.D., University of Pennsylvania, Philadelphia, Pennsylvania

Glycoprotein (GP) IIb/IIIa antagonists prevent binding of fibrinogen to its platelet receptor, thereby blocking the final, nonredundant pathway of platelet aggregate formation. Orally bioavailable GP IIb/IIIa antagonists were developed to extend the benefits observed with parenteral GP IIb/IIIa antagonists in acute cardiovascular interventions to more chronic atherothrombosis indications. These new agents needed to be differentiated from or add on to aspirin and/or ticlopidine. Five chemically distinct, orally bioavailable GP IIb/IIIa antagonists underwent Phase III clinical testing with rather disappointing results. At best the antagonists were noneffective, and at worst—based on large meta-analysis—they increased mortality. Here we discuss some of the reasons that may have lead to the failure of this compound class. We review the evolving biology of GP IIb/IIIa that was only incompletely developed at the time these antagonists underwent clinical testing, discuss the pharmacology and its implication for the PK/PD relationship, evaluate issues with clinical dose findings based on platelet biomarkers, and explore some of the potential explanations for the observed excess mortality. The lessons learned from the failed development of this compound class are discussed as it relates to general principles applicable to newer mechanisms. It should be noted that the lack of appropriate efficacy biomarkers to predict major adverse cardiovascular events is still hindering the development of new therapeutic entities for the treatment of atherothrombosis.

I. INTRODUCTION

Multi-cellular organisms require a patent vascular system for efficient transport of nutrients and metabolic by-products. The hemostatic system is designed to

Target Validation in Drug Discovery

maintain blood within the vascular system in a fluid state under physiological conditions. It is primed to react quickly to vascular injury in an explosive manner to minimize blood loss, mainly during injuries and child birth. As a consequence, evolutionary pressure favors efficient hemostatic mechanisms.

While a number of hemostatic mechanism can be separated, a synergy between all systems is required for efficient hemostasis. Hemostatic mechanisms include the vessel wall (vascular contraction to reduce blood flow/loss), exposure of tissue factor to initiate blood coagulation, platelets (primary hemostasis by adhesion to areas of injury and providing pro-coagulant surfaces), the coagulation system (leading to fibrin formation and platelet stimulation), the anti-coagulant system (shutting down coagulation enzymes), and the fibrinolytic system (removal of blood clots and initiation of wound healing).

Thrombosis is a pathological extension of hemostasis leading to blood clots in the vasculature. Thrombotic events mainly occur after the reproductive period and thus have a low evolutionary pressure. Thrombosis can be viewed as accident of nature with insufficient time to adapt through evolution to advances of modern medicine and longevity. Already in 1886, Rudolf Virchow identified the predisposing factors for thrombosis (also known as Virchow's triad), including alterations in the vessel wall, alterations in normal blood flow, and alterations in the composition of blood. Thrombotic diseases present clinically on the arterial site as coronary artery disease leading to acute coronary syndrome and sudden cardiac death, cerebrovascular diseases including transient ischemic attacks and strokes, peripheral arterial occlusive diseases, and on the venous site as deep vein thrombosis and pulmonary embolism.

A functional hemostatic system is essential for survival and alterations in either procoagulant-, anticoagulant-, and fibrinolytic systems or platelet number or functional responsiveness are associated with either bleeding or thrombosis. This conclusion is derived from mouse models or human monogenetic traits (for textbook coverage of Thrombosis and Haemostasis see Coleman et al., 2001; Loscalzo and Schafer, 1998; and Michelson, 2002).

Thrombosis is the main cause of morbidity and mortality in the western world, pointing to a large unmet medical need that is incompletely served by existing therapies. Parenteral GP IIb/IIIa antagonists show promise in the treatment of patients undergoing coronary revascularization procedures, including balloon angioplasty and stenting (see Chapter 8). To extend the benefits observed in acute treatment to chronic atherothrombotic diseases, a number of pharmaceutical companies pursued oral glycoprotein IIb/IIIa programs. This review focuses on antagonists that underwent extensive clinical evaluations in Phase III trials and is written by two scientists involved in the preclinical and clinical development of one of these programs.

II. ROLE OF GP IIb/IIIa IN PLATELET BIOLOGY: CLUES FOR TARGET VALIDATION?

Platelets receive activation and inhibitory signals (Figure 14.1). If activation signals exceed inhibitory responses, the conformationally labile GP IIb/IIIa is converted to a ligand binding competent active form. This process is known as inside-out signaling. GP IIb/IIIa is a member of the integrin family of cell surface receptors. The IIb subunit, also known as alpha IIb, is specific for the megakaryocytic lineage, whereas the IIIa subunit, also known as beta 3, is also a component of the vitronectin receptor found in various cell types.

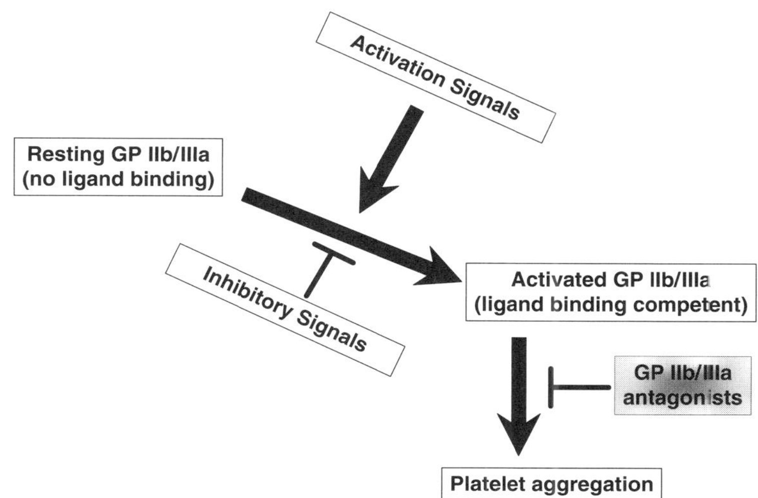

FIGURE 14.1 **GP IIb/IIIa antagonists block the final common pathway of platelet activation.** Platelets can be viewed as an environmental sensor for activation and inhibitory signals, regulating the affinity state of GP IIb/IIIa. Activation signals and respective receptor systems on platelets include thrombin (protease activated receptor 1; protease activated receptor 4; glycoprotein Ib/IX), adenosine diphosphate (P2Y1; P2Y12), thromboxane A2 (thromboxane/prostanoid receptors alpha and beta), serotonine (HT2A), collagen (integrin alpha 2 beta 1; GP VI), and vWF (GP Ib/IX). Inhibitory signals include prostacyclin and NO. All these receptors activate or inhibit specific signaling cascades. Excess of stimulatory signals ultimately result in GP IIb/IIIa activation, a process fundamental for a number of platelet responses, including adhesion and aggregation. GP IIb/IIIa binds multiple ligands including fibrinogen/fibrin, vWF, fibronectin, vitronectin, and soluble CD40 ligand. GP IIb/IIIa antagonists block ligand binding to GP IIb/IIIa without blocking platelet activation per se. Under conditions of high shear stress encountered in stenosed arteries, thrombus growth is mainly driven by the vWF/GP Ib/IX interaction and fibrinogen plays a secondary role in stabilizing larger platelet aggregates.

GP IIb/IIIa is present at approximately 40,000 to 80,000 copies per platelet (Niiya et al., 1987), translating to a blood concentration of 20 to 40 nM. In addition, an intracellular pool of GP IIb/IIIa becomes surface expressed upon stimulation with strong platelet agonists (Niiya et al., 1987; Woods et al., 1986). Active GP IIb/IIIa binds Arg-Gly-Asp (RGD)-containing adhesive glycoproteins (fibrinogen, von Willebrand factor (vWF), fibronectin, vitronectin) (Plow and Shattil, 2000). Cross-linking of platelets via fibrinogen/fibrin or other adhesive proteins is essential for platelet aggregation and the formation of a stable blood clot. Occupancy of GP IIb/IIIa leads to a series of intracellular responses, referred to as outside-in signaling, further amplifying platelet activation.

Important clues for GP IIb/IIIa target validation can be obtained from human defects known as Glanzmann's thrombasthenia and from mouse knockout models. Glanzmann's thrombasthenia is a heterogenous patient population with both quantitative and qualitative defects in the GP IIb/IIIa complex (for review, see Nurden and George, 2000). In general, thrombasthenia platelets lack fibrinogen in their secretory granules, fail to support clot retraction and fail to aggregate upon stimulation with platelet agonists (e.g., ADP, thrombin, collagen). Nonsense and missense mutations have been described both in the IIb and IIIa subunits of GP IIb/IIIa. The clinical bleeding manifestations include menorrhagia, easy bruising and purpura, epistaxis, and gastrointestinal hemorrhage. A poor correlation has been observed between the amount of residual GP IIb/IIIa and the severity of hemorrhagic disease. The GP IIb/IIIa density required for normal hemostasis, however, can be defined by clinical observations. Heterozygous individuals (50% reduction of GP IIb/IIIa) have no bleeding episodes, whereas 10% to 15% of functional GP IIb/IIIa is required for adequate hemostasis under severe challenge. Similarly, blocking $>80\%$ of GP IIb/IIIa receptors with a monoclonal antibody abolishes ADP-induced platelet aggregation. There are no reports of myocardial infarctions or stroke in thrombasthenic patients. Little atherothrombosis would be expected in this young patient group.

The phenotype of mouse knock-out and transgenic models is an important part of target validation. Knock-out models of GP IIb/IIIa, however, were described after all Phase III trials of the oral agents were either initiated or completed. In the case of the platelet/megakaryocyte specific integrin subunit alpha IIb, homozygous deficient mice presented with a moderate to severe bleeding phenotype, lacked platelet aggregation responses to ADP and collagen, showed drastically reduced platelet fibrinogen binding, and no clot retraction was observed (Tronik-Le Roux et al., 2000). There are no reports of *in vivo* antithrombotic effects in this mouse knock-out model. The phenotype of the heterozygous mice deserves mentioning because this better reflects the receptor blockade anticipated with small molecule antagonists for chronic therapy. Heterozygotes show a normal response to ADP-induced platelet aggregation, delayed reactivity toward collagen-induced platelet aggregation, and approximately 50% reduced fibrinogen binding. The interpretation of the phenotype of the beta III integrin subunit knock-out model is complicated by the fact that beta III is widely expressed outside the megakaryocytic lineage. Nevertheless, the phenotype of the knock-outs recapitulates all the cardinal features of human Glanzmann's thrombasthenia, including gastrointestinal and cutaneous hemorrhage, prolonged bleeding time, abnormal platelet aggregation and clot retraction (Hodivala-Dilke et al., 1999). Tail-bleeding times of the heterozygous animals were normal and GI bleeding was absent. Beta III null–mice are protected from $FeCl_3$-induced thrombosis of the carotid artery (Smyth et al., 2001). Heterozygotes presented a delayed occlusion of the vessel, suggesting that efficacy of small-molecule GP IIb/IIIa antagonists could be achieved at incomplete platelet blockade. Knock-outs were also protected from ADP or collagen- and epinephrine-induced thrombocytopenia, whereas no protection against tissue factor–induced systemic coagulation was observed. The main initiator of arterial thrombosis is believed to be tissue factor, and this observation could point to potential limitations of the GP IIb/IIIa mechanism.

In considering the extent of target validation for GP IIb/IIIa, both the antithrombotic efficacy (benefit) and hemorrhagic liability (risk) sides of the equation must be considered (Table 14.1). From the efficacy side, GP IIb/IIIa is clearly a biochemically plausible target based on the role of platelet fibrinogen binding in platelet function. This binding interaction can be blocked with RGD-containing peptides. Mouse genetic models were not available when oral GP IIb/IIIa antagonist programs started, but point now (retrospectively) to the need of a high degree of receptor blockade for efficacy. The human monogenetic trait is not informative for efficacy due to the young age of patients and insufficient case numbers. Clinical experience with parenteral agents and high level of platelet inhibition under high-risk situations (e.g., percutaneous coronary artery angioplasty) clearly showed a benefit for treatment with GP

TABLE 14.1 Target Validation for the GP IIb/IIIa Mechanism at Beginning of Phase III Studies.

Validation	Comment
Biochemical plausibility	Interaction of fibrinogen with GP IIb/IIIa is the final common pathway of platelet aggregation, and platelet aggregation, is believed to be essential for thrombus formation
Mouse genetics (KO, TG)	Not available
Human monogenetic proof of principle	Glanzmann's thrombasthenia: heterogenous defects in GP IIb or IIIa subunits; frequency too low to assess cardiovascular risk; bleeding signal
Pharmacological proof of concept	GP IIb/IIIa antagonists efficacious in preclinical thrombosis models
Clinical proof of concept (benefit/risk)	Acute efficacy of parenteral agents associated with increased but manageable risk of in hospital bleeding episodes.
Marketed drug	Parenteral agents approved for acute indications

IIb/IIIa antagonists (See chapter 8 of this volume). For the safety side, there is clearly a bleeding risk associated with this target based on the known importance of platelets for hemostasis, the bleedings observed in the knock-outs and patients with thrombasthenia, and clinical experience with the parenteral agents. Thus, it is obvious that exaggerated pharmacology will not be tolerated for this target. Two key questions, however, can not be answered from the available data sets. What is the minimal extent of platelet inhibition by GP IIb/IIIa antagonists required for improvement in antithrombotic efficacy with an acceptable long-term bleeding risk? What is the maximal extent of platelet inhibition tolerated during long-term dosing without a significant increase of bleeding risk? To obtain answers to these questions in view of the large unmet medical need for the treatment of atherothrombosis, several pharmaceutical companies discovered and developed orally bioavailable GP IIb/IIIa antagonists.

III. PHARMACOLOGY OF ORALLY BIOAVAILABLE GP IIb/IIIa ANTAGONISTS

Glycoprotein IIb/IIIa antagonists (see Figure 14.1) block the final common pathway of platelet activation. As a consequence, there is no redundant pathway that can bypass inhibition by GP IIb/IIIa antagonists. The *in vitro* and *in vivo* pharmacology of the five antagonists that progressed to Phase III trials is discussed in this section (Tables 14.2 and 14.3). The antagonists showed efficacy in *in vivo* preclinical thrombosis models and this subject is not further reviewed here. This review will emphasize specific features of the pharmacology that have implications for efficacy and safety.

A. *In Vitro* Receptor Pharmacology: Implications for PK and PD Relationships

The receptor pharmacology of the five antagonists that entered Phase III clinical trials is summarized in Table 14.2. Roxifiban is the most potent antagonist

TABLE 14.2 Summary of Platelet Binding Kinetic and Platelet Aggregation Inhibition (20 μM ADP) of GP IIb/IIIa Antagonists in Phase III Studies. The structures of the biological active (nonprodrug) GP IIb/IIIa antagonists are shown together with platelet aggregation and binding studies using radiolabeled inhibitors in platelet rich human plasma. The data are composed of previous publications (Anders et al., 2001; Seiffert et al., 2003). The target level of 10–20 μM ADP-induced platelet aggregation inhibition during clinical trials with the orally bioavailable agents (roxifiban, xemilofiban, orbofiban, lotrafiban, sibrafiban) was 30% to 70%. Thus, the IC_{50} values for platelet aggregation can be used as a guide for the clinically achieved plasma concentrations. *NA*, not available.

	Chemical Structure	Off-rate ($T_{1/2}$ in minutes)	Resting K_d (nmol/L)	Activated K_d (nmol/L)	IC_{50} Aggregation (nmol/L)
Roxifiban	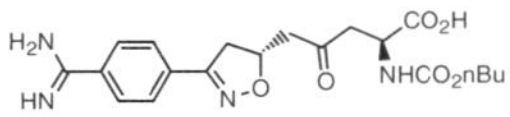 H_2N, HN, N-O, O, CO_2H, $NHCO_2nBu$	7	3.9	2.9	63
Xemilofiban	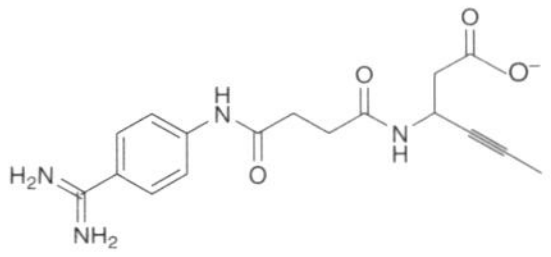 H_2N, NH_2, H N, O, O, N H, O, O^-	*NA*	*NA*	*NA*	80
Orbofiban	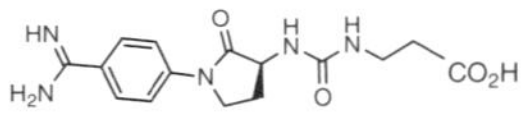 HN, H_2N, N, O, H N, H N, O, CO_2H	0.2	663	113	200
Lotrafiban	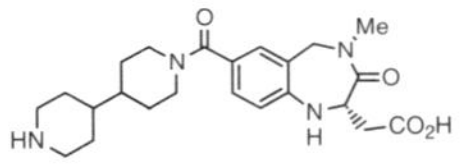 HN, N, O, Me, N, O, N H, CO_2H	0.2	60	5	60
Sibrafiban	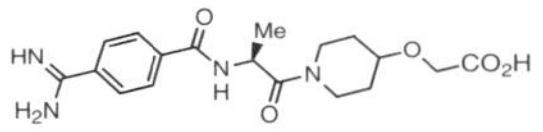 HN, H_2N, O, N H, Me, O, N, O, CO_2H	0.2	16	5	110

with similar affinity for resting and activated platelets and a slow off-rate. In contrast, orbofiban, lotrafiban, and sibrafiban are of lower affinity and showed preferential binding to activated platelets with rapid off-rates. These observations allow some predictions as to the expected clinical pharmacology profiles. For example, binding of roxifiban to resting platelets, combined with a slow off-rate suggest that platelets may provide a reservoir of roxifiban in the circulation, thereby resulting in a long half-life. This was indeed observed in clinical studies (see Table 14.3). The rapid clearance of the free, non–platelet bound form of roxifiban, however, suggests that newly synthesized platelets may not be protected (assuming that no redistribution occurs among platelets). In this respect, the survival time of human platelets in the circulation is seven to 10 days, indicating that 10% to 15% of platelets are replaced per day. In contrast, the lower affinity GP IIb/IIIa antagonists require a free plasma fraction during the dosing interval for efficacy and newly synthesized platelets are expected to be inhibited upon stimulation as well. Moreover, the fact that roxifiban is already bound to the target upon platelet activation suggest kinetic advantages in competing with fibrinogen binding to its receptor. This may be viewed as an advantage from an efficacy point of view, but may also increase bleeding liabilities. One other point of interest is the fact that the concentration of GP IIb/IIIa is 25 to 40 nM, far in excess of the affinity of the antagonists (besides orbofiban) for GP IIb/IIIa. Unlike the more common

TABLE 14.3 GP IIb/IIIa Antagonists That Have Undergone Phase III Clinical Trials. Abbreviations: ACS, acute coronary syndrome; PAD, peripheral artery disease; PCI, percutaneous coronary intervention; TIA, temporary ischemic attack.

	Phase III program	Primary Indication	Half-life (hours)	Dosing	Peak-Trough	Platelet aggregation inhibition	Anti-platelet co-medication
Orbofiban	OPUS-TIMI-16	ACS	4 to 6	Twice a day	Intermediate	40–60% low dose	Aspirin Ticlopidine (placebo group)
						50–80% high dose	Aspirin Ticlopidine (placebo group)
Xemilofiban	Excite	PCI	4	Three times a day	Intermediate	30–60% low dose	Aspirin Ticlopidine (placebo group)
						50–80% high dose	Aspirin Ticlopidine (placebo group)
Sibrafiban	Symphony I	ACS	11	Twice a day	Intermediate	>25% low dose	None
						>50% high dose	None
Sibrafiban	Symphony II	ACS	11	Twice a day	Intermediate	>25% low dose	Aspirin
						>50% high dose	None
Lotrafiban	Bravo	ACS, PAD, TIA, Stroke	4–8	Twice a day	Intermediate	50–90%	Aspirin
Roxifiban	Purpose	PAD	>24	Once a day	Low	40–75%	Aspirin

bimolecular receptor ligand interaction with ligand in vast excess of the receptor, the "titration" pharmacology where receptor is in excess of ligand results in steep concentration response curves with a Hill coefficient approaching two (as observed for the inhibition of platelet aggregation by roxifiban). The "titration" pharmacology suggests steep pharmacodynamic dose responses as well.

B. The Need for Pro-Drugs

The charged RGD sequence was used as a starting point by each of the GP IIb/IIIa antagonist medicinal chemistry programs. The resulting non–peptide antagonists depicted in the table are highly charged molecules that lack significant oral bioavailability. The antagonists were dosed in clinical trials as prodrugs that required conversion to the free, biologically active form shown in Table 14.2. Despite prodrugging, the human oral bioavailability of all five antagonists was relatively low and in general between 10% and 20%. In addition, the prodrug approach may have certain inherent liabilities that include inter-individual variability in drug delivery (for review of prodrug approaches, see Larsen and Osterguard, 2002). It should be noted that inter-individual variability could be a major issue with a narrow therapeutic index drug.

C. Peak/Trough

Some key clinical pharmacology characteristics for the five antagonists are summarized in Table 14.3. While roxifiban is a once a day drug candidate with a half-life of >24 hours, the remainder of the antagonists had to be dosed twice daily or three times a day. Half-lives in the range of 4 to11 hours combined with twice daily dosing predict peak to through variations in blood drug levels up to eight-fold. The narrow therapeutic index of GP IIb/IIIa antagonists precludes dosing to high peak levels to ensure therapeutic levels at trough. Accordingly, patients dosed with GP IIb/IIIa antagonists may be predicted to go through multiple daily cycles of overdosing (with the associated risk of bleeding) and underdosing (with associated risk of cardiovascular events). The reader is reminded that currently approved anti-platelet agents, aspirin, ticlopidine, and clopidogrel are irreversible inhibitors with minimal peak/trough variability. It should be noted that the free, not platelet-bound form of roxifiban is rapidly cleared. The lack of a free fraction, however, may also suggest a limitation for roxifiban. More specifically, stimulation of platelets with strong agonists exposes an internal pool of GP IIb/IIIa (up to 30% of total GP IIb/IIIa). We tried to address whether radiolabeled roxifiban has access to the internal pool in *in vitro* experiments. Platelets were preincubated with roxifiban for up to 6 hours (the longest time point to retain platelet reactivity *in vitro*), followed by stimulation with either ADP (no exposure of the internal pool) or thrombin-receptor activating peptide (exposure of the internal pool). We were unable to demonstrate equilibration of roxifiban with the internal pool during the prolonged *in vitro* incubation (unpublished observations). While we cannot exclude that equilibration occurs *in vivo*, this observation may point to the possibility that a significant number of GP IIb/IIIa binding sites remain unprotected during the once-daily dosing interval with roxifiban.

D. Clinical Dose-Finding Based on Platelet Biomarkers

The key issue encountered during the clinical development of GP IIb/IIIa antagonists, as with many other clinical development programs, is dose selection. The clinical event rate in chronic atherothrombosis, as a consequence of advancements in interventional cardiology, cardiac surgery, and drug therapy is rather low. Even extensive Phase II testing paradigms failed to provide reliable information on efficacious doses based on clinical endpoints. As a consequence, attempts were made to link drug doses to platelet biomarkers in order to predict efficacious doses based on drug-induced changes in platelet biomarkers. Inhibition of platelet aggregation was the most commonly used biomarker for exposure response modeling. The principle of platelet aggregation measurements is based on changes in physical-chemical properties of platelets upon activation. The most commonly used parameter is change in light transmission (turbidity). A blood sample is drawn from a subject/patient, platelet-rich plasma prepared by centrifugation, and after addition of a platelet agonist, the change in light transmission is monitored. As stated previously, platelet aggregation is dependent on the cross-linking of platelets to larger size aggregates by fibrinogen, a process concentration-dependently blocked by GP IIb/IIIa antagonists. Platelet-aggregation studies, however, can be performed under a number of different conditions, with variables including 1) choice of anticoagulant used to collect blood to ensure minimal preactivation of the platelets, 2) use of isolated platelets, platelet-rich plasma or whole blood, 3) kind and concentration of platelet agonist, and 4) methods of data analysis. It is well established that variations of these conditions will alter the apparent potency of anti-platelet agents. It is also unclear whether the extent of platelet aggregation inhibition for drugs with one mechanism of action will translate to other mechanisms of action. The most important questions however are 1) how relevant platelet aggregation tests are to assess the activation of platelets on atherosclerotic plaques *in vivo* and 2) how predictive aggregation studies are for efficacy and safety of specific anti-platelet agents.

ADP-induced platelet aggregation in citrated plasma was most commonly used to predict the human efficacious dose of GP IIb/IIIa antagonists. Experience with parenteral agents in acute settings suggested that high levels of platelet aggregation inhibition (>85%) were required for efficacy. The level of bleeding side effects associated with this high level of platelet aggregation inhibition was viewed as too high for chronic indications. Accordingly, Phase III studies were performed at significantly lower levels of platelet inhibition (see Table 14.3).

Human dose predictions were also based on preclinical pharmacology models. In general, high levels of *ex vivo* platelet aggregation inhibition, similar to those required for the parenteral agents in acute settings, were required to demonstrate efficacy in preclinical models. Thrombosis in the preclinical models is elicited by acute injury to a normal vessel wall, whereas atherothrombosis is viewed as an acute episode of a chronic disease process. This raises the question of how the acute injury inflicted on the vascular wall represents the atherosclerotic vessel wall. In addition, the reactivity of the GP IIb/IIIa antagonists with the platelet receptor in commonly used experimental animal species, with the exception of primates, is limited. As a result, the low affinity for the efficacy species resulted in the need of high concentrations of GP IIb/IIIa antagonists. The disparity in receptor pharmacology added an additional level of complexity to human dose predictions, as well as to the assessment of efficacy and safety.

The atherothrombosis field lacks reliable biomarkers for major adverse cardiovascular events. Thus, new therapeutic entities require morbidity and mortality endpoint trials for regulatory approval. The uncertainty around the human efficacious and safe dose, based on the lack of clinically validated biomarkers, combined with the lack of biomarkers modeling clinically relevant endpoints at early stages of drug development, led to multiple-arm, large-scale Phase III trials to properly test the GP IIb/IIIa mechanism.

IV. KEY RESULTS FROM PHASE III STUDIES

The Phase III clinical trials performed with orally bioavailable GP IIb/IIIa are summarized in Table 14.3.

The OPUS-TIMI 16 trial enrolled over 10,000 patients with acute coronary syndromes (Cannon et al., 2000). Patients were randomized within 72 hours of an acute event to high dose or low dose of orbofiban or placebo and dosing for at least 180 days was planned. All patients received aspirin and the placebo arm received ticlopidine in the event of stent placement. The trial was terminated prematurely because of an unexpected increase in 30-day mortality. Overall primary endpoints were not different among the treatment arms. Major or severe bleeding was higher with orbofiban (Cannon et al., 2000).

Xemilofiban was administered to more than 7000 patients undergoing percutaneous coronary revascularization for up to 182 days at two different doses (O'Neil et al., 2000). Death, myocardial infarction, or urgent revascularization rates were similar in all three trial groups. Moderate to severe bleeding was significantly higher in both dose groups of xemilofiban compared to placebo (O'Neil et al., 2000).

Sibrafiban was tested in two large Phase III programs. In Symphony I, more than 9000 patients that were stabilized after an acute coronary syndrome were dosed with low dose or high dose sibrafiban or aspirin for 90 days. Cardiovascular endpoints were not different among the study arms. Sibrafiban showed a dose-dependent increase in major bleeding compared to the aspirin group (The Symphony Investigators, 2000). The Symphony II trial was prematurely stopped after the results of Symphony I became available. Symphony II enrolled approximately 6500 stabilized ACS patients in three arms, low-dose sibrafiban plus aspirin, high-dose sibrafiban alone, or aspirin alone. Primary endpoints were increased in the high dose sibrafiban arm and major bleeding was increased in the low-dose sibrafiban plus aspirin arm (The Second Symphony Investigators, 2001).

The Bravo trial enrolled more than 9000 patients with vascular disease. Patients were randomized 1/1 to lotrafiban or placebo, with follow-up planned for 2 years. The dose of lotrafiban was adjusted based on age and renal function. All patients received aspirin at a dose selected at the discretion of the physician. The trial was prematurely stopped because of a statistically significant increase in mortality in the lotrafiban arm and the cause of excess death was vascular related. In addition, serious bleedings were more frequent in the lotrafiban arm compared to aspirin (Topol et al., 2003).

Roxifiban was tested in a patients with moderate to severe peripheral artery disease (PAD) on the background of aspirin with an intended treatment duration of 2 years. The study enrolled fewer than 400 patients and was stopped because an unacceptable bleeding rate combined with a lower than expected overall event rate. To the best of our knowledge, detailed results have not yet been published. The high bleeding signal may be due to the higher potency (binding to resting as well as activated platelets) and the long off-rate compared to first generation agents discussed above.

Some of the individual studies failed to detect the increased mortality associated with long-term GP IIb/IIIa inhibition. Meta-analysis of the first four Phase III trials with more than 30,000 patients followed for at least 30 days revealed a highly significant ($p = 0.001$) increase in mortality in patients dosed with GP IIb/IIIa antagonists. Consistent with the observed benefit of the parenteral agents, the rate of urgent revascularization was reduced in patients dosed with oral GP IIb/IIIa antagonists (Chew et al., 2001).

V. LACK OF EXPLANATION FOR EXCESS MORTALITY IN LARGE PHASE III TRIALS

In the following section, we will discuss several mechanisms that have been put forward to explain the excess mortality with oral GP IIb/IIIa antagonists. In our view, no single factor is sufficient to explain the clinical data and no general consensus on the reason for the excess mortality exists in the scientific community.

For example, in addition to bleeding risks associated with inhibition of platelet aggregation (vide supra), there is also an increased risk due to thrombocytopenia (TCP) associated with GP IIb/IIIa antagonist therapy. Studies with the parenteral administered GP IIb/IIIa antagonist, abciximab demonstrated an incidence of TCP between 2.4% and 5.4% (George et al., 2004; Merlini et al., 2004; Razakjr et al., 2005) when TCP is defined as a platelet count $<100 \times 109$ cell/L. The rate is much less if only severe TCP ($<20\,000 \times 109$ cell/L) is considered. The overall frequency of TCP was three-fold less with tirofiban (Merlini et al., 2004). The TCP itself can be life threatening due to increased bleeding risk. In addition, TCP may predict overall response to therapy. Analysis of a 30-day follow-up of the TARGET study revealed that individuals who became thrombocytopenic had increased mortality and increased target vessel revascularization versus patients who had not become thrombocytopenic (Merlini et al., 2004). This observation appears paradoxical because fewer platelets might be expected to decrease the risk of thrombosis or restenosis. Preactivation of platelets during TCP, however, may explain the increased thrombosis risk. A similar risk associated with TCP was not observed by George et al. (Hodivala-Dilke et al., 1999). The orally bioavailable GP IIb/IIIa antagonists also cause an increased incidence of TCP ranging from about 0 to 2% (Cannon et al., 2000; O'Neil et al., 2000; Seiffert et al., 2003; Symphony Investigators, 2000; Second Symphony Investigators, 2001). Patients given orally bioavailable GP IIb/IIIa antagonists may be at greater risk for sudden bleeding episodes than those receiving the acute therapy because the former will primarily be out-patients.

The mechanism underlying TCP has only recently been defined. Exposure of platelets to small-molecule GP IIb/IIIa antagonists induces alteration in the confirmation of GP IIb/IIIa giving rise to neo-epitopes referred to as ligand induced binding sites (LIBS) (Frelinger et al., 1990). These neo-epitopes may provide binding sites for preexisting antibodies or elicit a *de novo* immune response. Two mechanisms may contribute to the clearance of platelets from the circulation, ultimately resulting in TCP when the regenerative capacity of the bone marrow is exceeded by the clearance of platelets from the circulation. Platelets coated with antibody are targeted for clearance by the reticular endothelial system. Second, GP IIb/IIIa directed antibodies may activate platelets via an Fc-mediated mechanism. In the latter case, activated platelets are rapidly removed from the circulation. Initial support for an immune-mediated nature of the TCP came from studies with plasma from two patients who became thrombocytopenic on roxifiban (Billheimer et al., 2002). Using a differential (drug-dependent) enzyme linked immunosorbent assay (ELISA), which detects antibodies to neo-epitopes present upon binding of agonist to GPIIb/IIIa, it was shown that both patients underwent conversion to a highly positive drug-dependent antibody (DDAB) status temporally associated with TCP. In addition, discontinuation of roxifiban treatment reversed the fall in platelet count even though DDAB was present, demonstrating the exquisite drug dependency of the immune response. The DDAB titer itself depreciated over a 6-month period following discontinuation of the drug. The presence of DDAB has since been found in serum samples from patients who became thrombocytopenic on other GPIIb/IIIa antagonists.

From a clinical perspective, a decrease in agonist associated TCP can be achieved in two ways: by screening out those individuals that have DDAB or by development of a GP IIb/IIIa antagonist that does not elicit an immune response. Cases of roxifiban immune-mediated TCP fall into two categories: early-onset TCP, which occurs within the first 4 days of treatment and delayed-onset TCP, occurring during the second or third week of treatment. Using a DDAB ELISA, it was demonstrated that prescreening followed by a second test 8 days post dosing would decrease TCP by approximately 90% (Seiffert et al., 2003). Obtaining a nonimmunogenic GP IIb/IIIa antagonist would require screening the compound-receptor complex for the presence of neo-epitopes using a DDAB assay. Recent evidence, however, suggests that multiple drug-dependent epitopes are exposed in response to antagonist binding and that different conformational epitopes are exposed by different antagonists (Breth et al., 2005; Seiffert et al., 2003). Thus, screening for a nonimmunogenic antagonist would be a formidable task involving not one but a panel of DDAB to test for "conformational neutrality." The low rate of TCP observed with some of the oral GP IIb/IIIa antagonists makes it unlikely that TCP and associated sequelae are responsible for the increased mortality. Thus, TCP alone is insufficient to explain the excess mortality.

Although the GPIIb/IIIa antagonists inhibit aggregation, they do not inhibit platelet activation. Platelet activation causes the release of several factors that could affect the underlying vascular pathology, including thromboxane A2, P-selectin, ADP, platelet-derived growth factor and CD40L. For example, platelets are an important reservoir of soluble CD40L in the plasma, and elevated sCD40L, which has both pro-thrombotic and pro-inflammatory actions, and is associated with acute coronary syndromes (Garlichs et al., 2001; Viallard et al., 2002). In *in vitro* studies with activated platelets when GP IIb/IIIa antagonists yield a >80% blockade of aggregation sCD40L release is inhibited but with suboptimal concentration of the agonists the release of sCD40L is actually increased (Viallard et al., 2002). An increase in plasma sCD40L might help explain the increase in mortality observed in Phase III trials because the targeted platelet aggregation inhibition in the clinical trials was generally less than 80% (see Table 14.3). It should be noted that a causal role for elevated sCD40L in cardiovascular mortality has not been established.

In addition, it was reported that GP IIb/IIIa antagonists lead to a paradoxical activation of platelets. Methodological issues, however, severely questioned the validity of these conclusions, and this subject will not be further discussed here. Last, there is always a chance that other, not yet identified, off-target mechanisms common to the oral GP IIb/IIIa class are responsible for the excess mortality.

CONCLUSIONS AND LESSONS LEARNED

Could the failure of orally bioavailable GP IIb/IIIa antagonists have been predicted? Table 14.1 summarizes the level of target validation when large Phase III trials were initiated. Target validation based on human biology is hindered by the rare occurrence of Glanzmann's thrombasthenia, and animal models of GP IIb/IIIa deficiency were not available at the time the late-stage clinical programs were initiated. Furthermore, information on the risk and benefits of partial inhibition, the aim of the oral GP IIb/IIIa programs, cannot easily be obtained from genetic models. In contrast, atherothrombosis is an area of high unmet medical need. Thus, research to identify treatment modalities with improved benefit/risk was and still is clearly justified. The lack of validated biomarkers to model exposure response relationships led to a high uncertainty around the clinically beneficial dose. It remains unclear whether the lack of benefit in clinical trials is inherent to the mechanism targeted, or was a result of either insufficient clinical exploration or the pharmacokinetic properties of the antagonists. The postulated mechanism-based safety issues (e.g., paradoxical platelet activation) have not been generally accepted in the scientific community. The low event rate in the secondary atherothrombosis prevention setting underscores the need for extremely safe mechanisms of action. Targeting nonredundant final pathways precludes any compensation with the promise of high levels of efficacy. This is a risky approach for biological systems that are known to be essential for life.

REFERENCES

Anders, R., J. Kleinman, et al. (2001). Xemilofiban/orbofiban: Insights into drug development. *Cardiovasc. Drug Rev.* 19: 116–132.

Billheimer, J. T., I. B. Dicker, et al. (2002). Evidence that thrombocytopenia observed in humans treated with orally bioavailable glycoprotein IIb/IIIa antagonists is immune mediated. *Blood* 99: 3540–3546.

Breth, L., J. Kochie, et al. (2005). Identification and characterization of antibodies that bind GP IIb/IIIa: Antagonist complexes. *J. Immunol. Methods* 301: 11–20.

Cannon C. P., C. H. McCabe, et al. (2000). Oral glycoprotein IIb/IIIa inhibition by orbofiban in patients with unstable coronary syndromes (OPUS-TIMI16) trial. *Circulation* 102: 149–156.

Chew, D. P., D. L. Bhatt, et al. (2001). Increased mortality with oral platelet glycoprotein IIb/IIIa antagonists: A meta-analysis of phase III multicenter randomized trials. *Circulation* 103: 201–206.

Colman, R. W., J. Hirsh, et al. (2001). *Hemostasis and thrombosis.* Philadelphia: Lippincott Williams & Wilkins.

Frelinger, A. L., I. Cohen, et al. (1990). Selective inhibition of integrin function by antibodies specific for ligand-occupied receptor confomers. *J. Biol. Chem.* 265: 6346–6352.

Garlichs, C. D., S. Eskafi, et al. (2001). Patients with acute coronary syndrome express enhanced CD40 ligand/CD154 on platelets. *Heart* 86: 649–655.

George, B. J., R. E. Eckart, et al. (2004). Glycoprotein IIb/IIIa inhibitor-associated thrombocytopenia: Clinical predictors and outcome. *Cardiology* 102: 184–187.

Hodivala-Dilke, K. M., K. P. McHugh, et al. (1999). Beta 3 integrin-deficient mice are a model for Glanzmann thrombasthenia showing placental defects and reduced survival. *J. Clin. Invest.* 103: 229–238.

The Second Symphony Investigators. (2001). Randomized trial of aspirin, sibrafiban, or both for secondary prevention after acute coronary syndromes. *Circulation* 103: 1727–1733, 2001.

The Symphony Investigators. (2000). Comparison of sibrafiban with aspirin for prevention of cardiovascular events after acute coronary syndromes: A randomised trial. *Lancet* 355: 337–345.

Larsen, C. S. and J. Ostergaard (2002). Design and application of prodrugs. In: *Textbook of drug design and discovery,* P. Krogsgaard-Larsen, T. Liljefors, and U. Madsen (eds.). London & New York: Taylor & Francis, p. 410–458.

Loscalzo, J. and A. I. Schafer (1998) *Thrombosis and hemorrhage*. Baltimore: Williams & Wilkins.

Merlini, P. A., M. Rossi, et al. (2004). Thrombocytopenia caused by Abciximab or Tirofiban and its association with clinical outcome in patients undergoing coronary stenting. *Circulation* 109: 2203–2206.

Michelson, A. D. (2002). *Platelets*. Amsterdam: Academic Press.

Niiya, K., E. Hodson, et al. (1987). Increased surface expression of the membrane glycoprotein IIb/IIIa complex induced by platelet activation: Relationship to the binding of fibrinogen and platelet aggregation. *Blood* 70: 475–483.

Nurden, A. and J. N. George (2000). Inherited abnormalities of the platelet membrane: Glanzmann thrombasthenia, Bernard-Soulier syndrome, and other disorders. In: *Hemostasis and Thrombosis*, R. W. Coleman, J. Hirsh, V. J. Marder, A. W. Clowes, and J. N. George (eds.). Philadelphia: Lippincott Williams & Wilkins, 921–943.

O'Neil, W., P. Serruys, et al. (2000). Long-term treatment with a platelet glycoprotein-receptor antagonist after percutaneous coronary revascularization. *N. Engl. J. Med.* 342: 1316–1324.

Plow, E. F. and S. J. (2002). Shattil, Integrin IIb/IIIa and platelet aggregation. In: *Hemostasis and thrombosis* (4 ed.), R. W. Colman, J. Hirsh, V. J. Marder, Clowes AW and George JN. Philadelphia: Lippincott Williams & Wilkins, 479–492.

Razakjr, O. A., H. C. Tan, et al. (2005). Predictors of bleeding complications and thrombocytopenia with the use of abciximab during percutaneous coronary intervention. *J. Interven. Cardiol.* 18: 33–37.

Seiffert, D., D. L. Pedicord, et al. (2003). Regulation of clot retraction by glycoprotein IIb/IIIa antagonists. *Thromb. Res.* 108: 181–189.

Seiffert, D., A. M. Stern, et al. (2003). Prospective testing for drug-dependent antibodies reduces the incidence of thrombocytopenia observed with the small molecule glycoprotein IIb/IIIa antagonist roxifiban: Implications for the etiology of thrombocytopenia. *Blood* 101: 58–63.

Smyth, S. S., E. D. Reis, et al. (2001). Variable protection of beta 3 integrin-deficient mice from thrombosis initiated by different mechanism. *Blood* 98: 1055–1062.

Topol, E. J., D. Easton, et al. (2003). Randomized, double-blind, placebo-controlled international trial of the oral IIb/IIIa antagonist lotrafiban in coronary and cerebrovascular disease. *Circulation* 108: 399–406.

Tronik-Le Roux, D., V. Roullot, et al. (2000). Thrombasthenic mice generated by replacement of the integrin alpha IIb gene: Demonstration that transcriptional activation of this megakaryocytic locus precedes lineage commitment. *Blood* 96: 1399–1408.

Viallard, J. F., A. Solanilla, et al. (2002). Increased soluble and platelet-associated CD40 ligand in essential thrombocythemia and reactive thrombocytosis. *Blood* 99: 2612–2614.

Woods, V. L., L. E. Wolff, and D. M. Keller (1986). Resting platelets contain a substantial centrally located pool of glycoprotein IIb-IIIa complex which may be accessible to some but not other extracellular proteins. *J. Biol. Chem.* 261: 15, 242–15,251.

INDEX